LEAN SIX SIGMA in SERVICE

Applications and Case Studies

LEAN SIX SIGMA in SERVICE

Applications and Case Studies

Edited by
Sandra L. Furterer

CRC Press
Taylor & Francis Group
Boca Raton London New York

CRC Press is an imprint of the
Taylor & Francis Group, an **informa** business

CRC Press
Taylor & Francis Group
6000 Broken Sound Parkway NW, Suite 300
Boca Raton, FL 33487-2742

Printed in the United States of America on acid-free paper
10 9 8 7 6 5 4 3 2 1

International Standard Book Number-13: 978-1-4200-7888-6 (Hardcover)

Library of Congress Cataloging-in-Publication Data

Furterer, Sandra L.
Lean Six sigma in service : applications and case studies / Sandra L. Furterer.
p. cm.
Includes bibliographical references and index.
ISBN-13: 978-1-4200-7888-6
ISBN-10: 1-4200-7888-7
1. Process control. 2. Six sigma (Quality control standard) 3. Quality control--Statistical methods. I. Title.

TS156.8.F88 2009
658.5'62--dc22 2008048130

Visit the Taylor & Francis Web site at
http://www.taylorandfrancis.com

and the CRC Press Web site at
http://www.crcpress.com

Contents

Preface

This book grew out of the need for my students to better understand how to apply and integrate the many tools of the Lean and Six Sigma methodologies and toolkits. As the breadth of tools has increased across the integrated Lean Six Sigma methodology, I found that my students struggled not with applying individual tools, but how they would integrate the suite of tools to make sense of an unstructured problem, and ensure that they focused on what was critical to the customers. It is critical that the team that applies Lean Six Sigma is able to show improvement against the metrics that assess our customers' satisfaction.

This book would not be possible without the enthusiasm, dedication, commitment, energy, and quest for learning that all my Lean Six Sigma students exhibit.

My goal as author and editor of the Lean Six Sigma case book is to provide the learner with an understanding of how others applied Lean Six Sigma and a guide for how they might solve their organization's problems by applying Lean Six Sigma.

The case study data used in this book may be downloaded from the publisher's website at http://www.crcpress.com/e_products/downloads/download.asp?cat_no=78887. This data is an invaluable educational tool that will enhance the students' learning by working with the actual data that the Lean Six Sigma team members used to solve the real world problems discussed in this book.

Acknowledgments

My thanks and appreciation goes to my students who worked diligently on the Lean Six Sigma projects during the courses that I taught at the University of Central Florida (UCF), including Total Quality Improvement, and Seminar in Advanced Industrial Engineering Lean Six Sigma Application; and my students at East Carolina University in my Quality Assurance course. I am extremely grateful to my advisor, Ahmad Elshennawy, UCF, for his guidance in my PhD program and mentoring in the Quality field. I am also grateful to Grace Duffy and Frank Voehl who mentored me in the ASQ Community Good Works program, a joint venture with the American Society for Quality (Orlando chapter) and the UCF's Department of Industrial Engineering (spearheaded by Dr. Ahmad Elshennawy). I also thank Frank Voehl for his Master Black Belt assistance in providing Six Sigma certification through the Harrington Institute, Inc. for my students at UCF. Thanks also to David Collins, who embraced the concepts of Six Sigma in the city where he was finance director, allowing me to make significant change happen.

My special appreciation goes to my students who were so enthusiastically engaged in the Lean Six Sigma projects that became the case studies in this book: Amol Shah, Ariel Lazarus, Blake Hussion, Ethling Hernandez, Felix Martinez, Jose Saenz, Kevin Cochie, Khalid Buradha, Lawrence Lanos, Marcela Bernardinez, Matt Smith, Miguel Torrejon, Parker Rowe, Robert Beaver, Russ D'Angelo, Stefan McMurray, and Varshini Gopal.

My thanks and appreciation to the Six Sigma Black Belts who mentored my students on other Lean Six Sigma projects during my courses: Ala Battikhi, Alain Gaumier (late), Grace Duffy, Richard E. Biehl, Richard Matthews, Mike Kirchner, and Rosida Coowar.

To my Lean Six Sigma project champions and sponsors, whose commitment is so vital to making Lean Six Sigma happen: David Christiansen, David Collins, Debra Reinhart, Jose Murphy, Leslie Pagliari, Isa Nahmens, and Mark Angolia.

Special thanks go to several people for their contributions to the development and production of the book at Taylor & Francis Group, including Cindy Renee Carelli (senior editor) and Jill Jurgensen (project coordinator), Richard Tressider (project editor), and Srikanth Gopaalan (project manager) from Datapage.

Special thanks also go to my brother Dan Brumback who has, and who continues to provide mentoring and guidance to me throughout my life, and who provided feedback and editing of this book, along with his wife Elizabeth. I also want to thank my sister Kathy Bauman, and brothers Tim and Neil Brumback, who through their competitive natures spurred me to achieve.

I dedicate this book to my husband Dan and my children Kelly, Erik, and Zachary, who provide purpose in my life. My husband Dan has enabled me to reach so many of my dreams through his constant encouragement and support. My children provide the incentive for me to strive for excellence to be a guiding example for their lives. I also dedicate this book to my parents, Joan and the late Mel Brumback, who instilled in me the value of education and a lifelong love of books and learning.

Editor

Sandra L. Furterer, PhD, CSSBB, MBB, CQE, is currently an operational performance analyst with Holy Cross Hospital in Ft. Lauderdale, Florida, where she leads their Six Sigma process improvement effort. She also serves as adjunct faculty in the masters of science program at Southern Polytechnic State University in Marietta, Georgia. Previously, she was a business architect and Master Black Belt with the Information Systems Division, Enterprise Architecture Team at Wal-Mart Stores Inc. There she led the business architecture team in the retail systems development area, implementing best practices to achieve operational excellence in information systems application development processes.

Dr. Furterer received her PhD in industrial engineering with a specialization in quality engineering from the University of Central Florida in 2004. Dr. Furterer developed a state-of-the-art framework and roadmap for integrating Lean and Six Sigma methodologies for service industries and implemented the framework in a local government's financial services department. She received an MBA from Xavier University, and a bachelor and master of science in industrial and systems engineering from The Ohio State University.

She was an assistant professor in the Industrial Distribution and Logistics program at East Carolina University from 2006 to 2007. She was a visiting assistant professor and assistant department chair in the Industrial Engineering and Management Systems Department at the University of Central Florida from 2004 to 2006.

Dr. Furterer has more than 18 years of experience in business process and quality improvements. She is an ASQ-Certified Six Sigma Black Belt, Certified Quality Engineer, and a certified Harrington Institute Master Black Belt. An experienced consultant, Dr. Furterer has facilitated and implemented quality, statistics, and process improvement projects, using Six Sigma and Lean principles and tools. She has helped Fortune 100 companies, local government, and nonprofit organizations streamline their processes and implement information systems.

Dr. Furterer has published and/or presented 20 conference papers/proceedings in the areas of Lean Six Sigma, quality, operational excellence and engineering education. She is a senior member of the Institute of Industrial Engineers, a senior member in the American Society for Quality, a member of the American Society for Engineering Management and a member of the American Society for Engineering Education.

Dr. Furterer lives in Coral Springs, Florida with her husband and three children.

Contributors

Robert Beaver
Industrial Engineering and Management Systems
University of Central Florida
Orlando, Florida

Marcela Bernardinez
Industrial Engineering and Management Systems
University of Central Florida
Orlando, Florida

Khalid Buradha
Industrial Engineering and Management Systems
University of Central Florida
Orlando, Florida

Kevin Cochie
Industrial Engineering and Management Systems
University of Central Florida
Orlando, Florida

Russ D'Angelo
Industrial Engineering and Management Systems
University of Central Florida
Orlando, Florida

Sandra L. Furterer
Holy Cross Hospital
Fort Lauderdale, Florida

Varshini Gopal
Industrial Engineering and Management Systems
University of Central Florida
Orlando, Florida

Ethling Hernandez
Industrial Engineering and Management Systems
University of Central Florida
Orlando, Florida

Blake Hussion
Industrial Distribution and Logistics
East Carolina University
Greenville, North Carolina

Lawrence Lanos
Industrial Engineering and Management Systems
University of Central Florida
Orlando, Florida

Ariel Lazarus
Industrial Engineering and Management Systems
University of Central Florida
Orlando, Florida

Felix Martinez
Industrial Engineering and Management Systems
University of Central Florida
Orlando, Florida

Stefan McMurray
Industrial Distribution and Logistics
East Carolina University
Greenville, North Carolina

Parker Rowe
Industrial Distribution and Logistics
East Carolina University
Greenville, North Carolina

Jose Saenz
Industrial Engineering and
Management Systems
University of Central Florida
Orlando, Florida

Amol Shah
Industrial Engineering and
Management Systems
University of Central Florida
Orlando, Florida

Matt Smith
Industrial Distribution and Logistics
East Carolina University
Greenville, North Carolina

Miguel Torrejon
Industrial Engineering and
Management Systems
University of Central Florida
Orlando, Florida

1 Instructional Strategies for Using This Book

Sandra L. Furterer

CONTENTS

The purpose of this book is to provide a guide for learners and appliers of Lean Six Sigma methodologies and tools. The book is designed to engage the reader by enabling hands-on experience with real Lean Six Sigma project cases in a safe environment, where experienced Black Belt and Master Black Belts can help mentor the students in Lean Six Sigma. Case studies are designed to enable the student to work through the exercises and to provide sufficient background information so that they can apply the tools as if they collected the data themselves. The case discussions provide questions to allow students to compare their solutions with actual results realized by similar students struggling with learning and applying Lean Six Sigma. Another advantage is that the students are using real "messy" data that does not necessarily fit nicely into normal statistical distributions. This will help prepare them to touch actual data when they embark on real-world projects.

BUSINESS PROCESSES AND LEAN SIX SIGMA PROJECT BACKGROUNDS

The Lean Six Sigma projects consist of various service-oriented processes in academic and governmental environments. An overview of each process is provided for the students so that they understand the background of the project, as well as having sufficient information regarding the processes that need to be improved so that they can develop a project charter and scope the project. Data that were actually collected in the Lean Six Sigma projects are provided for application of Lean Six Sigma tools and appropriate statistical analysis. Case exercises are provided so that the students can solve the Lean Six Sigma or Design for Six Sigma projects for each phase of the Define–Measure–Analyze–Improve–Control (DMAIC) or Identify–Define–Design–Optimize–Validate (IDDOV) problem-solving methodology. Each phase provides the solution the students actually developed, that can be used as a guide to solve the next phase of the project.

LEAN SIX SIGMA CASE STUDY GOALS

To successfully complete the Lean Six Sigma case studies, participants must apply appropriate problem-solving methods and tools from the Lean Six Sigma toolkit to understand the problem, identify key customers and stakeholders, understand critical to satisfaction (CTS) characteristics, find critical factors and root causes of the problem, develop potential improvement recommendations, and develop a plan to control the new process.

LEAN SIX SIGMA TOOLS

During the case study, the class will use Lean Six Sigma, DMAIC and Minitab® tools that were most commonly used in the real project.

LEARNING DESIGN

Each exercise in the case study is designed so that the teams of students experience the factors listed below:

- Team interaction, definition of team ground rules, brainstorming, and consensus building, as well as the stages of team growth.
- Choosing how to apply Lean Six Sigma tools and problem-solving methods.
- Supporting their decisions and application of the tools with data.
- Reviewing information for relevant and irrelevant information and data, and reframing into what is important to solve the problem.
- Each exercise develops students' understanding and application of specific tools and problem-solving methods.
- Development of written reports and presentations, as well as the ability to present technical information.
- Application of project management tools to manage activities and complete tasks in a timely manner.
- Experience in solving an unstructured problem in a safe learning environment where mentoring is available.

The Instructor's Role

To facilitate the learning process, it is critical for the instructor to act as a coach or mentor to the student teams. It can also be helpful to have Six Sigma Black Belts and/or Master Black Belts experienced in applying Lean Six Sigma tools and methods assigned to each student team to mentor them in the application of Lean Six Sigma problem-solving. Local sections of the American Society for Quality can be a great resource for providing experienced Six Sigma Black Belts and Master Black Belt volunteers.

The instructor could organize the students into teams of 4–6 students. Most Six Sigma programs solve complex problems with problem-solving teams. There is a

Rate yourself on a scale of 1 to 5 in the following areas:

Rating Scale

Element	1 No Experience	2 Little Experience	3 Some Experience	4 Fairly Extensive Experience	5 Extensive Experience
Project Team					
Project Team Leadership					
Lean Six Sigma Tools and Methods					

FIGURE 1.1 Project team assessment.

great deal of value in having students work together as a team to solve the problems. They can learn how to work more effectively as a team, and team members can transfer learning across the team members because students grasp the difficult concepts of Lean Six Sigma at different paces. An effective way to organize the teams is to ask the students the questions provided in Figure 1.1, and try to distribute the experienced team leaders, problem-solvers, and team members across the teams.

REQUIRED KNOWLEDGE LEVELS BY LEAN SIX SIGMA PROJECTS

The Lean Six Sigma projects included in this book include different knowledge levels and depth of understanding to best apply the Lean Six Sigma tools. Figures 1.2 through 1.7 show the student level and tools applied by project, so that the instructor can select the appropriate cases for their students.

There are three different student levels defined as follows:

- Beginner: Early (up to Junior) undergraduate student with no exposure to Lean Six Sigma, and little statistical background.
- Intermediate: Senior undergraduate or master's graduate student with some exposure (theoretical knowledge) to Lean Six Sigma tools and some statistical background.
- Advanced: master's or PhD graduate students with theoretical learning of Lean Six Sigma tools and some statistical background, as well as having worked on a Lean Six Sigma project.

The chapter objectives are detailed below.

Chapter 1: provides an overview of the text and the instructional strategies to best use this book.

Chapter 2: provides an overview of Lean Six Sigma and the DMAIC problem-solving methodology and tools as applied to services and transaction-based processes.

Chapter 3: provides an overview of Design for Six Sigma and the IDDOV design methodology as applied to services and transaction-based processes.

Chapters 4 through 9: provide detailed projects, case exercises and discussions to enable the student to perform Lean Six Sigma or Design for Six Sigma projects, and learn and apply these methodologies and tools.

Chapter 10: understand and be able to carry out a Lean Six Sigma project assessment to determine what the team did well and areas for improving the Lean Six Sigma program and future projects.

Chapter 11: provides some insight into the future of Lean Six Sigma and some challenges that organizations may face in their Lean Six Sigma journey.

Lean Six Sigma Project: Sunshine High School Discipline Process Improvement

Team Members: Marcela Bernardinez, Khalid Buradha, Kevin Cochie, Jose Saenz, Master Black Belt: Dr. Sandy Furterer

Book chapter	Methodology applied	Tools applied	Student level
Chapter 4	Lean Six Sigma DMAIC	▪ Project chartering ▪ Stakeholder analysis ▪ Project planning ▪ SIPOC, process maps ▪ Operational definitions ▪ CTS ▪ VOC, VOP ▪ VOC surveys ▪ VOP matrix ▪ Measurement system analysis (Gage R&R) ▪ Benchmarking ▪ Cost of poor quality ▪ Cause & effect analysis ▪ Process and waste analysis ▪ Histogram, graphical and data analysis ▪ Correlation analysis ▪ Regression analysis ▪ Statistics and confidence intervals ▪ Hypothesis testing, ANOVA ▪ Attribute survey analysis ▪ DPPM/DPMO ▪ Process capability ▪ QFD ▪ Recommendations for improvement; action plans ▪ Training plans; procedures ▪ Mistake proofing ▪ Control plan ▪ Control charts ▪ Replication opportunities ▪ Standardize work ▪ Dashboards, scorecards	Advanced

FIGURE 1.2 Methodology, tool, student level mapping for Chapter 4.

Lean Six Sigma Project: Financial Services Improvement in a City Government

Team Members: Author—Sandy Furterer

Book chapter	Methodology applied	Tools applied	Student level
Chapter 5	Lean Six Sigma DMAIC	▪ Project chartering ▪ Stakeholder analysis ▪ Project planning ▪ Responsibilities matrix ▪ SIPOC, process maps ▪ Operational definitions ▪ CTS ▪ Pareto chart ▪ VOC, VOP ▪ VOC surveys ▪ VOP matrix ▪ Statistical analysis ▪ Cost of poor quality ▪ Cause & effect analysis ▪ Process and waste analysis ▪ Histogram, graphical and data analysis ▪ Correlation analysis ▪ Regression analysis ▪ Hypothesis testing ▪ Attribute survey analysis ▪ DPPM/DPMO ▪ Process capability ▪ QFD ▪ Recommendations for improvement; action plans ▪ Cost/benefit analysis ▪ Training plans; procedures ▪ Mistake proofing ▪ Control plan ▪ Control charts ▪ Replication opportunities ▪ Standard work, kaizen ▪ One-piece flow ▪ Visual control, kanban ▪ Dashboards, scorecards	Intermediate

FIGURE 1.3 Methodology, tool, student level mapping for Chapter 5.

Lean Six Sigma Project: Industrial Distribution and Logistics (IDIS) Program Recruiting Process Design

Team Members: Blake Hussion, Stefan McMurray, Parker Rowe, Matt Smith
Master Black Belt: Dr. Sandy Furterer

Book chapter	Methodology applied	Tools applied	Student level
Chapter 6	Lean Six Sigma DMAIC	▪ Project chartering ▪ Stakeholder analysis ▪ Project planning ▪ Responsibilities matrix ▪ SIPOC, process maps ▪ Operational definitions ▪ CTS ▪ Pareto chart ▪ VOC, VOP ▪ VOC surveys ▪ VOP matrix ▪ Cost of poor quality ▪ Cause & effect analysis ▪ Process and waste analysis ▪ Failure mode and effect analysis ▪ 5S ▪ Hypothesis testing ▪ Attribute survey analysis ▪ DPPM/DPMO ▪ Recommendations for improvement; action plans ▪ Training plans; procedures ▪ Control plan ▪ Replication opportunities ▪ Standard work, Kaizen ▪ Dashboards, scorecards	Beginner

FIGURE 1.4 Methodology, tool, student level mapping for Chapter 6.

Lean Six Sigma Project: CECS Inventory and Asset Management Process Improvement

Team Members: Felix Martinez, Varshini Gopal, Amol Shah, Robert Beaver, Russ D'Angelo, Miguel Torrejon; Master Black Belt: Dr. Sandy Furterer

Book chapter	Methodology applied	Tools applied	Student level
Chapter 7	Lean Six Sigma DMAIC	▪ Project chartering ▪ Stakeholder analysis ▪ Project planning ▪ Responsibilities matrix ▪ SIPOC, process maps ▪ Operational definitions ▪ CTS ▪ Pareto chart ▪ VOC, VOP ▪ VOC surveys ▪ VOP matrix ▪ Benchmarking ▪ Cost of poor quality ▪ Statistical analysis ▪ Cause & effect analysis ▪ Process and waste analysis ▪ Histogram, graphical and data analysis ▪ Failure mode and effect analysis ▪ 5S ▪ Attribute survey analysis ▪ DPPM/DPMO ▪ Recommendations for improvement; action plans ▪ QFD ▪ Cost/benefit analysis ▪ Training plans; procedures ▪ Control plan ▪ Dashboards, scorecards	Intermediate

FIGURE 1.5 Methodology, tool, student level mapping for Chapter 7.

Lean Six Sigma Project: High School Advanced Placement Open Access Process Assessment

Team Members: Marcela Bernardinez, Ethling Hernandez, Lawrence Lanos
Ariel Lazarus, Felix Martinez, Master Black Belt: Dr. Sandy Furterer

Book chapter	Methodology applied	Tools applied	Student level
Chapter 8	Lean Six Sigma DMAIC	▪ Project chartering ▪ Stakeholder analysis ▪ Project planning ▪ Responsibilities matrix ▪ SIPOC, process maps ▪ Operational definitions ▪ CTS ▪ Pareto chart ▪ VOC, VOP ▪ Statistical analysis ▪ VOP matrix ▪ Cost of Poor Quality ▪ Statistical analysis ▪ Cause & effect analysis ▪ Waste analysis ▪ Correlation analysis ▪ Regression analysis ▪ Histogram, graphical and data analysis ▪ Hypothesis testing, ANOVA ▪ DPPM/DPMO ▪ Recommendations for improvement; action plans ▪ QFD ▪ Training plans; procedures ▪ Control plan ▪ Control charts ▪ Replication opportunities ▪ Dashboards, scorecards	Advanced

FIGURE 1.6 Methodology, tool, student level mapping for Chapter 8.

Lean Six Sigma Project: Project Charter Review Process Design—A Design for Six Sigma Case Study

Team Members: Carrie Harris, Emily McKenzie, Bridget Corp
Master Black Belt and Author: Dr. Sandy Furterer

Book chapter	Methodology applied	Tools applied	Student level
Chapter 9	Design for Six Sigma IDDOV	▪ Project chartering ▪ Stakeholder analysis ▪ Project planning ▪ Data collection plan ▪ VOC ▪ QFD ▪ Process map ▪ Operational definitions ▪ CTS ▪ Failure mode and effect analysis ▪ Process and waste analysis ▪ VOP matrix ▪ Implementation plan ▪ Statistical process control ▪ Process capability analysis ▪ Training plans; procedures ▪ Dashboards, scorecards ▪ Mistake proofing ▪ Hypothesis testing	Beginner

FIGURE 1.7 Methodology, tool, student level mapping for Chapter 9.

2 Lean Six Sigma Roadmap Overview

Sandra L. Furterer

CONTENTS

LEAN SIX SIGMA OVERVIEW

Lean Six Sigma is an approach focused on improving quality, reducing variation, and eliminating waste in an organization. It is the combination of two improvement programs: Six Sigma and Lean Enterprise. The former is a quality management philosophy and methodology that focuses on reducing variation; measuring defects (per million output/opportunities); and improving the quality of products, processes, and services. The concept of Six Sigma was developed in the early 1980s at Motorola Corporation. Six Sigma was popularized in the late 1990s by the General Electric Corporation and their former CEO, Jack Welch. Lean Enterprise is a methodology that focuses on reducing cycle time and waste in processes. Lean Enterprise originated from the Toyota Motor Corporation as the Toyota production system (TPS), and increased in popularity after the 1973 energy crisis. The term "lean thinking" was coined by James P. Womack and Daniel T. Jones in their book *Lean Thinking* (Womack and Jones 1996). The term "lean enterprise" is used to broaden the scope of a Lean program from manufacturing to embrace the enterprise or entire organization (Alukal 2003). Figure 2.1 shows the evolution to the combined methods of Lean and Six Sigma.

The concepts of control charts and statistical process control (SPC) were developed by Walter Shewhart at Western Electric in the 1920s. Dr. W. Edwards Deming installed SPC in Japanese manufacturing as he assisted Japan in their rebuilding efforts after World War II. Japan's successes in the 1970s repopularized SPC in

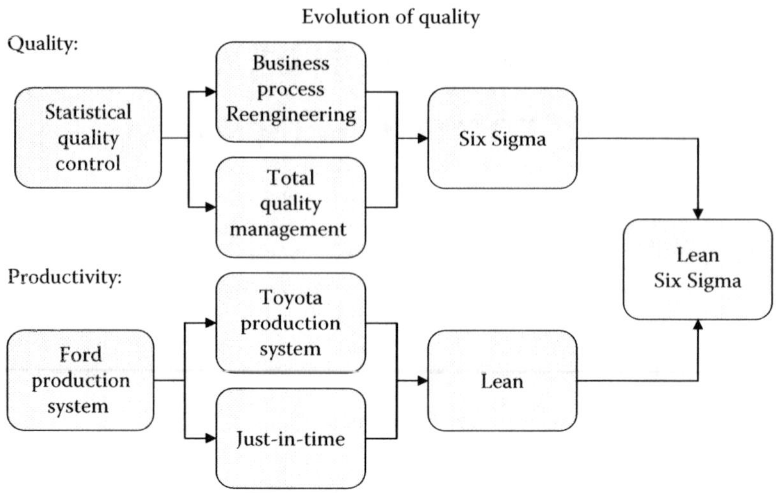

FIGURE 2.1 Evolution of quality and productivity to Lean Six Sigma. (From Furterer, S.L., ASQ Conference on Quality in the Space and Defense Industries, Critical Quality Skills of Our Future Engineers. March 2006.)

U.S. businesses. Total quality management (TQM) was a natural outgrowth of SPC, adding a process improvement methodology. In the 1980s, Business process reengineering (BPR) and TQM became popular. BPR encouraged completely throwing out the old process and starting over, many times within the context of implementing changes in major information systems. TQM focused on a less structured approach with the principles of quality and process improvement. These methodologies evolved into Six Sigma.

On the productivity side, the Ford production system was used to assemble cars, which was the basis for the TPS. Just-in-time (JIT) production philosophies joined with TPS, which evolved into Lean. Now Lean and Six Sigma are merging to capitalize on the best of both improvement philosophies and methodologies.

Six Sigma uses the Define, Measure, Analyze, Improve, and Control (DMAIC) problem-solving approach, and a wide array of quality problem-solving tools. Use of these tools is based on the type of process studied and the problems encountered. There are many powerful tools in the Lean tool set that help to eliminate waste, organize, and simplify work processes.

LEAN SIX SIGMA APPLICATIONS IN PRIVATE INDUSTRY

The concept of combining Lean manufacturing and Six Sigma principles began in the middle to late 1990s, and quickly took hold as companies recognized the synergies. There are many examples of manufacturing companies implementing a combined effort of Lean and Six Sigma. An early example, starting in 1997, was by an aircraft-engine-controls firm, BAE Systems Controls, in Fort Wayne, Indiana.

They blended Lean manufacturing principles with Six Sigma Quality tools. Their "Lean Sigma" strategy was "designed to increase velocity, eliminate waste, minimize process variation, and secure its future in the evolving aerospace market" (Sheridan 2000). They started with implementing Lean initiatives and then identified a synergy between Lean and the Six Sigma quality program that had been launched while the company was a part of General Electric. BAE Systems Controls implemented the following Lean initiatives: (1) kaizen events, (2) takt time driven one-piece-flow product cells, (3) kanban pull system and point-of-use storage bins on the plant floor, (4) lean production cells, (5) mistake proofing, and (6) use of a multiskilled workforce. As part of the Six Sigma program, they implemented statistical methods and team leadership with the use of Black Belts. In BAE Systems Control's implementation of Lean Six Sigma, they improved productivity by 97% and customer lead time by 90%. Their value-added productivity increased 112% in five years, work in process was reduced by 70%, product reliability improved by 300%, and there were zero lost workdays in 1999 (Sheridan 2000).

Another early innovator combining Lean and Six Sigma was the Maytag Corporation, which implemented Lean Sigma® in 1999. They designed a new production line using the concepts of Lean and Six Sigma. Maytag reduced utilized floor space to one-third of that used by Maytag's other product lines. Maytag also cut production costs by 55%. Their Lean Sigma effort helped them to achieve savings worth millions of dollars (Dubai Quality Group 2003).

Lean Six Sigma has been implemented at Northrop Grumman, an aerospace company. They had already started to implement Lean Thinking when they embarked upon their Six Sigma program. Northrop integrated the WorkOut® events (problem-solving process developed at GE) with Lean Thinking methods and kaizen events. They used the strategies and methods of Six Sigma within their product teams, not as a stand-alone program. Their formal process integrated WorkOut, kaizen, and DMAIC into the Six Sigma Breakthrough WorkOut. Subject matter experts and a Black Belt were used on their project team. They carried out a 4–5 day Define/Measure phase. They then did the Measure, Analyze, and Improve phases for about 30 days each. The final activities included a post WorkOut phase as the Control, Integrate, and Realize phase (McIlroy and Silverstein 2002).

Lockheed Martin Aeronautical Systems reduced costs and improved competitiveness, customer satisfaction, and the first-time quality of all its manufactured goods. They had separate Lean and Six Sigma projects, depending on the objective of the project and the problem that needed to be solved (Kandebo 1999).

The Six Sigma DMAIC problem-solving methodology is used to improve processes. DMAIC phases are well defined and standardized, but the steps carried out in each phase can vary based on the reference used. The Define phase is where the scope of the project charter is developed. The goal of the Measure phase is to understand and baseline the current process. In the Analyze phase, we analyze the data collected in the Measure phase to identify the root causes of the problems identified. In the Improve phase, the improvement recommendations

Define	Measure	Analyze	Improve	Control
1. Develop project charter 2. Identify stakeholders 3. Perform initial VOC and identify CTS 4. Select team and launch the project 5. Create project plan	6. Define the current process 7. Define detailed VOC 8. Define the VOP and current performance 9. Validate Measurement System 10. Define COPQ and Cost/Benefit	11. Develop cause and effect relationships 12. Determin and validate root causes 13. Develop process capability	14. Identify breakthrough & select solutions 15. Perform cost/benefit analysis 16. Design future state 17. Establish performance targets, project scorecard 18. Gain approval to implement and implement 19. Train and execute	20. Measure results & manage change 21. Report scorecard data & create process control plan. 22. Apply P–D–C–A process. 23. Identify replication opportunities 24. Develop future plans

FIGURE 2.2 DMAIC activities.

are developed and implemented. The goal of the Control phase is to ensure that improvements had a positive impact and that they will be sustained and controlled. Figure 2.2 is a description of the activities that can be carried out within each phase of the DMAIC problem-solving methodology (adapted from Brassard and Ritter 2001).

The DMAIC approach (the detailed steps and most frequently used tools applied within each phase shown in Figure 2.2) is described as follows (Brassard and Ritter, LLC 2001).

PHASE I: DEFINE

The purpose of the Define phase is to delineate the business problem and scope of the project and the process to be improved. The following steps can be applied to meet the objectives of the Define phase:

1. Develop project charter
2. Identify customers and stakeholders
3. Define initial voice of customer (VOC) and critical to satisfaction (CTS) criteria
4. Form the team and launch the project
5. Create project plan

Figure 2.3 shows the main activities mapped to the tools or deliverables most typically used during each step of the Define phase.

	Define Activities	Tools/Deliverables
1	Develop project charter	▪ Project charter ▪ SIPOC (Supplier–Inputs–Process–Output–Customer) ▪ High-level process map
2	Identify customers and stakeholders and perform stakeholder analysis	▪ Stakeholder analysis definition ▪ Stakeholder commitment scale ▪ Communication planning worksheet
3	Perform Initial Voice of Customer (VOC) and identify Critical to Satisfaction (CTS)	▪ Critical to Satisfaction (CTS) summary
4	Select team and launch the project	▪ Responsibilities matrix ▪ Ground rule ▪ IFR (Items for Resolution)
5	Create project plan	▪ Work plan

FIGURE 2.3 Define activities and tools/deliverables.

1. Develop Project Charter

The first step in the Define phase is to identify and delineate the problem. The project charter can help to identify the elements that help to scope the project, and identify the project goals.

A project charter template is provided in Figure 2.4.

The elements of the project charter that help to scope and define the business problem are described as follows.

Project name: Describes the process to be improved, along with the project goal.

Project overview: Provides a project background and describes basic assumptions related to your project.

Problem statement: A clear description of the business problem. What is the challenge or the problem that the business is facing? The problem statement should consider the process that is affected. Define the measurable impact of the problem. The team should be specific as to what is happening, when it is occurring, and what the impact or consequences are to the business problem.

Customers/stakeholders: Define the customers, both internal and external, and the stakeholders that are being affected by the problem or process to be improved.

CTS: Identify what is important to each customer/stakeholder group. They can be identified by what is critical to quality (defects), delivery (time), and cost.

Goal of the project: What is the quantifiable goal of the project? It may be too early in the problem-solving method to identify a clear target, but at least

Project Name: Name of the Lean Six Sigma project.

Project Overview: Background of the project.

Problem Statement: Business problem: describe what, impact, consequences.

Customer/Stakeholders: (Internal/External) Key groups impacted by the project.

What is important to these customers–CTS: Critical to satisfaction, the key business drivers.

Goal of the Project: Describe the improvement goal of the project.

Scope Statement: The scope of the project; what is in the scope and what is out of scope?

Finacial and Other Benefit(s): Estimated benefits to the business, tangible and intangible.

Potential Risks: Risks that could impact the success of the project. Can assess risk by probability of occurrence and potential impact to the project.

Milestones: DMAIC phase and estimate completion dates.

Project Resources: Champion, Black Belt Mentor, Process Owner, Team Members.

FIGURE 2.4 Project charter template (adapted from Wal-Mart Global Continuous Improvement Training 2008).

a placeholder should be identified relating to what should be measured and improved.

Scope statement: The scope should clearly identify the process to be improved, and what is included or excluded from the scope for the Lean Six Sigma project. The scope can also address the organizational boundaries to be included and, possibly more importantly, which should be excluded. It can also include a temporal scope of the timing of the process and data collection activities. The deliverable scope includes what specifics should be delivered from the project, such as improvement recommendations and the implementation plan.

Projected financial and other benefits: Describes potential savings, revenue growth, cost avoidance, cost reduction, cost of poor quality (COPQ), as well as less tangible benefits such as impact to morale, elimination of waste, and inefficiencies.

Potential risks: Brainstorm the potential risks that could affect the success of the project. Identify the probability that the risk could occur, on a high, medium, or low scale. Identify the potential impact to the project if the risk does occur, on a high, medium, or low scale. The risk mitigation strategy identifies how you would potentially mitigate the impact of the potential risk if it does occur.

Project resources: Identify the project leader who is in charge of the overall project. Identify the division and department of the project leader or project team. Identify the process owner, the person who is ultimately responsible for implementing the improvement recommendations. The project champion is at the director (or above) level who can remove the barriers to successful project implementation. The project sponsor is the executive-level person who sponsors the project initiative and is the visible representative of the project and improvements. Continuous Improvement Mentor or the Master Black Belt is the team's coach who helps mentor the team members in applying the tools and DMAIC methodology. Finance is the financial representative who approves the financial benefits or savings established during the project. Team members or support resources are people who are part of the project team, or who provide support, information, or data to the project team.

Milestones: The milestones are the estimated key dates when each phase will be completed, and when the project improvements will be approved.

Suppliers–Input–Process–Output–Customer (SIPOC)

The SIPOC (Pyzdek 2003) is a useful tool in the Define phase to help scope the project and understand the process. SIPOC shows the interrelationships between the customers and suppliers, and how they interact with the process. It also identifies the inputs used in the process steps and the outputs of the process. The process steps transform the inputs into the outputs. The best way to construct the SIPOC is to identify the five to seven high-level process steps that bound the process. For each process step, identify the inputs to the process and who supplies the inputs. Next identify the outputs of each process step and the customer of the output. An example of an SIPOC for creating a circular advertisement is shown in Figure 2.5.

High-Level Process Map

A process is a description of activities that transforms inputs to outputs. A process map is a graphical representation of the process, interrelationships, and sequence of steps. The high-level or level-1 process map utilized in the Define phase can be derived from the process steps identified in the SIPOC. The process steps can be simply turned 90° and be displayed horizontally instead of vertically. Process

Suppliers	Inputs	Process	Outputs	Customers
Merchandising	Strategies, market information	Identify items to advertise	Items to advertise	Marketing
Marketing	Items to advertise	Identify price and discounts	Prices, discounts	Ad firm
Ad firm	Prices, discounts, items	Design circular	Circular design	Ad firm
Ad firm	Circular design	Develop circular	Draft circular	Marketing
Marketing	Draft circular	Approve circular	Approval	Ad firm
Ad firm	Approval	Finalize circular	Final circular	Marketing
Marketing	Final circular	Distribute circular	Circular	Customers

FIGURE 2.5 SIPOC example.

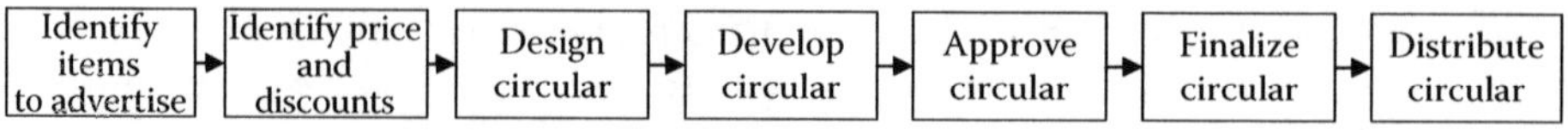

FIGURE 2.6 Level-1 process map.

maps are a valuable tool in helping to understand the current process, identifying the inefficiencies and nonvalue-added activities, and then creating the future state process during the Improve phase. If there is sufficient knowledge of the process, a more detailed, level-2 process can be created in the Define phase, but additional interviews must usually be held to collect the information. A level-1 process map is therefore usually sufficient, as shown in Figure 2.6, which is a process map for an advertising circular process for a company.

2. Identify Customers and Stakeholders

It is critical to clearly identify the customers and stakeholders that are affected by the process because the quality of the process is defined by the customers. Quality is measured by first understanding, then exceeding, the customers' requirements and expectations. There is a high cost of an unhappy customer: Ninety-six percent of unhappy customers never complain; 90% of those who are dissatisfied will not buy again; and each unhappy customer will tell his or her story to as many as 14 people (Pyzdek 2003).

Customers and stakeholders can be my peers, people who report to me, my boss, other groups within the organization, suppliers, and external customers. The customers can include internal and external customers of the process. Each process does not always interface directly with an external customer of the company, but will have internal customers. The latter are people who receive some output from the process, such as information, materials, products, or a service step. It is ultimately the boundary of the process that is being improved that determines who the customer is.

The stakeholder analysis definition identifies the stakeholder groups, their role, and how they are impacted, as well as their concerns related to the process. There is an additional column that provides a quick view of whether the impact is positive (+), such as reducing variation, or negative (–), such as resistant to change. This is a high-level view that will be further detailed in the Measure phase. Figure 2.7 is an example of a stakeholder definition.

The next step in the stakeholder analysis is to understand the stakeholders' attitudes toward change, as well as potential reasons for resistance. Additionally, the team should understand the barriers to change as a result of the resistance. Activities, plans, and actions should then be developed that can help the team overcome the resistance and barriers to change. A definition of how and when each stakeholder group should participate in the change effort should be developed in the Define phase, and then updated throughout the DMAIC project. Figure 2.8 shows a stakeholder commitment.

The stakeholder commitment scale can be used to summarize where the stakeholders are regarding their acceptance or resistance to change. The team should determine, based on initial interviews and prior knowledge of the stakeholder groups, the current level of support or resistance to the project. "Strongly supportive" indicates

Stakeholder Analysis Definition			
Stakeholders	**Role description**	**Impact/concern**	**+ / −**
External customer	Customers who receive our marketing efforts related to marketing programs, including advertising circulars and commercials.	• Timely information • Accurate information • Coupons	+ + +
Marketing	Internal marketing department who plan, develop and deploy marketing programs.	• Timely deployment • Ability to reach and impact customers	+ +
Information technology	Information technology department that provides technology	• Clear requirements • Accurate data	+ +

FIGURE 2.7 Stakeholder analysis definition.

Stakeholder commitment scale							
Stakeholders	**Strongly against**	**Moderate against**	**Neutral**	**Moderate support**	**Strongly against**	**Communication plan**	**Action plan**
External customer		X →	→	→	→ O	• Surveys • Market research	Test pilot
Marketing			X →	→	→ O	• Email • Meetings	Engage on project
Information technology	X →	→	→	→ O		• Intranet • Email • Meeting	Engage on project Communicate process and requirements
X = At beginning of project							O = At end of project

FIGURE 2.8 Stakeholder commitment scale.

that these stakeholders are supportive of advocating for and making change happen. "Moderately supportive" indicates that the stakeholders will help, but they will not strongly. "Neutral stakeholders" will allow the change and not stand in the way, but they will not go out of their way to advocate for the change. "Moderately against stakeholders" do not comply with the change, and have some resistance to the project. "Strongly against stakeholders" will not comply with the change and will actively and vocally lobby against the change. A strategy to move the stakeholders from their current state to where the team needs them to be by the end of the project should be developed. This change strategy should include how the team will communicate with the stakeholders and activities in their action plan to gain support and implement change.

3. Define Initial VOC and Identify CTS

In the Define phase, the team can carry out an initial VOC data collection to understand the CTS criteria, which are the elements of a process that significantly affect the output of the process. It is critical to focus on the CTS throughout the phases of the DMAIC problem-solving process and the Six Sigma project.

In the Define and Measure phases, the focus is on collecting information from the customer to understand what is important to them regarding the process, product, or service. In the Measure phase, the team should identify the metrics to measure the processes that are directly related to the CTS criteria. In the Analyze phase, the team should analyze the root causes related to the CTS. The improvement recommendations implemented are aligned with eliminating the root causes related to the CTS in the Improve phase. The variability to be controlled by implementing control mechanisms in the Control phase should reduce the variability related to the CTS.

Some references refer to identifying the CTQ, but the CTS broadens the elements of itself by including CTQ, critical to delivery (CTD), and critical to cost (CTC). There may also be critical elements in the process to measure that are related not only to quality, delivery, and cost, but also to time. For a Six Sigma project, not everything should be a CTS. The CTS should be specific to the scope of the project and the process to be improved. If there are more than a few CTS measures identified for the project, the scope is probably too large for a reasonable Six Sigma project to be completed in 3–6 months. The CTS should describe the customer need or requirement, not how to solve the problem.

The steps to identify the CTS are shown as follows. (George, Rowlands, Price, and Maxey 2005):

1. Gather appropriate VOC data from market research, surveys, focus groups, interviews, etc
2. Extract key verbatims from the VOC data collections, identifying why a customer would do business with your organization
3. Sort ideas and find themes, develop an Affinity or Tree Diagram
4. Be specific and follow up with customers where needed
5. Extract CTS measures and specifications from customer information
6. Identify where you are missing data and fill in the gaps

The VOC is a term used to "talk to the customer" to hear their needs and requirements or their "voice." Many mechanisms can be used to collect VOCs, including interviews, focus groups, surveys, customer complaints and warranty data, market research, competitive information, and customer buying patterns. We will further discuss VOCs during the Measure phase, where more detailed and extensive VOCs can best be done. The initial VOC is used to identify the CTS. In the Define phase, the CTS summary is a listing of the CTS measures based on knowledge of the process and the customer to this point.

4. Form Team and Launch the Project

The Six Sigma project team should be selected based on those team members who have knowledge of the process, and have the commitment to work on the project. The roles and responsibilities of the project team members should be clearly defined.

A team is a group of people working together to achieve a common purpose. Teams need a clearly defined purpose and goals that are provided through the Six

Sigma project charter. They also need well-defined roles, and responsibilities, which can be provided through developing a responsibilities matrix (Figure 2.9) (Scholtes, Joiner, and Streibel 2003). The responsibilities matrix identifies the team members, their roles, and high-level responsibilities on the Six Sigma project. Another important component of forming the team is to brainstorm and identify ground rules. Ground rules identify how the members of the team will interact with each other and ensure that behavioral expectations are clearly defined at the start of the project. The team's common set of values and ethics can be established during the development of the team ground rules.

Sample ground rules for a team:

- Treat everyone with respect
- Listen to everyone's ideas
- When brainstorming, do not evaluate ideas
- Contribute fully and actively participate
- Come to team meetings prepared
- Make decisions by consensus
- Identify a back-up resource to complete tasks when not available

Role / Responsibility	Team leader	Black belt	Champion	Process owner	Team members
Facilitate meetings	X				
Manage project	X				
Mentor team members	X	X			
Transfer knowledge of Six Sigma tools		X			
Remove roadblocks			X		
Monitor project progress			X		
Approve project			X		
Implement improvements				X	
Subject matter expertise				X	
Apply Six Sigma tools					X
Statistical Analysis					X
Data collection					X

FIGURE 2.9 Responsibility matrix.

5. Create Project Plan

The Lean Six Sigma project plan is developed in this last step of the Define phase. The resources, time, and effort for the project are planned. A project plan template is provided in Figure 2.10. Additional tasks can be identified within each phase and major activity.

Excel® or a project planning software such as Microsoft® Project can be used to track tasks completed against the project plan. An important part of project planning is to carry out a risk analysis to identify potential risks that could impact the successful completion of the project.

The team can brainstorm potential risks to the project. They can also assess the probability that each risk would occur on a scale of high, medium, or low occurrence. The impact of the risk should also be assessed, i.e., if the risk were to occur, what level of impact would it have on the successful completion of the project (high, medium, or low)? It is also important to develop a risk mitigation strategy that identifies that if the risk occurs, how will the team mitigate the impact of the risk to reduce or eliminate the impact of the risk? Figure 2.11 shows a simple risk matrix.

Another tool that is useful while planning and managing the project is an item for resolution (IFR) form. This helps the team to document and track items that need to be resolved. It enables the team to complete the planned agendas in meetings, by allowing a place to "park" items that arise that cannot be resolved in the meeting due to time constraints, or lack of data or access to appropriate decision makers. Figure 2.12 shows an IFR form and includes a description of the item to be resolved. A priority (high, medium, low) should be assigned to each item. The status of the item, open (newly opened), closed (resolved), or hold (on hold—not being actively worked on), should be identified. The owner who is responsible for resolving the issue, as well as the dates that the item was opened and resolved, should be completed on the IFR form. A description of the resolution should also be included. This helps the team keep track of key decisions and ensures that the items are resolved to the satisfaction of all team members. The log of IFRs can also be used during the lessons learned activity after the project is complete to identify where problems arose and how they were resolved, so that these items can be incorporated into the risk mitigation strategies for follow-on projects.

Another helpful tool that should be developed in the Define phase, but should be used throughout the Lean Six Sigma project, is a communication plan. The communication plan can be used to identify strategies for how the team will communicate with all key stakeholders. It can be useful to help overcome resistance to change by planning how frequently and the manner in which the team will communicate with the stakeholders. Each key stakeholder or audience of a communicated message should be identified. The objectives or messages that will be communicated are then developed. The media or mechanism of how to communicate with the audience is then identified (e.g., face-to-face, email, websites). The frequency of the communication is important, especially for those more resistant to change, because they have more frequent communication. The last element of the communication plan is to clearly identify who is responsible for developing and delivering the communication to the audience. A communication plan is shown in Figure 2.13.

Task name	Duration	Start date	End date	Resources	Predecessor
Define					
1. Develop project charter					
2. Identify stakeholders					1
3. Perform initial VOC and identify CTS					2
4. Select team and launch the project					3
5. Create project plan					4
Measure					
6. Define the current process					5
7. Define the detailed VOC					6
8. Define the VOP and current performance					7
9. Validate measurement system					8
10. Define COPQ and Cost/benefit					9
Analyze					
11. Develop cause and effect relationships					10
12. Determine and validate root causes					11
13. Develop process capability					12
Improve					
14. Identify breakthrough & slect practical approaches					13
15. Perform cost/benefit analysis					14
16. Design future state					15
17. Establish performance targets, project scorecard					16
18. Gain approval to implement, and implement					17
19. Train and execute					18
Control					
20. Measure results & manage change					19
21. Report scorecard data and create process control plan					20
22. Apply P–D–C–A process					21
23. Identify replication opportunities					22
24. Develop future plans					23

FIGURE 2.10 DMAIC project plan.

Potential risks	Probability of risk occurring (High/ Medium/Low)	Impact of risk (High/Medium/Low)	Risk mitigation strategy

FIGURE 2.11 Risk matrix.

ITEMS FOR RESOLUTION

#	Issue	Priority	Status	Owner	Open date	Resolved date	Resolution

FIGURE 2.12 Item for Resolution (IFR) form.

Audience	Objectives/Message	Media/Mechanism	Frequency	Responsible

FIGURE 2.13 Communication plan.

Much of the work of the team is performed within meetings. It is crucial to effectively manage meetings during the Lean Six Sigma project work. Following are some tips for effective team meetings.

Team meeting management:
Some best practices for team meeting management are:

- Respect people and their time
- Determine critical/required participants for emails, meetings, and decisions
- Cancel or schedule meetings ahead of time
- Always create a meeting agenda and send it out in advance of the meeting. The agenda should include required and optional participants
- Recap action items and meeting minutes
- Use voting in emails to make easy decisions, or agree upon a meeting time
- Track meeting attendance, and resolve habitual lack of attendance

The planned meeting agenda should include the following (Scholtes, Joiner, and Streibel 2003):

1. Date, time, and proposed length of the meeting
2. Name of meeting facilitator

3. Meeting location
4. Required and optional attendees
5. Purpose of the meeting
6. Desired outcomes
7. Topic with time and proposed outcome for each topic

Some tips that the meeting facilitator can use to keep the meeting productive are (Scholtes, Joiner, and Streibel 2003):

- Listen and restate what you think you heard
- Ask for clarification and examples
- Encourage equal participation, circle the group
- Summarize ideas and discussion
- Corral digressions, get back to the agenda
- Close the discussion

Summary

The Define phase is a critical phase of the project. It is important to spend ample time in the Define phase developing the project charter and getting the buy-in of the project champion, the team members, and all stakeholders. The time spent clearly defining the scope of the project will reap dividends by reducing issues during the remaining phases of the project. A process or a problem poorly defined will require the team to revisit the Define phase when improvement efforts bog down or lose focus in subsequent phases.

PHASE II: MEASURE

The purpose of the Measure phase is to understand and document the current state of the processes to be improved, collect the detailed VOC information, baseline the current state, and validate the measurement system. The activities done and tools applied during the Measure phase are as follows:

6. Define the current process
7. Define the detailed VOC
8. Define the voice of the process (VOP) and current performance
9. Validate the measurement system
10. Define the COPQ

Figure 2.14 shows the main activities mapped to the tools or deliverables most typically used during that step.

6. Define the Current Process

The first step of the Measure phase is to profile the current state. SIPOC and process mapping are excellent tools to document the current process steps, the information that is used, the people who perform the work, and the internal and external

	Measure activities	Tools/Deliverables
6	Define the current process	▪ Process map ▪ Operational definitions ▪ Metrics ▪ Baseline ▪ Data collection plan
7	Define the detailed Voice of the Customer (VOC)	▪ Surveys, interviews, focus groups ▪ Affinity diagram ▪ Quality function deployment
8	Define the Voice of the Process (VOP) and current performance	▪ Pareto charts ▪ VOP matrix ▪ Benchmarking, check sheets, histograms ▪ Statistics
9	Validate the measurement system	▪ Measurement system validation ▪ Gage R&R (Repeatability & Reproducibility)
10	Define the Cost of Poor Quality (COPQ)	▪ Cost of Poor Quality

FIGURE 2.14 Measure phase activities and tools/deliverables.

Level	Type/Name	Purpose
Level 1	Macro or high level	Scope the improvement project Provide project and process boundaries Provide a high-level view of the process
Level 2	Process map	Identify process improvement areas Identify process inefficiencies Identify waste
Level 3	Process map or process flow chart	Identify improvement area Identify value vs. nonvalue-added activities Provide detailed how-to (almost procedural level)

FIGURE 2.15 Process map level and purpose.

customers of the services. In a process improvement effort there are typically three levels of process maps that are used to help with documenting the current or AS-IS process. Figure 2.15 shows the three levels and where they should be applied. An example of a level-2 process map for making a peanut butter and jelly sandwich is shown in Figure 2.16.

It is also important to identify process measures and related metrics that are used to measure the quality and productivity of the processes. The current profile of the people and cultural state should be understood, including the level of skills and training of employees, as well as their levels of resistance or acceptance to change.

The steps to completing a process map are:

1. Identify level (1, 2 or 3) to map and document
2. Define the process boundaries
3. Identify the major activities within the process

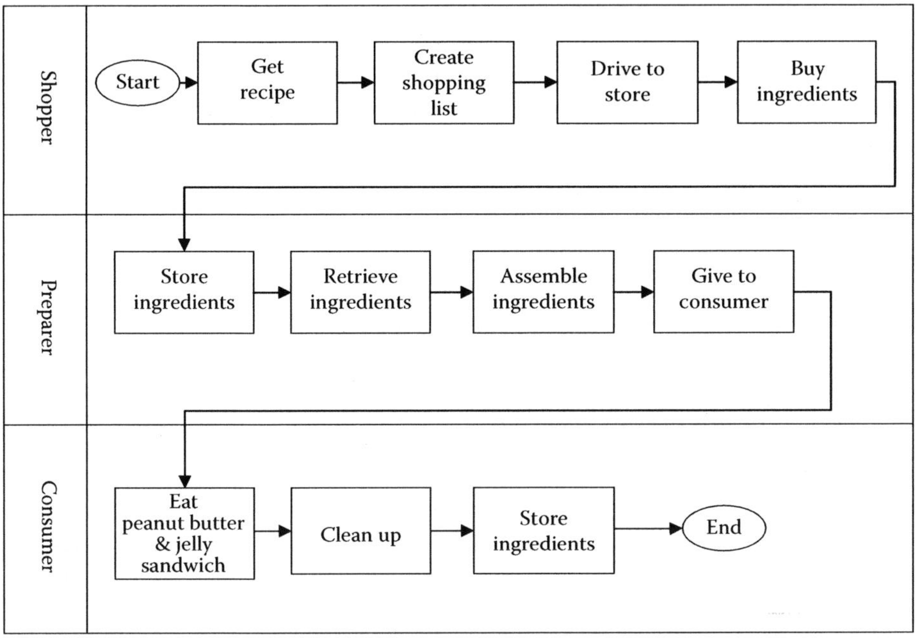

FIGURE 2.16 Process map for making a peanut butter and jelly sandwich.

4. Identify the process steps and uncover complexities using brainstorming and storyboarding techniques
5. Arrange the steps in a time sequence and differentiate operations by symbol
6. Validate the Process map by a "walkthrough" of the actual process and by having other process experts review it for consistency

7. Define Detailed VOC Information

In the Measure phase, the VOC information should be collected to define the customers' expectations and requirements with respect to the service delivery process. VOC is an expression for listening to external customers and understanding their requirements for your product or service. Examples of requirements are their expectations for responsiveness, such as turnaround time on vendor (customer) invoices, or error rates, such as employee (customer) expectations of no errors on their paycheck. The VOC can be captured through interviewing, surveys, focus groups with the customers, complaint cards, warranty information, and competitive shopping. Quality function deployment (QFD) can be used to organize the VOC information.

Personal interviews are an effective way to gain the VOC, but it can be expensive and training of interviewers is important to avoid interviewer bias. However, additional questioning can occur to eliminate misunderstanding. The objectives of the interview should be clearly defined before the interviews are held.

Customer surveys are a typical way to collect VOC data. The response rate on surveys tends to be low; 20% is a "good" response rate. It can also be extremely difficult to develop a survey that avoids and asks the questions that are desired. Customer survey collection can be quite expensive. The steps to create a customer survey are as follows (Malone 2005):

1. Conceptualization: Identify the survey objective and develop the concept of the survey, and what questions you are trying to answer from the survey.
2. Construction: Develop the survey questions. A focus group can be used to develop and/or test the questions to see if they are easily understood.
3. Pilot (try out): Pilot the questions by having a focus group of representative people from your population. You would have them review the questions, identify any unclear or confusing questions, and tell you what they think that the questions are asking. You would not use the data collected during the pilot in the actual results of the surveys.
4. Item analysis: Item analysis provides a statistical analysis to determine which questions answer the same objectives, as a way to reduce the number of questions. It is important to minimize the number of questions and the total time required to take the survey. Typically, the survey time should be 10 minutes or less.
5. Revision: Revise the survey questions and roll out the customer survey, or pilot again if necessary

Focus groups are an effective way to collect VOC data. A small representative group, typically 7–10 people, are brought together and asked to respond to predetermined questions. The focus group objective should be developed and the questions should support the objective. The participants should be selected by a common set of characteristics. The goal of a focus group is to gather a common set of themes related to the focus group objective. There is no set sample size for focus groups. Multiple focus groups are typically run until no additional themes are derived. Advantages of focus groups are (Pyzdek 2003):

- They tend to have good face validity (i.e., responses are in the words of the focus group participants)
- Typically more comments are derived than in an interview with one person at a time
- The facilitator can probe for additional information and clarification
- Information is obtained relatively inexpensively

Some of the disadvantages of focus groups are (Pyzdek 2003):

- The facilitator skills dictate the quality of the responses
- They can be difficult to schedule
- It can be difficult to analyze the dialogue due to participant interactions

Affinity diagrams organize interview, survey, and focus group data after collection. The affinity diagram organizes the data into themes or categories (Scholtes, Joiner, and Streibel 2003). The themes can first be generated, and then the data organized into the themes, or the detailed data can be grouped into the themes. An example of a simple affinity diagram for ways to study for a Six Sigma Black Belt exam is shown in Figure 2.17.

A data collection plan should be developed to identify the data to be collected that relate to the CTS criteria.

The data collection plan ensures:

- Measurement of CTS metrics
- Identification of the right mechanisms to carry out data collection
- Collection and analysis of data
- Definition of how and who is responsible to collect the data

Figure 2.18 shows a data collection plan.

The steps for creating a data collection plan in the Measure phase are

1. Define the CTS criteria
2. Develop metrics
3. Identify data collection mechanism(s)

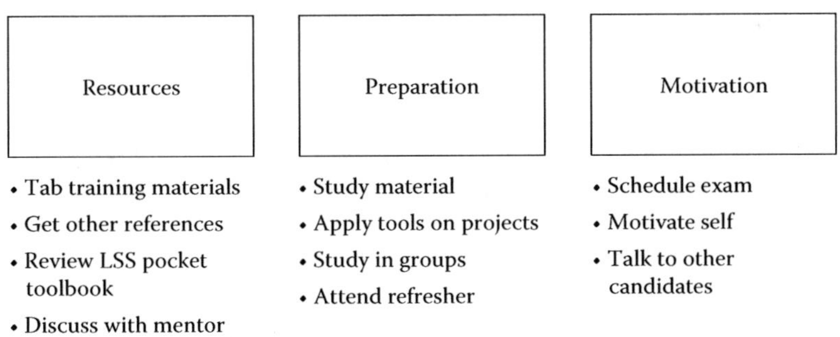

FIGURE 2.17 Affinity diagram for Six Sigma Black Belt exam preparation.

Critical to Satisfaction (CTS)	Metric	Data collection mechanism (survey, interview, focus group, etc.)	Analysis mechanism (statistics, statistical tests, etc.)	Sampling plan (sample size, sample frequency)	Sampling instructions (who, where, when, how)
Speed to market	Cycle time	Project management tool	Statistics (mean, variance); *t*-test	One year of projects	Collect data from project management system for last year
	Functionality delivered	Requirements traceability tool	Count	50 projects (30 development, 20 support)	Extract data based on sampling plan

FIGURE 2.18 Data collection plan for software application development Six Sigma Project.

4. Identify analysis mechanism(s)
5. Develop sampling plans
6. Develop sampling instructions

A description of each step in the development of data collection plan is given in the following:

1. *Define the CTS criteria*: (George, Rowlands, Price, and Maxey 2005): CTS is a characteristic of a product or service that fulfills a critical customer requirement or a customer process requirement. CTS measures are the basic elements in driving process measurement, improvement, and control.
2. *Develop metrics*: In this step, metrics are identified that help to measure and assess improvements related to the identified CTS measures. Some rules-of-thumb for selecting metrics are to (Evans and Lindsey 2007):

    - Consider the vital few vs. the trivial many
    - Focus on the past, present, and future
    - Link metrics to meet the needs of shareholders, customers, and employees

It is vital to develop an operational definition for each metric, so it is clearly understood how the data will be collected by anyone who collects it. The operational definition should include a clear description of a measurement, including the process of collection. Include the purpose and metric measurement. It should identify what to measure, how to measure it, and how the consistency of the measure will be ensured. A summary of an operational definition is given in the following section.

Operational Definition

Defining the Measure: Definition

A clear, concise description of a measurement and the process by which it is to be collected (George, Rowlands, Price, and Maxey 2005).

1. Purpose: Provides the meaning of the operational definition, to provide a common understanding of how it will be measured.
2. Clear way to measure the process

    - Identifies what to measure
    - Identifies how to measure
    - Makes sure the measuring is consistent

3. Identify data collection mechanism(s): Next you can identify how you will collect the data for the metrics. Data collection mechanisms include customer surveys, observation, work sampling, time studies, customer complaint data, emails, websites, and focus groups.
4. Identify analysis mechanism(s): Before collecting data, consider how you will analyze the data to ensure that you collect the data in a manner that enables the analysis. Analysis mechanisms can include the types of statistical

tests or graphical analysis that will be performed. The analysis mechanisms can dictate the factors and levels for which you may collect the data.

5. Develop sampling plans: You should determine how you will sample the data, and the sample size for your samples. Several types of sampling are (Gitlow and Levine 2005):
    - Simple random sample: Each unit has an equal chance of being sampled.
    - Stratified sample: The *N* (population size) items are divided into subpopulations or strata, and then a simple random sample is taken from each stratum. This is used to decrease the sample size and cost of sampling.
    - Systematic sample: *N* (population size) items are placed into *k* groups. The first item is chosen at random, the rest of the samples selected every *k*th item.
    - Cluster sample: *N* items are divided into clusters. This is used for wide geographic regions.
6. Develop sampling instructions: Clearly identify who will be sampled, where you will sample, and when and how you will take your sample data.

QFD and the house of quality are excellent tools to help to translate the customer requirements from the VOC into the technical requirements of your product, process, or service. They can also be used to relate the customer requirements to potential improvement recommendations developed during the Improve phase. Figure 2.19 shows the format for the house of quality.

The steps for creating a house of quality are (Evans and Lindsey 2007).

1. Define the customer requirements or CTS characteristics from VOC data. The customer can provide an importance rating for each CTS.
2. Develop the technical requirements with the organization's design team.

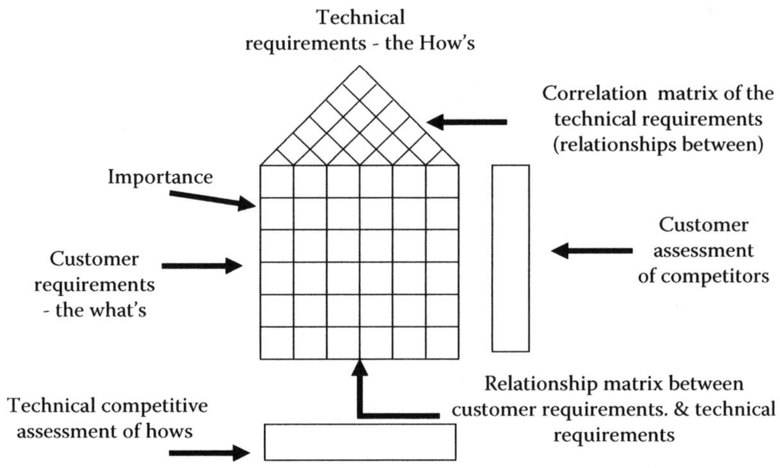

FIGURE 2.19 Quality function deployment house of quality.

3. Perform a competitive analysis, having the customers rank your product, process, or service against each CTS to each of your competitors.
4. Develop the relationship correlation matrix by identifying the strength of the relationship between each CTS and each technical requirement. Typically a numerical scale of 9 (high strength of relationship), 3 (medium strength of relationship), 1 (low strength of relationship), and blank (no relationship) is used.
5. Develop the trade-offs or relationships between the technical requirements in the roof of the house of quality. You can identify a positive (+) relationship between the technical requirements—as one requirement increases the other also increases; no relationship (blank); or a negative (–) relationship—there is an inverse relationship between the two technical requirements. An example of a positive relationship can be illustrated in the design of a fishing pole. The line gauge and tensile strength both increase as the other increases. A negative relationship can be illustrated by line buoyancy and tensile strength. As the tensile strength of the line increases, the buoyancy will decrease.
6. The priorities of the technical requirements can be summarized by multiplying the importance weightings of the customer requirements by the strength of the relationships in the correlation matrix. This helps to identify which of the technical requirements should be incorporated into the design of the product, process, or service first.

8. Define the VOP and Current Performance

There are many tools that can be used to assess the VOP and current performance. We shall discuss the VOP matrix, Pareto charts, benchmarking, check sheets, and histograms.

VOP Matrix

A VOP matrix, developed by the author, can be used to achieve integration and synergy between the DMAIC phases and the critical components of the process to enhance problem solving. The VOP matrix includes the CTS, the related process factors that impact the CTS, the operational definition that describes how the CTS will be measured, the metric, and the target for the metric. An example of a VOP matrix for the inventory asset management process for a college in a university is shown in Figure 2.20 (Furterer 2004).

Pareto Chart

A Pareto chart helps to identify critical areas causing most of the problems. It provides a summary of the vital few rather than the trivial many. It demonstrates the Pareto principle that 80% of the problems are created by 20% of the causes, so that these root causes can be investigated in the Analyze phase. It helps us to arrange the problems in order of importance and focus on eliminating the problems in the order of highest frequency of occurrence.

Following are the steps for creating a Pareto chart:

Identify assets	Process factors	Operational definition	Metric	Target
Faculty/staff awareness of process	Procedures exist	Procedures exist and are auditable	Number of departments with procedures	100% of departments have procedures by Jan. 1
	Training in procedures	All faculty will take 1 hour training session within 3 months of hire	Number of faculty trained	100% of faculty are trained within 3 months of hire or Jan. 1
Documented location of assets	Procedures exist	Procedures exist and are auditable	Number of departments with procedures	100% of departments have procedures by Jan. 1
	Training in procedures	All faculty will take 1 hour training session within 3 months of hire	Number of faculty trained	100% of faculty are trained within 3 months of hire or Jan. 1
Identify assets	Description on PO	All purchasers will input detailed description of asset on PO	Number of POs with detailed description	95% of POs sampled have detailed descriptions
	Description in system	PO description will transfer to asset management system	Number of asset descriptions in asset mgt.	95% of POs sampled have detailed descriptions
Efficiency of yearly scanning	Training	All property managers will be trained in process	Number property managers trained	100% of property managers trained within 3 months
	Process	Quality of process	Proportion of items found on first try	95% of items found on first scan

FIGURE 2.20 VOP matrix for inventory asset management process (Furterer 2004).

Step 1: Define the data categories, defects, or problem types
Step 2: Determine how the relative importance is defined (dollars, number of occurrences)
Step 3: Collect the data and compute the cumulative frequency of the data categories
Step 4: Plot a bar graph, showing the relative importance of each problem area in descending order. Identify the vital few to focus on

An example of a Pareto chart that identifies the resolution categories for problems reported to an information systems help desk for a financial application is shown in Figure 2.21.

Benchmarking

Benchmarking is a tool that provides a review of the best practices to be potentially applied to improve your processes. In a Six Sigma project, process benchmarking is typically carried out. The organization should document the process that they will

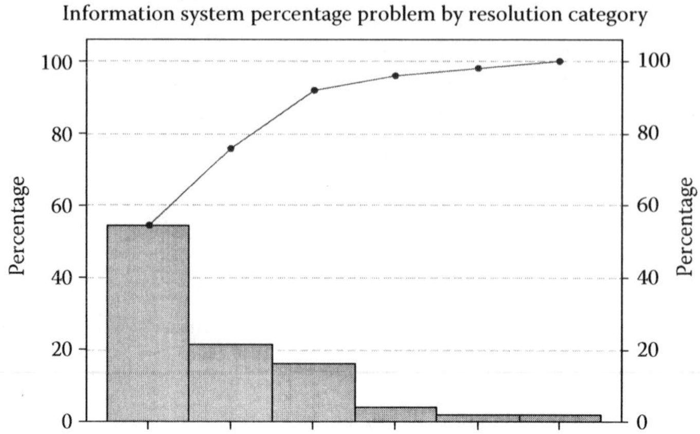

Resolution category	Training	Conversion	Software	Printer	Set up	User security
Percentage	54	21	16	4	2	2
Percent	54.5	21.2	16.2	4.0	2.0	2.0
Cum %	54.5	75.8	91.9	96.0	98.0	100.0

FIGURE 2.21 Pareto chart of resolution categories to an information systems help desk.

benchmark, then select who they will benchmark. It is not necessary to benchmark a company in the same industry, but to focus on the process to be benchmarked, and select an organization that is known for having world class or best practice processes. The next step is to work with the organization to collect the data and understand how the data can be used to identify ways to improve your processes and identify potential improvement opportunities to be implemented in the Improve phase. This is similar to the benchmarking processes of Motorola (Evans and Lindsey 2007). It is important to be careful when processing a benchmark to ensure that you are comparing apples with apples, i.e., the organization's characteristics are similar to your own, so that the benchmarked process applies to your process.

Check Sheet

A check sheet is a graphical tool that can be used to collect data on the process and the types of defects so that root causes can be analyzed in the Analyze phase. The steps to create a check sheet are:

Step 1: Choose a characteristic to track, i.e., defect types
Step 2: Set up the data collection check sheet
Step 3: Collect data using the check sheet

An example of a check sheet for potential errors when loading data for an on-line research system is shown in Figure 2.22.

A Pareto chart can then be created from the data collected on a check sheet.

Histogram

A histogram is a graphical tool that provides a picture of the centering, shape, and variance of the distribution of data. Minitab or Excel is commonly used to create a

Defect type	Tally	Total
Incorrect case name		5
Incorrect docket number		2
Incorrect court name		6
Incorrect cite segment		2
Incorrect decided date		7
Incorrect segment coding		2
Missing text		10
Incorrect primary embedded citations		2
Copyright material included		3

FIGURE 2.22 Check sheet for errors loading data.

histogram. It is important to graph the data in a histogram as the first step to understanding the data.

Statistics

Statistics can also be used to assess the VOP related to the metrics that are measured. Once the data are collected, they can be tested to see if the data distribution follows a normal distribution using a test for normality. The null hypothesis says that the data are normal. If the null hypothesis is not rejected, then the statistics that would describe the normal distribution are the mean and the standard deviation. The mean is the average of the sample data. The mean describes the central location of a normal distribution. The sample standard deviation is the square root of the sum of the differences between each data value and the mean divided by the sample size less one. Standard deviation (Sigma) is a measure of variation of the data; 99.997% of all data points within the normal distribution are within Six Sigma.

9. Validate the Measurement System

It is important to validate the measurement system to ensure that you are capturing the correct data and that the data reflect what is happening. It is also important to be able to assess a change in the process with our measuring system as well as the measurement system error. We must ensure that the measurement system is stable over time and that we are collecting the data that will allow us to make appropriate decisions.

Measurement Systems Analysis

A measurement systems analysis includes the following steps (Gitlow and Levine 2005):

1. Prepare a flow chart of the ideal measurement system
2. Prepare a flow chart of the current measurement system
3. Identify the gaps between the ideal and current measurement systems
4. Perform a gage repeatability & reproducibility (R&R) study

The measurement process variation is due to two main types of variations:

- Repeatability related to the gage
- Reproducibility related to the operator

The gage R&R study assesses repeatability and reproducibility. Minitab or other statistical software can be used to assess the measurement system error and improve the measurement system if necessary.

10. Define the COPQ

The last step in the Measure phase can be to assess the COPQ related to your Six Sigma project. The COPQ identifies the cost related to poor quality or not doing things right the first time. The COPQ translates defects, errors, and waste into the language of management (cost or dollars). There are four categories of COPQ: (1) prevention; (2) appraisal; (3) internal failures; and (4) external failures.

Prevention costs are all the costs expended to prevent errors from being made or the costs involved in helping the employee do the job correctly every time. Appraisal costs are the results of evaluating already completed output and auditing the process to measure conformance to established criteria and procedures. Internal failure cost is defined as the cost incurred by the company as a result of errors detected before the output is accepted by the company's customer. External failure cost is incurred by the producer because the external customer is supplied with an unacceptable product or service (Harrington Group 2004).

Examples of prevention costs are:

- Methods improvements
- Training
- Planning for improvement
- Procedures
- Quality improvement projects
- Quality reporting
- Data gathering and analysis
- Preventive maintenance
- SPC training costs
- ISO 9000 training costs

Examples of appraisal costs are:

- Inspections
- Process audits (SPC, ISO)
- Testing activity and equipment depreciation allowances
- Product audits and reviews
- Receiving inspections and testing
- Reviews (meeting time)
- Data collection
- Outside endorsements and certifications

Examples of internal failure costs are:

- Reaudit, retest, and rework
- Defects and their impact
- Unscheduled lost time
- Unscheduled overtime
- Excess inventory
- Obsolescence
- Scrap
- White-collar mistakes

Examples of external failures are:

- Warranty
- Technical support
- Customer complaints
- Customer bad-will costs
- Customer appeasement costs
- Lost business (margin only) due to poor quality
- Product liability
- Return/refunds
- White-collar mistakes

The COPQ can help to identify potential categories of waste embedded in the process.

PHASE III: ANALYZE

The purpose of the Analyze phase is to analyze the data collected related to the VOC and the VOP to identify the root causes of the process problems, and to develop the capability of the process. The activities performed and tools applied during the Analyze phase are as follows.

11. Develop cause and effect relationships
12. Determine and validate root causes
13. Develop process capability

Figure 2.23 shows the main activities mapped to the tools or deliverables most typically used during this step.

11. Develop Cause and Effect Relationships

There are several tools that can be used to generate the root causes of the problems identified in the Measure phase.

Cause and Effect Diagram

The cause and effect diagram can be used to brainstorm and document the root causes of an effect or problem. It is helpful to group the causes into categories, or

	Analyze activities	Deliverables
11.	Develop cause and effect relationships	Cause and effect diagrams Cause and effect matrix Why-why diagram
12.	Determine and validate root causes	Process analysis, histograms, and graphical analysis, waste elimination and summary of wastes, 5S, kaizen, FMEA, correlation analysis, regression analysis, basic statistics, confidence intervals, hypothesis testing, ANOVA, survey analysis.
13.	Develop process capability	DPPM/DPMO, process capability

FIGURE 2.23 Analyze phase activities and tools/deliverables.

use categories to brainstorm the causes. Typical categories are people, machine, materials, methods, measurements, and environment. Transactional categories are places, procedures, policies, people, and information systems. The steps for creating a cause and effect diagram are:

1. Define problem
2. Brainstorm all possible types of causes
3. Brainstorm and organize causes by groups: people, machines, materials, methods, measurement, and environment. Can also add information systems
4. Brainstorm/identify subcauses for each main cause

An example cause and effect diagram is shown in Figure 2.24.

Cause and Effect Matrix

The cause and effect matrix (George, Rowlands, Price, and Maxey 2003) can be used to understand if the same root causes contribute to multiple effects. It is helpful to use the cause and effect matrix if you have multiple CTS characteristics or effects. The matrix establishes the relationship Y = F(X), where Y equals the output variables, and X represents the input/process variables or root causes. To create the cause and effect matrix, brainstorm the potential causes for the multiple CTS measures or problems. The cause and effect matrix helps to relate the CTS measures or output variables (Ys) to the process or input variables (Xs). The team can rate the strength of the relationship between the CTS (effects) and the causes. A scale of 9 can be used for a high relationship, 3 for a medium relationship, 1 for a low relationship, and blank as no relationship. The customer should rate the importance of each CTS on a scale of 1 to 10, with 10 being the highest importance. This importance can be multiplied by the relationship number to gain a total priority of the effects to understand where process improvement recommendations should be focused in the Improve phase. The relative weightings provide the order of importance, with 1 being the first effect to focus on related to the highest total score. A template for a cause and effect matrix is shown in Figure 2.25.

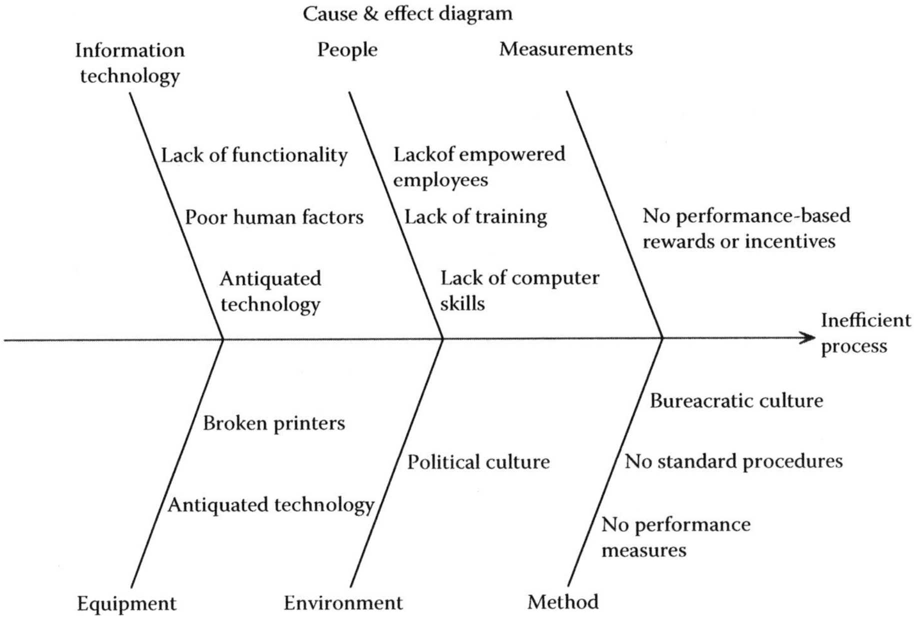

FIGURE 2.24 Cause and effect diagram.

	Effects					
	Y1	Y2	Y3	Y4	Total	Relative Weighting
Causes/Importance						
X1						
X2						
X3						
X4						
X5						

FIGURE 2.25 Cause and effect matrix template.

Why-Why Diagram

The Why-Why diagram is also a powerful tool to generate root causes. It uses the concept of the 5 "whys", where you ask the question "why?" several times until the root cause is revealed. Following are the steps to create a Why-Why diagram (Summers 2006):

1. Start on left with problem statement
2. State causes for the problem
3. State causes of each problem

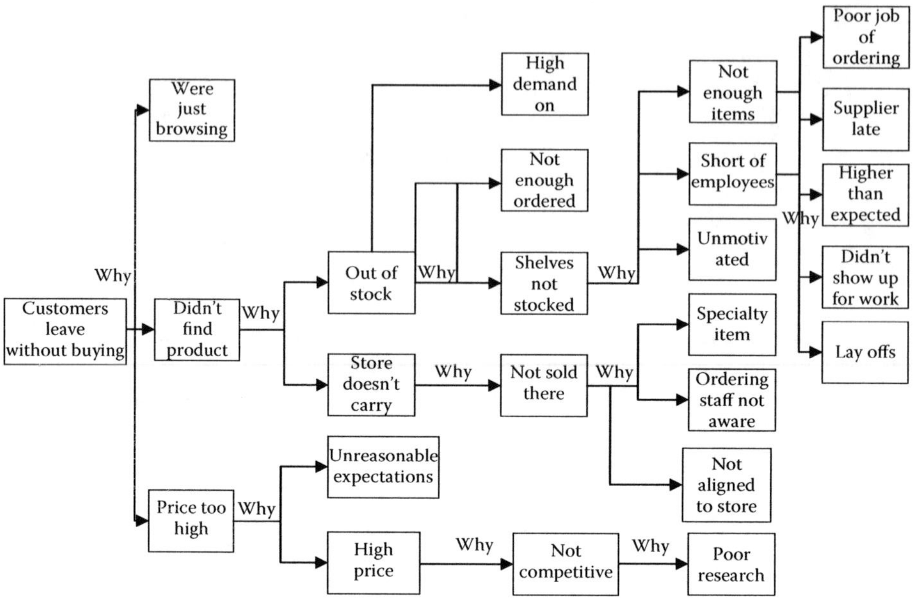

FIGURE 2.26 Why-Why diagram.

4. Keep asking "why?" five times
5. Try to substantiate the causes with data
6. Draw the diagram

Figure 2.26 shows a sample Why-Why diagram for why potential customers leave a store without making a purchase. It is critical that, once you brainstorm the potential root causes of the problems, you collect additional data to substantiate the causes.

12. Determine and Validate Root Causes

Process Analysis

To determine and validate root causes, the Six Sigma team can perform a process analysis coupled with waste elimination. A process analysis consists of the following steps:

1. Document the process (using process maps from the Measure phase)
2. Identify nonvalue-added activities and waste
3. Consider eliminating nonvalue-added activities and waste
4. Identify and validate (collect more data if necessary) root causes of nonvalue-added activities and waste
5. Begin generating improvement opportunities

Value-added activities are activities that the customer would pay for, that add value for the customer. Nonvalue-added activities are those that the customer would not want to pay for, or do not add value for the customer. Some are necessary (e.g., for legal, financial reporting, or documentation reasons) whereas others are unnecessary and should be reduced or eliminated. You can assess the percent of value-added activities as:

100 × (number of value-added activities/number of total activities)%

Value-added activities include operations that add value for the customer, whereas nonvalue-added activities include delays, storage of materials, movement of materials, and inspections. The number of total activities includes the value-added activities and the nonvalue-added activities.

You can also calculate the percent of value-added time as:

100 × (total time spent in value-added activities/total time for process)%

Typically, the percentage of value-added time is about 1–5%, with total nonvalue-added time equal to 95–99%.

During the process analysis, the team can focus on areas to identify inefficiencies in the following areas (*Process Flow Analysis Training Manual*):

- Can labor-intensive process be reduced, eliminated, or combined?
- Can delays be eliminated?
- Are all reviews and approvals necessary and value-added?
- Are decisions necessary?
- Why is rework required?
- Is all of the documentation, tracking, and reporting necessary?
- Are there duplicated process across the organization?
- What is slipping through the cracks and causing customer dissatisfaction?
- What activities require accessing multiple information systems
- Travel—look at the layout requiring the travel
- Is it necessary to store and retrieve all of that information, do we need that many copies?
- Are inspections necessary?
- Is the sequence of activities or flow logical?
- Are standardization, training, and documentation needed?
- Are all of the inputs and outputs of a process necessary?
- How are the data and information stored and used?
- Are systems slow?
- Are systems usable?
- Are systems user-friendly?
- Can you combine tasks?
- Is the responsible person at too high or too low of a level?

Waste Analysis

Waste analysis is a Lean tool that identifies waste into eight categories to help brainstorm and eliminate different types of wastes. The eight waste categories are all considered nonvalue-added activities and should be reduced or eliminated when possible. Waste is defined as anything that adds cost to the product without adding value. The eight wastes are:

- Transportation: Moving people, equipment, materials, and tools
- Over production: Producing more product or material than is necessary to satisfy the customers' orders (or faster than is needed)

- Motion: Unnecessary motion, usually at a micro or workplace level
- Defects: Errors in not making the product or delivering the service correctly the first-time
- Delay: Wait or delay for equipment or people
- Inventory: Storing product or materials
- Processing: Effort that adds no value to a product or service, incorporating requirements not requested by the customer
- People: Not using people's skills or mental, creative, and physical abilities

5S Analysis

5S is a Lean tool that helps to organize a workplace. The 5S are:

- *Simplify*: Clearly distinguish between what is necessary and what is unnecessary, disposing of the unnecessary. A red tag is used to identify items that should be reviewed for disposal
- *Straighten*: Organize the necessary items so they can be used and returned easily
- *Scrub*: Fix the root cause of the dirt or disorganization
- *Stabilize*: Maintain and improve the standards of the first three S's
- *Sustain*: Achieving the discipline or habit of properly maintaining the correct 5S procedures

Kaizen

Kaizen is a Lean tool that stands for "kai" (means "change") and "zen" (means "for the good"). It represents the continuous incremental improvement of an activity to constantly create more value for the customer by eliminating waste. A kaizen consists of short-term activities that focus on redesigning a particular process. A kaizen event can be incorporated into the Analyze or Improve phase of the Six Sigma project to help design and/or implement a focused improvement recommendation.

The kaizen event follows the Plan–Do–Check–Act (PDCA) cycle, including the following steps (George, Rowlands, Price, and Maxey 2003):

Plan:

1. Identify need: Determine the purpose of the kaizen
2. Form kaizen team: Typically 6–8 team members
3. Develop kaizen objectives: To document the scope of the project. The objectives should be Specific, Measurable, Attainable, Realistic, and Time-based (SMART)
4. Collect current state baseline data: From the Measure phase or additional data as needed
5. Develop schedule and kaizen event agenda: Typically one week or less

Do:

6. Hold kaizen event using DMAIC
   Sample kaizen event agenda:
   - Review kaizen event agenda
   - Review kaizen objectives and approach

- Develop kaizen event ground rules with team
- Present baseline measure and background information
- Hold event:
    - Define: Problem (derived from objectives), agree on scope for the event
    - Measure: Review measure baseline collected
    - Analyze: Identify root causes, wastes, and inefficiencies
    - Improve: Create action item list and improvement recommendations
    - Control: Create standard operating procedures to document and sustain improvements. Prepare summary report and present to sponsor
- Identify and assign action items
- Document findings and results
- Discuss next steps and close meeting

7. Implement: Implement recommendations, fine tune, and train.
   Check/act:

8. Summarize: Summarize results
   Kaizen summary report items:
   - Team members
   - Project scope
   - Project goals
   - Before kaizen description
   - Pictures (with captions)
   - Key kaizen breakthroughs
   - After kaizen description
   - Results
   - Summary
   - Lessons learned
   - Kaizen report card with follow-up date

9. Control: If targets are met, standardize the process. If targets are not met, or the process is not stabilized, restart kaizen event PDCA cycle.

Failure Mode and Effect Analysis (FMEA)

A FMEA is a systemized group of activities intended to recognize and evaluate the potential failure of a product or process, identify actions that could eliminate or reduce the likelihood of the potential failure occurring, and document the entire process (Pyzdek 2005).

The FMEA process includes the following steps:

1. Document process, define functions
2. Identify potential failure modes
3. List effects of each failure mode and causes
4. Quantify effects: severity, occurrence, detection
5. Define controls

6. Calculate risk and loss
7. Prioritize failure modes
8. Take action
9. Assess results

A simple FMEA form is shown in Figure 2.27 (Pyzdek 2005; George, Rowlands, Price, and Maxey 2005).

The risk priority number (RPN) is calculated by multiplying the Severity times the Occurrence times the Detection. The Severity is estimated for the failure, and given a numerical rating on a scale of one (low severity) to ten (high severity). The Occurrence is given a numerical rating on a scale of 1 (low probability of occurrence) to 10 (high probability of occurrence). The detection scale is reversed, where a numerical rating is given on a scale of 1 (failure is easily detected) to 10 (failure is difficult to detect).

A Pareto chart can be created based on the RPN values to identify the potential failures with the highest RPN values. Recommendations should be developed for the highest value RPN failures to ensure that they are incorporated into the improvement recommendations in the Improve phase.

Correlation Analysis

Correlation analysis measures the linear relationship between two quantitative variables that provides the relationship Y = F(X), or the dependent variable (CTS) is a function of the independent variable(s). A correlation analysis can be run between any two variables, regardless of whether they are both independent, or one is dependent and the other is independent. The correlation coefficient (r) is a measure of the strength of the relationship between the two variables. The r value falls between +1.0 and –1.0. The +1.0 signifies a positive strong correlation or positive linear relationship between the two variables. As one variable increases, so does the other. For example, as the number of customers increases, the sales increase. The –1.0 signifies a negative strong correlation or negative linear relationship between two variables. As one variable increases, the other variable decreases. A general rule of thumb is that an r value of +.80 or greater, or –.80 or less, signifies a significant correlation between the two variables.

Process step	Potential failure mode	Potential effects of failure	SEVERITY	Potential causes of failure	OCCURRENCE	Current process controls	DETECTION	RPN	Recommended action

FIGURE 2.27 Failure mode and effect analysis form.

If we establish a correlation between x and y, that does not necessarily mean a variation in x caused a variation in y. There may be a third variable that is causing the variation that we have not accounted for. To conclude that there is a relationship between two variables does not mean that there is a cause and effect relationship.

Linear Regression Analysis

Linear regression analysis is a statistical tool that generates a prediction equation that allows us to relate independent factors to our response variable or dependent variable. We use the coefficient of determination, R^2, which allows us to identify the fit of the prediction equation. A simple linear regression relates a single x variable to a single y variable. A simple linear regression equation attempts to fit a line with an equation of $Y = a + b\,x$, where Y is the dependent or response variable, a is the Y-intercept value, and b is the slope of the line. A multiple linear regression relates multiple x values to a single y value. The linear regression requires at least one independent variable and at least one dependent variable. The R^2 value will be between 0 and 1.0. Ideally, the R^2 value should be greater than 0.64 to represent a significant model. This means that our model accounts for 64% of the variation in the output.

Confidence Intervals

Confidence interval estimation or confidence intervals provide a range where there is some desired level of probability that the true parameter value is contained within it. It helps us to determine how well the subgroup average approximates the population mean. Confidence intervals help us understand the parameters between which the true population parameter, the mean or standard deviation, lies. As the sample size increases, the confidence interval width decreases. As the confidence level increases, say from 95 to 99%, the confidence interval will increase in width.

Hypothesis Testing

The purpose of hypothesis testing is to:

- Determine if claims on process parameters are valid
- Understand the variables of interest, the CTS measures

The hypothesis test begins with a theory, claim, or assertion about a particular characteristic (CTS) of one or more populations or levels of the X (independent variable).

The null hypothesis is designated as H_0 (pronounced "H-O") and defined as there is no difference between a parameter and a specific value. The alternative hypothesis is designated as there is a difference between a parameter and a specific value. The null hypothesis is assumed to be true, unless proven otherwise. If you fail to reject the null hypothesis, it is not proof that the null hypothesis is true.

In hypothesis testing there are two types of errors. A type I error (alpha risk) is the risk of rejecting the null hypothesis when you should not. The probability of a type I error is referred to as alpha. A type II error (beta risk) is the risk of not rejecting the null hypothesis when you should. The probability of a type II error is referred to as beta.

When performing a hypothesis test, you select a level of significance. The level of significance is the probability of committing a type I error, and is typically .05 or .01. Figure 2.28 shows the type I and type II errors.

The steps for performing a hypothesis test are:

1. Formulate the null and alternative hypotheses
2. Choose the level of significance (alpha) and the sample size (n)
3. Determine the test statistic
4. Collect the data and compute the sample value of the test statistic
5. Run the hypothesis test in Minitab or some other statistical package
6. Make the decision. If the *p*-value is less than our significance level (alpha), reject the null hypothesis; if not, then there is no data to support rejecting the null hypothesis. Remember, if *p* is low, H_0 must go!

Some of the most common hypothesis tests are summarized in Figure 2.29.

Analysis of Variance (ANOVA)

ANOVA is another hypothesis test, that is used for when you are testing more than two variables of populations. Following are the steps for running ANOVA:

1. Formulate the null and alternative hypotheses
2. Choose the level of significance (alpha) and the sample size (n)

		Conclusion drawn	
		Do not reject H_0	Reject H_0
Actual or True State	H_0 True	Correct conclusion	Type I error
	H_0 False	Type II error	Correct conclusion

FIGURE 2.28 Type I and type II errors.

Test statistics	Number of variables	Test	Parameters
Mean	1	1 sample Z	Variance
Mean	1	1 sample *t*	Variance unknown
Mean	2	2 sample *t*	Variance unknown, assume equal variances
Mean	2	2 sample *t*	Variance unknown, do not assume equal variances
Mean	2	Paired *t*-test	Paired by subject (before and after)
Proportion	1	1 proportion	
Proportion	2	2 proportion	
Variance	1	1 variance (chi-square)	
Variance	2	Variance (F-test)	

FIGURE 2.29 Summary of hypothesis tests.

3. Collect the data
4. Check for normality of the data, using a test for normality
5. Check for equal variances using an F-test
6. Run the ANOVA in Minitab or another statistical package
7. Make the decision

Common types of ANOVA:

- One-way ANOVA: testing one variable at different levels, such as testing average grade-point averages for high-school students for different ethnicities.
- Two-way ANOVA: testing two variables at different levels, such as testing the average grade-point averages for high-school students for different ethnicities and by grade level.

Customer Survey Analysis

Most surveys are attribute or qualitative data, where you are asking the respondent to answer questions using some type of Likert scale, asking importance, the level of agreement, or, perhaps, the level of excellence (Malone 2005).

Ways to analyze survey data are:

1. Summarize the percentage or frequency of responses in each rating category using tables, histograms, or Pareto charts.
2. Perform attribute hypothesis testing using chi-square analysis.

Unlike hypothesis testing with variable data, with attribute data we are testing for dependence, not a difference, but you can think "makes a difference".

We formulate our hypothesis as (Malone 2005):

- H_0: "{factor A} is independent of {factor B}"
- H_a: "{factor A} is dependent on {factor B}"

In addition to the *p*-value, we use contingency tables to help understand where the dependencies (differences) exist.

The customer survey analysis steps include (Malone 2005):

1. State the practical problem
2. Formulate the hypotheses
3. Enter your data in Minitab or another statistical package
4. Run the chi-square test
5. Translate the statistical conclusion into practical terms

If *p*, the significance level, is low, then reject H_0 (if *p* is low, H-O must go). If you fail to reject the null hypothesis, H_0, that means that you fail to reject the hypothesis that the values are independent. If you reject H_0, that means that they are dependent, or that dependencies or differences exist.

13. Develop Process Capability

To develop the process capability, you can calculate the defects per million opportunities (DPMO) and related sigma level, or you can calculate the capability indices. We will discuss DPMO first.

DPMO

Six Sigma represents a stretch goal of six standard deviations from the process mean to the specification limits when the process is centered, but also allows for a 1.5 sigma shift toward either specification limit. This represents a quality level of 3.4 defects per million. This is represented in Figure 2.30. LSL represents the lower specification limit and USL represents the upper specification limit.

The greater the number of σ values, the smaller the variation (the tighter the distribution) around the average. DPMO provides a single measure to compare the performances of very different operations, giving an apples-to-apples, comparison, not apples-to-oranges. Figure 2.31 shows a Sigma-to-DPMO conversion.

DPMO is calculated as (Brassard and Ritter 2001):

$$\text{DPMO} = \frac{\text{Defects} \times 1{,}000{,}000}{\text{Units} \times \text{Opportunities}}$$

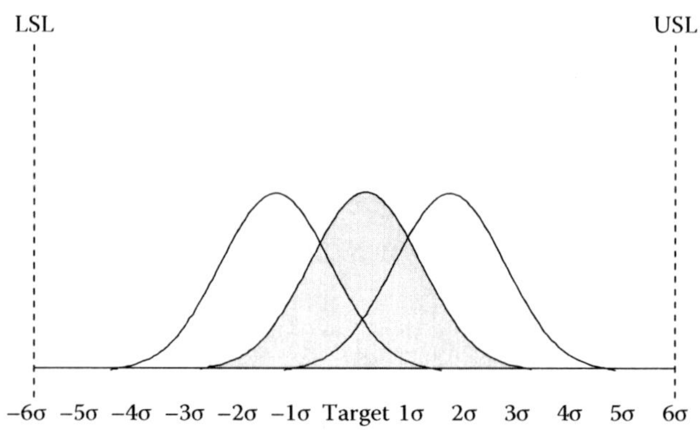

FIGURE 2.30 3.4 DPMO representing a Six Sigma quality level, allowing for a 1.5 sigma shift in the average.

Sigma level	DPMO
6 σ	3.4 DPMO
5 σ	233 DPMO
4 σ	6,210 DPMO
3 σ	66,810 DPMO
2 σ	308,770 DPMO
1 σ	691,462 DPMO

FIGURE 2.31 Sigma to DPMO-conversion (assuming 1.5 sigma shift).

where Defects is the number of defects in the sample, Units is the number of units in the sample, and Opportunities is the number of opportunities for error. For example, for taking a sample of 100 purchase orders, finding 5 defects, with 30 fields on the purchase order (opportunities for errors), we calculate a DPMO of 1667 or about 4.4 sigma.

Process Capability Study

Process capability is the ability of a process to produce products or provide services capable of meeting the specifications set by the customer or designer. You should conduct a process capability study only when the process is in a state of statistical control. Process capability is based on the performance of individual products or services against specifications. According to the central limit theorem, the spread or variation of the individual values will be greater than the spread of the averages of the values. Average values smooth out the highs and lows associated with individuals.

The steps for performing a process capability study are

1. Define the metric or quality characteristic. Perform your process capability study for the metrics that measure your CTS characteristics defined in the Define and Measure phases
2. Collect data on the process for the metric; take 25–50 samples
3. Perform a graphical analysis (histogram)
4. Perform a test for normality
5. Determine if the process is in control and stable, using control charts. When the process is stable, continue to step 6
6. Estimate the process mean and standard deviation
7. Calculate the capability indices, usually Cp, Cpk (Summers 2006):

$$\text{Cp} = \frac{\text{Upper specification limit} - \text{Process mean}}{6\sigma}$$

$$\text{Cpk} = \text{Minimum of } \{\text{CPU, CPL}\}$$

where:

$$\text{CPU} = \frac{\text{Upper specification limit} - \text{Process mean}}{3\sigma}$$

$$\text{CPL} = \frac{\text{Process mean} - \text{Lower specification limit}}{3\sigma}$$

A process can be in control but may not necessarily meet the specifications established by the customer or engineering. You can be in control and not capable. You can be out of control or unstable but still meet specifications. There is no relationship

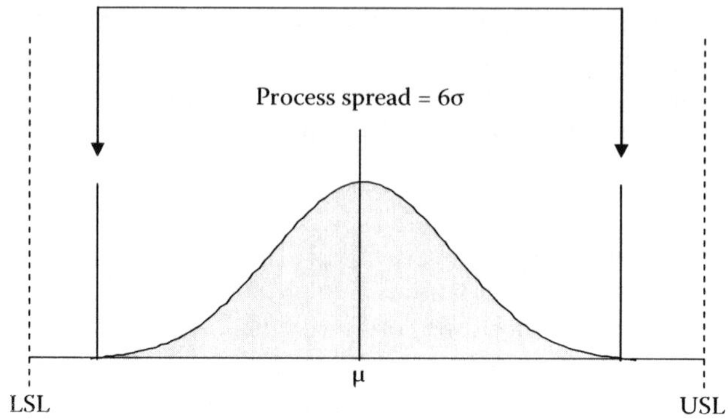

FIGURE 2.32 Process is quite capable.

between control limits and specification limits. However, you must be in control before you use the estimates of standard deviation from your process to calculate process capability and your capability indices.

There are typically three scenarios regarding process capability (Summers 2006):

1. Process spread is less than the specification spread. The process is quite capable. Figure 2.32 shows this scenario. Cp and Cpk are >1.33.
2. Process spread is equal to the specification spread, an acceptable situation, but there is no room for error. If the mean shifts, or the variation increases, there will be a nonconforming product. Figure 2.33 shows this scenario. where Cp = Cpk = 1.
3. Process spread is greater than the specification spread. The process is NOT capable. Figure 2.34 shows this scenario. where Cp and Cpk are <1.

PHASE IV: IMPROVE

The purpose of the Improve phase is to identify improvement recommendations, design the future state, implement pilot projects, train, and document the new processes. The activities performed and tools applied during the Improve phase are discussed below.

14. Identify improvement recommendations
15. Perform cost/benefit analysis
16. Design future state
17. Establish performance targets and project scorecard
18. Gain approval to implement, then implement
19. Train and execute

Figure 2.35 shows the main Improve activities mapped to the tools or deliverables most typically used during that step.

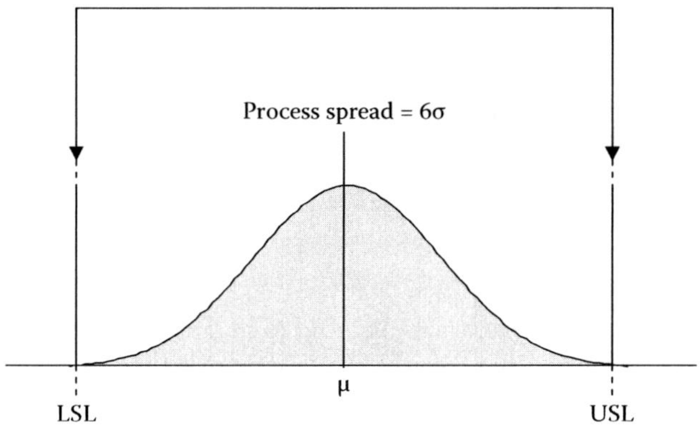

FIGURE 2.33 Process is just capable.

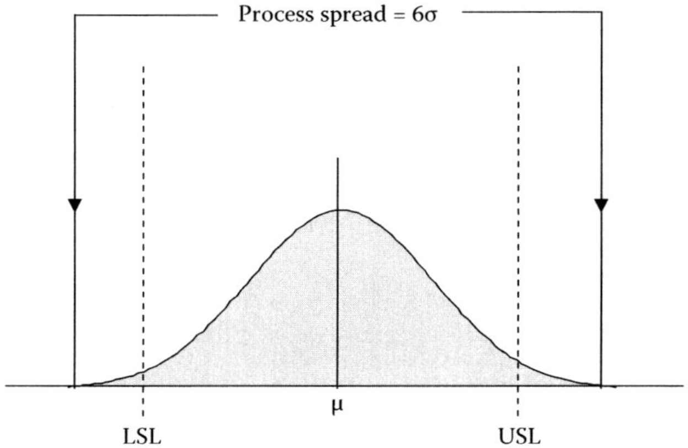

FIGURE 2.34 Process is not capable.

	Improve activities	Deliverables
14	Identify improvement recommendations	▪ Revised QFD ▪ Recommendations for improvement ▪ Action Plan
15	Perform cost/benefit analysis	▪ Cost/benefit analysis ▪ Cost of poor quality
16	Design future state	▪ Future state process map ▪ Future state Value Stream Map ▪ Design of Experiments
17	Establish performance targets and project scorecard	▪ Dashboards/scorecards ▪ Revised VOP Matrix
18	Gain approval to implement, and implement	▪ Project presentations
19	Train and execute	▪ Training plans, procedures

FIGURE 2.35 Improve phase activities and tools/deliverables.

14. Identify Improvement Recommendations

The Six Sigma team should use the data collected in the first three phases of the DMAIC to identify improvement recommendations. The recommendations should be designed to eliminate the root causes. The team should anticipate things that can go wrong or the type of resistance that you may have, and design plans to work around these barriers

QFD (Revised)

The QFD and the house of quality can be used to map the CTS to the improvement recommendations to ensure that the recommendations help to satisfy the CTS characteristics. The customer requirements on the left of the QFD matrix become the CTS measures, and the technical requirements become the improvements.

The house of quality can also be used to prioritize the recommendations, by weighting the CTS measures and multiplying the weighting by the strength of the relationship in the relationship matrix of the QFD matrix.

Recommendations for Improvement

The recommendations for improvement can be a list, with descriptions of the recommendation ideas. During the design of the recommendations, input should be obtained from the process owners to assess the reasonableness of the solutions. Information from the following previously used tools can be used to develop the improvement recommendations:

- QFD
- Waste analysis
- Process analysis
- Cause and effect analysis (diagram and matrix)
- Why-why diagram
- 5S
- Kaizen
- Process map or value stream map

Action Plan

The action plan for the improvement recommendations can be divided into short-term and long-term recommendations. Short-term is three months or less, and long-term recommendations are those that can be implemented in more than three months.

Action planning should include project plans, with resources, timelines, and risk analysis.

Change management is important when planning the improvement recommendations pilot projects. The following steps can be used to enhance the probability of success for the pilot projects:

1. Create a vision of the future state and communicate with the stakeholders
2. Understand what is in it for each stakeholder; include them in the solution
3. Identify who will be resistant to change, and who will be receptive. Gain some early adopters, and engage them in the change process

4. Implement the change, incorporating training and appropriate resources
5. Monitor improvement, measure and assess results
6. Implement new processes, systems, and organizational structures if needed

15. Perform Cost/Benefit Analysis

Cost/Benefit Analysis

You can perform a cost/benefit analysis or a COPQ assessment to determine the return on investment or savings that are estimated by implementing the improvement recommendations.

In a cost/benefit analysis, we need to decide whether the benefits of the project outweigh the costs. We can use the benefit/cost (B/C) ratio:

$$\text{B/C ratio} = \frac{\text{Present value of benefits}}{\text{Present value of costs}}$$

If the B/C ratio is >1, than you can accept the project; the benefits outweigh the costs.

COPQ

The COPQ categories can be used to assess the costs that can be eliminated by implementing the improvement recommendations. The team can use this in the cost/benefit analysis, or compare the improvement in the COPQ before and after the recommendations are implemented.

16. Design Future State

It is important to design the new future state by developing a future state process map. The team should challenge the boundaries and incorporate quality and Lean principles.

Future State Process Map

The future state process map is simply a process map of the new process incorporating the improvement recommendations.

Design of Experiments (DOE)

DOE can be used to identify key variables and levels that optimize process performance and improvement quality. It helps to design a robust process that is insensitive to uncontrollable factors. It allows you to look at many factors simultaneously and to assess the interaction of variables. It enables the identification of the critical factors and the associated levels for the process design.

The steps for performing a DOE are:

1. Set experimental objectives
2. Select process variables
3. Select an experimental design and identify hypotheses

4. Execute the design
5. Check that the data are consistent with the experimental assumptions
6. Analyze and interpret the results
7. Use/present the results, incorporate them into your future state design

There are many types of experimental designs, the most common are

- One-factor experiments: allow for the manipulation of one factor
- Two-factor experiments: allow for the manipulation of two factors
- Full factorial experiments: consist of all possible combinations of all selected levels of the factors to be investigated
- Fractional factorial experiments: study only a fraction or subset of all the possible combinations of factors

17. Establish Performance Targets and Project Scorecard

In this step, the team should identify the performance targets for the metrics identified in the Measure phase. They also should track pilot project status using project scorecards.

Dashboards/Scorecards

You should also create dashboards or scorecards to assess the performance of your process after trying out the improvement recommendations in the pilot projects.

The project scorecards or dashboards can be used to identify improvements in your process against the metrics that you have identified. The metrics should relate to your CTS characteristics.

Scorecards should include the following ways to present your metrics (Pyzdek 2003):

- Assessment of improvement of central tendency and variation overtime; SPC average and range charts can be used to meet this objective
- Graphical distribution using a histogram for the most recent time period
- Assessment of quality or number of defects, using SPC percent defective chart
- Outliers showing a distribution of individual defectives using a dot plot.

Revised VOP Matrix

You can revise your VOP matrix from the Measure phase to relate your CTS measures, process factors, operational definition, metric, and more realistic targets for your metrics.

18. Gain Approval to Implement, then Implement

The Six Sigma project team should create a presentation and deliver it to the project sponsor, champion, and other management that must approve the improvement recommendations. The champion presentation should be a high-level executive summary.

The team can use the PDCA cycle to implement the pilot recommendation projects. Plan: you can plan your improvements; Do: implement your improvements, usually on a pilot scale; Check: verify that the improvements improved the process based on your metrics; and Act: if the improvements made a positive difference, implement them on a broader scale; if not, refine the improvements and try again. You may go through the PDCA cycle several times.

19. Train and Execute

The team should develop detailed procedures as necessary to ensure consistency of the new process. They should develop and roll out training. The "train the trainer" concept is sometimes used to reduce the resources needed to train. A core group of people are trained on the new process and then they train others in the organization, and become subject matter experts. The process owners should be included in the change process, and changes should be communicated to the appropriate stakeholders. The team can use the future state process map as a training guide. They should assess the effectiveness of the training as part of the control plan in the next phase.

PHASE V: CONTROL

The purpose of the Control phase is to measure the results of the pilot projects, and manage the change on a broader scale; report scorecard data and the control plan; identify replication opportunities; and develop future plans for improvement. The activities performed and tools applied during the Control phase are discussed below.

20. Measure results and manage change
21. Report scorecard data and create process control plan
22. Apply the P–D–C–A process
23. Identify replication opportunities
24. Develop future plans

Figure 2.36 shows the main Control activities mapped to the tools or deliverables most typically used during that step.

	Control activities	**Deliverables**
20	Measure results and manage change	Hypothesis tests, design of experiments
21	Report scorecard data and create process control plan	Basic statistics, graphical analysis, sampling, mistake proofing, FMEA, control plan, process capability, DPMO; control charts
22	Apply P–D–C–A process	Replication opportunities
23	Identify replication opportunities	Standard work, kaizen
24	Develop future plans	Dashboards/scorecards, action plans

FIGURE 2.36 Control phase activities and tools/deliverables.

20. Measure Results and Manage Change

In the Control phase, the team should verify that training and implementation were carried out correctly. They need to collect and analyze data to ensure process performance and improvements were made. The teams must further manage the change for roll-out of the pilot recommendations on a broader scale. The team needs to keep all of the stakeholders in the loop by developing and implementing a communications plan. We will collect data after we improve the process, for the same CTS measures and metrics identified in the Measure phase. We will then assess if the changes implemented made a "statistically" significant difference, using

- Hypothesis testing

or

- DOE

21. Report Scorecard Data and Create Process Control Plan

In this step, the team should demonstrate the impact of the project's metrics, and create or revise the process control plan. The plan helps to deploy the Six Sigma approach across large areas and to coach groups through the major quality processes.

The purpose of the control plan is to maintain the gains. If a conscious plan and effort are not made to ensure that people continue to use the new process, the gains can slip, and when push comes to shove, and people get pressured and busy, they can very easily slip back to their old ways and old processes. The control plan can include (Pyzdek 2005):

- Deploying new policies, and removing outdated policies
- Implementing new standards
- Modifying procedures
- Modifying quality appraisal and audit criteria
- Updating prices and contract bid models
- Changing information systems
- Revising budgets
- Revising forecasts
- Modifying training

Useful tools that can be used to derive the information to create a control plan include:

- Project planning for creating the control plan
- Brainstorming
- FMEA
- SPC
- Process map
- Training
- Procedures

- Mistake proofing
- Statistics, graphical tools, sampling, FMEA, process capability, DPPM/DPMO

A control plan format is provided in Figure 2.37 (Brissard and Ritter 2001).

For each major process step on the future process map, the control plan should identify how you will control the process step (control mechanism), how you will measure the process step, how critical it is to ensure control for that step, actions to be taken if problems occur, and who is responsible for monitoring control for each process step.

Mistake Proofing

Mistake proofing is a tool that helps to prevent errors in your process. Errors are inadvertent, unintentional, accidental mistakes made by people because of the human sensitivity designed into our products and processes.

Mistake proofing (also called Poka Yoke) is the activity of awareness, detection, and prevention of errors that adversely affect our customers, and our people and result in waste.

Some of the underlying mistake-proofing concepts are:

- You should have to think to do it wrong, instead of right
- Easy-to-perform inspection at the source
- Reduces the need for rework and prevents further work (and cost) on a process step that is already defective
- Simplifies prevention and repair of defects by placing responsibility on the responsible worker

SPC Charts

SPC charts are an effective tool to monitor and control the process, and ensure that the process is not out of control. SPC control charts are a graphical tool for monitoring the activity of an ongoing process. The most commonly used control charts are also referred to as Shewhart control charts, because Walter A. Shewhart first proposed the general theory in the 1920s at AT&T Western Electric. Figure 2.38 identifies the most commonly used control charts.

Process step	Control mechanism	Measure/ metric	Criticality (High, Medium, Low)	Action to be taken if problems occur	Responsibility

FIGURE 2.37 Control plan.

Most common variables charts	Most common attributes charts
▪ X-bar and R-charts (average and range) ▪ X-bar and s-charts (average and standard deviation) ▪ X (individual) and moving range	▪ P-charts (proportion nonconforming) ▪ NP-charts (number nonconforming items) ▪ C-charts (number nonconformities) ▪ U-charts (number nonconforming per unit)

FIGURE 2.38 Commonly used control charts.

The following steps can be used to implement control charts:

1. Determine the type of chart, quality characteristic, sample size and frequency, and data collection mechanism
2. Select the rules for out of control conditions
3. Collect the data (10–25 subgroups)
4. Order data based on time order
5. Calculate the trial control limits and create charts (Minitab)
6. Identify out of control conditions
7. Remove points where you can assign causes
8. Recompute the control limits

22. Apply P–D–C–A Process

Apply the P–D–C–A to help people continually improve the process. There is the need to focus on:

- What are we trying to accomplish?
- How will we know that change is an improvement?
- What change can we make that will result in improvement?
- If the process is performing to plan, then standardize the activities; if not, then study why not and develop a new plan for improvement.
- Focus on the next most important root cause and implement additional improvements.

23. Identify Replication Opportunities

In this step, it is important to identify opportunities where you can replicate the same process in the organization. This will leverage the improvement effort across the organization, and potentially save additional money for the company. Identifying replication opportunities can help to support organizational learning.

24. Develop Future Plans

The purpose of developing future plans is to recognize the time and effort that went into the Lean Six Sigma project by reflecting on the lessons learned and incorporating these into future projects. Some important questions are:

- Have you identified lessons learned?
- Have you identified the next opportunity for improvement?
- Have you shared the learnings with others?
- Have you documented the new procedures?
- Has everyone been trained that needs to be?
- Have you taken the time to celebrate?

Dashboards and scorecards can be used to assess where you need to focus improvement efforts in the future. Also, the cause and effect analysis can be used to identify the next root cause to focus improvements on.

SUMMARY

This chapter provided a comprehensive overview of the DMAIC problem-solving approach along with key tools of each phase.

REFERENCES

Alukal, G., Create a lean, mean machine, *Quality Progress*, 36 (4), 29–36, 2003.

Brassard, M., and Ritter, D., *Sailing through Six Sigma*, Brassard and Ritter, LLC., Marietta, GA, 2001.

Dubai Quality Group, *The Birth of Lean Sigma*, The Manage Mentor, Dubai, 2003.

Evans, J. and Lindsay, M. *The Management and Control of Quality*, 5th ed., Thomson South-Western, Mason, OH, 2002.

Furterer, S. L., Critical quality skills of our future engineers, ASQ Conference on Quality in the Space and Defense Industries, Cape Canaveral, Florida, March 2006.

George, M., Rowlands, D., Price, M., Maxey, J., *Lean Six Sigma Pocket Toolbook*, McGraw-Hill, New York, 2005.

Gitlow, H., and Levine, D., *Six Sigma for Green Belts and Champions*. Prentice-Hall, Englewood Cliffs, NJ., 2005.

Harrington Group, Inc., IEMS Conference, Cocoa Beach, Florida, 2004.

Kandebo, S., Lean, Six Sigma yield dividends for C-130J, *Aviation Week & Space Technology*, July 12, 1999.

Malone, L., Class notes from a guest lecture, ESI 5227, University of Central Florida, Department of Industrial Engineering and Management System, Orlando, FL, 2005.

McIlroy, J. and Silverstein, D., Six Sigma deployment in one aerospace company. Six Sigma Forum website, www.sixsigmaforum.com, 2002.

Process Flow Analysis Training Manual, Control Data, 1982.

Pyzdek, T., *The Six Sigma Handbook*, McGraw-Hill, New York, 2003.

Sheridan, J., Lean Sigma synergy, *Industry Week*, October 16, 2000.

Summers, D. *Quality*, 4th ed., Pearson Prentice-Hall, Upper Saddle River, NJ, 2006.

U.S. Census Bureau website, North American Product Classification System. http://www.census.gov/eos/www/napcs/faqs.htm.

Womack, J. and Jones, D., *Lean Thinking: Banish Waste and Create Wealth in Your Corporation*. Simon & Schuster, New York, 1996.

Wal-Mart Global Continuous Improvement Training Course, Six Sigma Black Belt training, Bentonville, AR, 2008.

BIBLIOGRAPHY

Brue, G., *Six Sigma for Managers*, McGraw-Hill, New York, 2002.

Certified Six Sigma Black Belt Primer, Quality Council of Indiana, West Terre Haute, 2001.

Chowdhury, S., *Design for Six Sigma*, Dearborn Trade Publishing, Chicago, 2002.

Evans, J. and Lindsay, M., *An Introduction to Six Sigma of Process Improvement*, Thomson South-Western, Mason, OH, 2005.

Furterer, S., East Carolina University, Industrial Distribution and Logistics program, ITEC 4300, Quality Assurance course lecture notes, Greenville, NC, 2006.

Furterer, S., University of Central Florida, Department of Industrial Engineering and Management Systems, ESI 5227, Total Quality Improvement course lecture notes, Orlando, FL, 2005.

George, M., *Lean Six Sigma*, McGraw-Hill, NewYork, 2004.

Harry, M. and Schroeder, R. *Six Sigma: The Breakthrough Management Strategy Revolutionizing the World's Top Corporations*. Doubleday, New York, 1999.

Louis, R. *Integrating Kanban with MRP II: Automating a Pull System for Enhanced JIT Inventory Management*, Productivity Press, New York, 2005.

Malavé, C., Lecture notes, Texas A&M University, 2004.

Scholtes, P., Joiner, B., Streibel, B., *The Team Handbook*, 3rd ed., Oriel Incorporated, Madison, WI, 2003.

Summers, D. C. S., *Six Sigma Basic Tools and Techniques*, Prentice-Hall, Englewood Cliffs, NJ, 2007.

Wal-Mart Global Continuous Improvement training material, Six Sigma Black Belt Course, Bentonville, AR, 2007.

3 Design for Six Sigma Roadmap Overview

Sandra L. Furterer

CONTENTS

DFSS OVERVIEW

Design for Six Sigma (DFSS) is a methodology that can be used to systematically design new products, services, or processes. It embeds the underlying management philosophies, principles, and the Six Sigma stretch goal of the Six Sigma and DMAIC methodology. DFSS focuses on designing a product, service, or process correctly the first time, so less time needs to be spent downstream in improving the product, service, or process. We will discuss DFSS and the Identify, Define, Design, Optimize, Validate (IDDOV) methodology as it relates to service-oriented and transaction-based settings, instead of designing products.

Subir Chowdhury believes that Six Sigma can take an organization only so far, and that organizations must focus on designing good products and processes, so there is less need to improve them, which can prevent errors from occurring (Chowdhury 2005). From Deming's quality and profitability cycle, improved quality of design can lead to higher perceived value by the customer, which can contribute to increased market share, margins, revenue, and profitability (Deming 1986).

Unlike Lean Six Sigma, which typically uses the DMAIC problem-solving methodology, DFSS literature discusses applying many different methodologies to design the new products or services, such as Design, Measure, Analyze, Design, Validate (DMADV), IDDOV, Identify, Design, Optimize, Validate (IDOV), and Design, Measure, Analyze, Design, Optimize, Verify (DMADOV) (Proseanic et al. 2009).The author adapted the IDDOV methodology discussed by Chowdhury (2005), and developed the roadmap for applying DFSS using IDDOV to service-oriented processes.

The IDDOV process could be used when you are creating a brand new process that has never been done before in your organization, or to make a major redesign of an existing process. This existing process may be too broken to provide guidance for the redesign.

The benefits of applying Design for Six Sigma and IDDOV compared with Six Sigma and the DMAIC are that you are not constrained by an existing process, and you do not need to collect large amounts of VOP data, or spend time baselining a nonexistent or seriously broken process.

Figure 3.1 illustrates the steps that are part of each phase of the IDDOV methodology. Figure 3.2 maps the tools most typically used to each phase of the IDDOV methodology (Chowdhury 2005).

Following is a roadmap of how to apply IDDOV and the main tools that could be applied when designing a service process.

IDENTIFY

The purpose of the Identify phase is to define the business problem or opportunity, to scope the project by developing a project charter, and to identify the stakeholders impacted by the project. The main activities to be performed in the Identify phase are as follows:

1. Develop project charter
2. Perform stakeholder analysis
3. Develop project plan

Figure 3.3 shows the main activities mapped to the tools or deliverables most typically used during that step.

The tools applied in the Identify phase of the IDDOV are the same that are used in the DMAIC Define phase, and already discussed in Chapter 2. The team structure, with Black Belt and Master Black Belt mentors, project champions and sponsors, process owners, and working team members would be applied to the DFSS IDDOV similar to the Six Sigma DMAIC. The project charter elements would be similar, except the scope can be somewhat more difficult to define because we do not have an existing process to use as a scope, nor a process that can be documented

IDENTIFY	DEFINE	DESIGN	OPTIMIZE	VALIDATE
1. Develop project charter 2. Perform stakeholder analysis 3. Develop project plan	4. Collect VOC 5. Identify CTS measures and targets 6. Translate VOC into technical requirements	7. Identify process elements 8. Design process 9. Identify potential risks and inefficiencies	10. Implement pilot process 11. Assess process capabilities 12. Optimize design	13. Validate process 14. Assess performance, failure modes, and risks 15. Iterate design and finalize

FIGURE 3.1 DFSS IDDOV activities for service-oriented process design.

IDENTIFY	DEFINE	DESIGN	OPTIMIZE	VALIDATE
▪ Project charter ▪ Stakeholder analysis ▪ Project plan ▪ Risk matrix ▪ Responsibilities matrix ▪ Items for resolution (IFR) ▪ Ground rules ▪ Communication plan	▪ Critical to Satisfaction (CTS) ▪ Summary & targets ▪ Data collection plan ▪ VOC ▪ QFD ▪ Benchmarking ▪ Operational definitions ▪ Interviewing ▪ Focus groups ▪ Surveys ▪ Affinity diagram ▪ Market research ▪ SWOT ▪ VOP matrix	▪ Process element summary ▪ Process map ▪ Basic statistics ▪ Failure mode and effect Analysis ▪ Risk assessment ▪ Simulation ▪ Prototyping ▪ DOE ▪ Process analysis ▪ Multivoting ▪ Criteria-based matrix ▪ Pugh concept selection technique ▪ Waste analysis ▪ VOP matrix	▪ Process capability ▪ Simulation ▪ Implementation Plan ▪ Process map ▪ Communication plan ▪ Process analysis ▪ Waste analysis ▪ Cost/benefit analysis ▪ Statistical Process Control ▪ Training plans ▪ Procedures ▪ Mistake ▪ Proofing ▪ Design of experiment ▪ Pilot	▪ Prototyping ▪ Testing ▪ Pilot ▪ Mistake proofing ▪ Dashboards ▪ Scorecards ▪ Statistical process Control ▪ Statistical analysis ▪ Hypothesis tests ▪ ANOVA ▪ Design of Experiments ▪ Replication Opportunities

FIGURE 3.2 DFSS IDDOV tools and deliverables for service-oriented process design.

	Identify activities	Tools/Deliverables
1	Develop project charter	▪ Project charter ▪ Risk matrix
2	Perform Stakeholder Analysis	▪ Stakeholder analysis definition ▪ Stakeholder commitment scale ▪ Communication plan
3	Develop project plan	▪ Project plan ▪ Responsibilities matrix ▪ Items for resolution (IFR) ▪ Ground rules

FIGURE 3.3 Identify phase activities and tools/deliverables.

using a SIPOC, which helps to define the process, inputs, and outputs, as well as the stakeholders of the process. However, thinking through what the potential process steps would be and who would supply inputs and transform these into outputs, and who would receive those outputs would still be helpful to consider. A stakeholder analysis (including defining the stakeholders for the project and identifying their potential acceptance or resistance to change) would be performed. Project planning activities would include developing a detailed work breakdown structure (WBS) and project plan with roles, responsibilities, estimated durations, and prerequisite relationships of the activities. A responsibilities matrix identifies who is responsible for what during the project, and is an important part of the Identify phase to clearly set expectations of team members. The ground rules also help to clarify expectations of behavior and how the team will operate. A communication plan can help to clearly identify how the team will communicate and interact with the stakeholders. A risk plan, often part of the project charter, can be used to identify potential risks that could impede project progress, as well as identify mitigation and control strategies to avoid and control the risks should they occur. A sample project plan for the IDDOV methodology is shown in Figure 3.4.

DEFINE

The purpose of the Define phase is to understand the voice of the customer (VOC), what is important to the customers as defined by the critical to satisfaction (CTS)

Activity number	Phase/Activity	Duration	Predecessor	Resources
1.0	Identify			
1.1	Develop project charter			
1.2	Perform stakeholder analysis		1.1	
1.3	Develop project plan		1.2	
2.0	Define		1.0	
2.1	Collect voice of customer (VOC)			
2.2	Identify CTS measures and targets		2.1	
2.3	Translate VOC into technical requirements		2.2	
3.0	Design		2.0	
3.1	Identify process elements			
3.2	Design process		3.1	
3.3	Identify potential risks and inefficiencies		3.2	
4.0	Optimize		3.0	
4.1	Implement process			
4.2	Assess process capabilities		4.1	
4.3	Optimize design		4.2	
5.0	Validate		4.0	
5.1	Validate process			
5.2	Assess performance, failure modes, and risks		5.1	
5.3	Iterate design and finalize		5.2	

FIGURE 3.4 Project plan.

measures and to translate the customer's requirements into the technical elements of the process to be designed. The following activities can be applied to meet the objectives of the Define phase:

4. Collect VOC
5. Identify CTS measures and targets
6. Translate VOC into technical requirements

Figure 3.5 shows the main activities mapped to the tools or deliverables most typically used during that step.

Collecting the VOC for DFSS is very similar to the VOC data collection discussed in the DMAIC VOC collection. The data collection plan would be used to identify the data to be collected that would support the assessment of the proposed CTSs and to validate these CTS from the customers' perspective. Interviews, focus groups, surveys, and market research are some of the most common ways to collect VOC. The main difference between DFSS and Six Sigma would be that existing customer complaints, warranty information, and other data from an existing process would not be available or would not apply to our new process that we are designing. Benchmarking can be powerful in the DFSS arena so that the organization looks outside of itself to understand industry and even outside of industry best practices that can be used as a model for our process design.

It is important to clearly summarize the CTS so that we can operationally define the metrics and then translate these into the process elements that form the technical requirements of our new process. Quality function deployment (QFD) can be used to relate the customer requirements and CTS to the process elements and the technical requirements. The customer requirements would be prioritized by the customers through market research techniques. The strength of the relationship between the customer requirements and the technical requirements would be identified by the process design team. These relationship strengths would be multiplied by the CTS priorities to derive a relative weighting of the technical requirements. We will use these technical requirements as the process elements as input to the Design phase.

	Define activities	Tools/Deliverables
4	Collect VOC	▪ Data collection plan ▪ VOC ▪ Interviewing, surveying, focus groups, market research
5	Identify CTS measures and targets	▪ Critical to satisfaction (CTS) summary & targets ▪ Affinity diagram ▪ QFD ▪ Operational definitions ▪ SWOT ▪ VOP matrix
6	Translate VOC into technical requirements	▪ QFD ▪ Benchmarking

FIGURE 3.5 Define phase activities and tools/deliverables.

DESIGN

The purpose of the Design phase is to understand the elements of the process that can ensure the CTS of the customers and stakeholders are met, to design the new process, and to identify potential risks, failures and inefficiencies that could occur in the new process. The main activities to be performed in the Design phase are:

7. Identify process elements
8. Design process
9. Identify potential risks and inefficiencies

Figure 3.6 shows the main activities mapped to the tools or deliverables most typically used during the Design phase.

The first step in the Design phase is to analyze the VOC data that was collected in the Define phase. Attribute survey analysis using chi-square statistical analysis would be used to analyze attribute survey data. Data collected from the VOC would be used to generate the elements that would be incorporated into a process, or potential alternate process concepts. Potential elements could be categorized by people, process and technology. The people aspects would be which organizations and roles would be involved in owning and contributing to the process; the cultural and political aspects, resistance to change, training and skill sets available, and organizational structure. The process elements could pertain to any policies and procedures that may impact the process, understanding the activities needed to be performed, as well as how to measure and assess performance. The technology elements would pertain to what technologies would be needed, such as using a SharePoint® site or perhaps an off-the-shelf or internally developed information system.

There are many techniques that are part of the DFSS tool kit that can help to generate and brainstorm process elements and concepts, such as traditional brainstorming and Nominal Group Technique, channel and analogy brainstorm-

	Design activities	Tools/Deliverables
7	Identify process elements	▪ Process element summary
8	Design process	▪ Basic statistics ▪ Simulation ▪ Prototyping ▪ DOE ▪ Process Analysis ▪ Multivoting ▪ Criteria-based matrix ▪ Pugh concept selection technique ▪ VOP matrix
9	Identify potential risks and inefficiencies	▪ Failure mode and effect analysis ▪ Risk assessment ▪ Process analysis ▪ Waste analysis

FIGURE 3.6 Design phase activities and tools/deliverables.

ing, antisolution brainstorming and brainwriting, assumption busting, and TRIZ (Chowdhury 2005).

Traditional brainstorming includes sharing ideas in a group and writing them on a flip chart or white board. Nominal Group Technique structures the brainstorming into first a silent generation then a round-robin idea sharing. Important in any brainstorming activity is to hold the criticism and evaluation until after ideas are generated. Channel brainstorming allows a group to focus on a subcategory of a task to make the brainstorming more manageable. Analogy brainstorming allows participants to focus on a similar or parallel issue to generate ideas, and then link it back to the original issue. Antisolution brainstorming asks the participants to generate ideas of how they could make the process even worse, punching holes in your own argument. In brainwriting, each participant writes down an idea, and then passes it to the person next to them who then builds on the idea or concept. Assumption busting is when the brainstorming group, instead of asking "why?", they ask "why not?" (Chowdhury 2005).

The Theory of Inventive Problem Solving (pronounced "trees;" TRIZ) was developed by Genrick Altshuller and his colleagues. He developed the TRIZ principles by reviewing thousands of patent applications and extracting the key principles. He developed a set of principles that can be used to cultivate inventions to eliminate corporate contradictions and problems while generating creative solutions. A TRIZ principle encourages the team to look at the past, present and future of the process when designing the process. The following steps describe the TRIZ process (Chowdhury 2005):

1. Think of the ideal vision, process, or system
2. Think of ways to improve the process or function
3. Think of ways to eliminate or reduce undesired functions
4. Think of ways to segment the process
5. Think of ways to copy existing ideas or processes
6. Think of a disposable concept

There is a great deal of depth and richness in the TRIZ concept related to a tangible product design. Presented here are the elements of TRIZ that could apply to designing intangible service processes. A TRIZ case study reference is given in the References section (www.triz-journal.com/archives/2000/06/c/index.htm).

The Pugh concept selection technique is a technique for evaluating and selecting concepts. If you have several different process elements or concepts to choose from, you could use this technique. You would first brainstorm potential solutions or concepts, and generate criteria upon which to compare the concepts. Then you would select one of the concepts as the "candidate" concept. It does not matter which concept you select as the candidate concept. You then compare each of the other (new) concepts with the candidate for each comparison criteria. If the new concept is better than the candidate for those criteria, you would place a plus sign in the cell where the new concept intersects the criteria. If the new concept is worse than the candidate concept for the criteria, a minus sign is placed in the cell. If the new concept is the same as the candidate on those criteria, a zero is placed in the cell. Figure 3.7 shows

	CONCEPTS						
CRITERIA	1	2	3	4	5	6	7
A	–	–	–	0	Candidate	0	–
B	–	0	–	–	Concept	0	–
C	+	+	–	–	"	–	–
D	+	–	–	+	"	–	+
E	+	+	–	–	"	–	–
PLUSES	3	2	0	1		0	1
MINUSES	2	2	5	3		3	4
ZEROS	0	1	0	1		2	0

FIGURE 3.7 Pugh concept selection technique.

a Pugh matrix. You would select the few concepts with the most pluses and the fewest minuses. You could also attack the weaknesses of the few concepts and enhance them with the strengths of the surviving alternatives (Chowdhury 2005).

After you identify the process elements or concepts, the team can then design the process. A process map is a great tool to communicate the steps of the new process. It helps to think through sequencing, who does what in the process, as well as the information that is needed to perform each step of the process and what output is transformed by each process step.

A FMEA is a great tool to help think through the potential risks in a process, or where the failures can occur. By thinking of potential failure modes for each process step, identify the probability of occurrence, the impact or severity to the stakeholders if the failure occurs and the ability to detect the failure, we can develop recommendations to incorporate into the process to reduce the probability for failure, reduce the impact if the failure occurs, and improve the ability to detect the failure.

Process and waste analyses can be performed to identify potential process inefficiencies and wasteful activities. These were discussed in Chapter 2 in the DMAIC methodology. The process analyses helps to identify which activities you have defined in the process that do not add value, which could be further eliminated, combined or reduced. The waste analyses identifies activities that do not add value and which are wasteful.

A description of the process metrics that will be embedded in the process should be defined. An operational definition includes the purpose of the measure, as well as a specific and detailed description of how you would measure the metric.

Some other tools, beyond the scope of this text, could include performing simulations, prototypes, and design of experiments to help in designing the process.

OPTIMIZE

The purpose of the Optimize phase is to understand the elements of the process that can ensure the CTSs of the customers and stakeholders are met, to pilot the new process, to assess process capabilities and to identify potential risks, failures and inefficiencies that could occur in the new process. The main activities to be performed in the Optimize phase are:

10. Implement pilot process
11. Assess process capabilities
12. Optimize design

Figure 3.8 shows the Optimize activities mapped to the tools and deliverables typically used in the Optimize phase.

The team should gain the appropriate approvals to pilot the process from the process owners and stakeholders. A presentation of the project to this point may help to communicate the value of the project and the new process. To implement the process, the team who to the team would develop an implementation plan that include each implementation activity, who would be responsible for implementing each step, the stakeholders the activity would impact and the due date for when the activity would be complete. Figure 3.9 shows an implementation plan template.

Statistical process control is an effective tool to help ensure your process performance is being attained. It can highlight trends and identify when something goes

	Optimize activities	Tools/Deliverables
10	Implement pilot process	▪ Implementation plan ▪ Communication plan ▪ Training plan ▪ Procedures
11	Assess process capabilities	▪ Process capability ▪ Simulation
12	Optimize design	▪ Process map ▪ Process analysis ▪ Waste analysis ▪ Cost/Benefit analysis ▪ Statistical process control ▪ Mistake proofing ▪ Design of experiments

FIGURE 3.8 Optimize phase activities and tools/deliverables.

Activity	Responsible	Due date	Stakeholders impacted

FIGURE 3.9 Implementation plan template.

wrong in the process (assignable cause), that would encourage us to investigate the root cause of the problem.

The process capability would be assessed by collecting data for the metrics previously identified. For service processes, the data does not necessarily follow a normal distribution, so nonnormal capability analysis should be used. If attribute control charts are used to control the process, the process capability is the average value or center line of the control chart when the process is in control. If the process is not meeting the target metrics and expectations of the customers and stakeholders, further redesign of the process can be performed. Further process and waste analysis would be helpful for the redesign. Also, if training was not implemented during the pilot process, it should be considered first to ensure the new process is being consistently understood and practised, and skill transfer is occurring. Training plans would include the topics to be covered, as well as the targeted training audience, the expected length of the topic, any expected prerequisite knowledge, and the instructional strategies to be applied. Figure 3.10 shows a training template example.

Detailed procedures also help to train stakeholders in the process to ensure consistency and repeatability of the process. We cannot improve a process if it is not first consistent, stable and repeatable. This provides a baseline upon which to further optimize and improve the process.

VALIDATE

The purpose of the Validate phase is to validate the process, assess the performance, failure modes and risks, and iterate through a revised process until you are ready to finalize and stabilize the new process. The main activities that are performed in the Validate phase are:

13. Validate process
14. Assess performance, failure modes, and risks
15. Iterate design and finalize

The activities and related tools and deliverables of each activity are shown in Figure 3.11 for the Validate phase.

Training topic	Target audience	Expected length of topic	Prerequisite knowledge	Instructional strategy
Process mapping	Process analysts	four hours	Concepts of Six Sigma or design for Six Sigma	Workshop with hands-on exercises building process maps
Design for Six Sigma IDDOV methodology	Business analysts, process engineers	12 days across three separate weeks, three months apart	None	Workshops with hands-on case; mentored work projects

FIGURE 3.10 Training plan template.

	Validate activities	Tools/Deliverables
13	Validate process	▪ Design of experiments ▪ Pilot ▪ Statistical analysis
14	Assess performance, failure modes, and risks	▪ Mistake proofing ▪ Dashboards ▪ Scorecards ▪ Hypothesis tests ▪ ANOVA
15	Iterate design and finalize	▪ Replication opportunities ▪ Statistical process Control

FIGURE 3.11 Validate phase activities and tools/deliverables.

The first activity in the Validate phase is to validate the process is meeting the CTS metric targets. Developing a dashboard or scorecard to display the key metrics to management is helpful to ensure the process is performing to expectations and specifications.

The process should be piloted for some time to assess the performance of the process. The appropriate statistical or analysis of variance tests can be performed to assess the performance. If the process is not meeting expectations, further mistake proofing can be applied to reduce errors and to maintain consistency. Mistake proofing focuses on raising awareness, vigilance, and the ability to prevent errors from occurring.

When using the statistical tests, care must be taken to check if the data follows a normal distribution and, if it does not, to use the appropriate nonnormal statistical test. Replication opportunities also should be assessed to determine if the same process or similar concepts can be applied elsewhere in the organization. Future plans for further improving the process should also be developed.

REFERENCES

Chowdhury, S. *Design for Six Sigma: The Revolutionary Process for Achieving Extraordinary Profits*, Kaplan Press, 2005.

Deming, W.E. *Out of the Crisis*, The W. Edwards Deming Institute, Palos Verdes Estates, CA, 1986.

Simon, K. What is design for Six Sigma and how does design for Six Sigma compare to DMAIC? iSixSigma, 2000. Accessed Feb. 25, 2009 at www.isixsigma.com/library/content/c020722a.asp

Proseanic, V., Tananko, D. Visnepolschi, S., The experience of the anticipatory failure determination (AFD) method applied to an engine concern, *TRIZ Journal*, accessed on Feb 25, 2009 at www.triz-journal.com/archives/2000/06/c/index.htm

4 Sunshine High School Discipline Process Improvement—A Lean Six Sigma Case Study

Marcela Bernardinez, Khalid Buradha, Kevin Cochie, Jose Saenz, and Sandra L. Furterer

CONTENTS

PROCESS OVERVIEW

The Sunshine High School (SHS)* is one of the largest high schools in the Orange County Public School system, with more than 3400 students and 340 faculty mem-

* Used to generalize the high school.

bers. The student population is very diverse, comprising many nationalities and students from various socio-economic backgrounds. The campus is divided by an East Campus that consists exclusively of freshman students and a West Campus comprising sophomores through seniors. The leadership team consists of a principal, three assistant principals, and nine deans. The discipline program is charged with the responsibility of providing a safe and effective learning environment. The discipline system is overseen by one assistant principal, and three deans. This program is affected by many factors, including student attendance, student adherence to code of conduct, and classroom management, and discipline. The discipline program is a system of subprocesses that work together to achieve an environment conducive to quality learning.

The mission of SHS is to advance achievement for all students with the education necessary to be responsible, successful citizens.

To ensure that all students succeed, they are committed to the following:

- Encourage students to develop pride in their school and community
- Recognize all students, faculty, staff, and community for their achievements
- Create a culture of academic rigor and relevance
- Use data to identify what is essential to know
- Set high expectations that hold students and adults accountable for improvement
- Create a curriculum framework that drives instruction
- Provide students with real-world application of skills and knowledge
- Create multiple pathways to rigor and relevance based on students individual strengths
- Provide sustained professional development focused on improving instruction
- Obtain parental and community involvement
- Establish and maintain safe and orderly schools
- Offer effective leadership development for administrators, teachers, parents, and community

One of the substitute teachers at SHS has noticed a lack of standardization in the discipline process across different classrooms. The administration is also concerned that the students who get referred to the office for discipline problems miss one to several class periods, while the paperwork is sent from the teacher to the office.

This Lean Six Sigma project will look at the discipline system as a whole initially. It will then focus on key subprocesses to make recommendations for system-wide improvement, thus improving the overall academic environment.

The discipline system has been divided into the following subprocesses described below.

Classroom discipline and referral initiation

This subprocess consists of the actions that occur in the classroom when a faculty member observes a student infraction of the code of conduct. This subprocess is initiated with a student infraction and ends with classroom discipline imposed or the initiation of a referral to the dean's office.

Dean's office discipline actions

This process begins from student referral, including processing of that student, data entry of the completed referral, and ends with feedback to the faculty member who initiated the referral.

Attendance contract

This process is initiated when a student is put on an attendance contract and includes the process to track the contract and subsequent discipline penalties if the contract is violated.

It is important to administration that the Lean Six Sigma project provides a statistically based analysis of student discipline data in the student information system to understand which students have the highest percentage of discipline referrals by class, race/ethnicity, and socio-economic level, as well as the percentages by type of discipline referral.

It is assumed that the Lean Six Sigma team will have access to SHS process owners' information and database information. The Lean Six Sigma team will execute the DMAIC process. However, the team's primary goal is to recommend appropriate implementation improvements, as well as a control plan that ultimately institutionalizes changes that are implemented.

SHS process owners and administrators will assist in deployment of faculty, student, and parent surveys, as well as interviews of key stakeholders within the administrative staff and leadership team. The SHS principal is new to Lean Six Sigma but is convinced of its value for improving the discipline process at SHS.

The Lean Six Sigma mentors can meet with the Six Sigma team to provide background and process information on the discipline process, having worked with SHS in the past, as well as guide the participants with coaching on applying the Lean Six Sigma tools. A sample discipline referral form is included in Figure 4.1.

DEFINE PHASE EXERCISES

It is recommended that the students work in project teams of 4–6 students throughout the Lean Six Sigma Case Study.

1. **Define Phase Written Report**
   Prepare a written report from the case study exercises that describes the Define phase activities and key findings.

2. **Lean Six Sigma Project Charter**
   Use the information provided in the Process Overview section above, in addition to the project charter format to develop a project charter for the Lean Six Sigma project.

3. **Stakeholder Analysis**
   Use the information provided in the Process Overview section above, in addition to the stakeholder analysis format to develop a stakeholder analysis, including stakeholder analysis roles, an impact definition, and stakeholder resistance to change.

County Public Schools
Safety/Discipline Referral Form

Student Number: ___________ Incident No.: ______________ Student Name: ___________
Sex: ____ Race: ____ Grade: ______ Date of Infraction: __________ Parent/Guardian Name: _______________ Home Phone: _________ Work Phone: ___________ Referred By: _____________ Instructor/Staff #: ____ Bus Trip #: ____ Period ______ Time: _______ Location of Infraction: ___________________ Details of Offense: ____________________________
Administrator #: ________________ * Must be reported to Law Enforcement Offense(s) pertaining to this referral

	1A Cheating		2A Destroy prop/ Vand < $10		3A battery*		4A alcohol*
	1B classroom disruption		2B Disrespect		3B Breaking & Entering*		4B Arson*
	1C disorderly conduct		2C Fighting		3C Destroy prop/ vand ($10–$100)		4C Assault of emp/vol/stdts*
	1D disrespect for others		2D Forgery		3D Disrespect		4D Battery of emp/vol/stdts*
	1E dress code		2E Gambling		3E Extortion/threats		4E bomb threats/ explosions*
	1F failure to report detention		2F Insubordination /def		3F Fighting*		4F drugs*
	1G false/mislead information		2G Intimidation/ threats		3G Firecrackers/ works		4G false Fire alarm*
	1H insubordination		2H Misconduct on sch bus		3H Gross insubordination/Def		4H firearms*
	1I misconduct on school bus		2I Repeat misc/ Less Serious		3I Illegal organization		4I incite/lead/ participate*
	1J Profane/obs/ abusive lang		2J Stealing under $10		3J Possess of contraband material		4J Larceny/theft*
	1K repeated misconduct		2K Unauthorized assembly		3K Repeated misc/ more serious		4K other weapons*
	1L Tardiness		2L Bullying		3L Smoking/other use tobacco*		4L Repeat Misc/ more serious
	1M Unauth abs school/class		2M Other serious miscond.		3M Stealing over $10		4M Robbery*
	1N Bullying				3N Trespassing*		4N Sexual battery*
	1O Other				3O Violation of curfew		4O Sexual Harassment*
					3P Bullying		4P Sexual Offenses*
					3Q Other serious misconduct		4Q Violation early reentry

FIGURE 4.1 Safety/discipline referral form.

							4R Motor vehicle theft*
							4S Motor vehicle theft*
							4T other

Action(s) taken for this referral

	A Parental Contact		F Return of Prop/ Pay/Restit.		L Referral to Intervention Program		R Suspend from School
	B Counseling & Direction		G Retention		M Confiscate Unauthor. Material		S Suspend 10 Days Exp/Removal
	C Verbal Reprimand		H Saturday School		N Special Program/ School		X Probationary Plan (KG–05)
	D Special Work Assignment		I Behavior Contract/Plan		P In-school Suspension		Z Peer Mediation
	E Withdrawal of Privileges		K Alt Class Assignment		Q Suspension from Bus		

---------Suspension Information--------- ----------Other Information----------

From: ___ # Days: __ To: ____ Return: ______ Detention: ______ Sat Schl: ______ Other: _____

Ex Ed Student: Yes/No Sum Sch Susp: Yes/No Early Reentry: ______ From: _______ To: _______

Administrator's Comments: ____

____

School Name; ____________________ ________________ Time departed from office

Administrator's Signature: ____________________ Student's Signature: ____________________

Parent's Signature: ____________________

FIGURE 4.1 (Continued)

4. **Team Ground Rules and Roles**
   Develop the project team's ground rules and team members' roles.

5. **Project Plan and Responsibilities Matrix**
   Develop your team's project plan for the DMAIC project. Develop a responsibilities matrix to identify the team members who will be responsible for completing each of the project activities.

6. **SIPOC**
   Use the information provided in the Process Overview section above, to develop a SIPOC of the high-level process.

7. **Team Member Biographies**
   Each team member should create a short biography (bio) of themselves so that the key customers, stakeholders, project champion, sponsor, Black Belt, and/or Master Black Belt, can get to know them, and understand the skills and achievements that they bring to the project.

8. **Define Phase Presentation**
   Prepare a presentation (PowerPoint) from the case study exercises that provides a short (10–15 minutes) oral presentation of the Define Phase deliverables and findings.

DEFINE PHASE

1. Define Phase Report

A written report of the Define phase for the SHS Discipline Process Improvement project, including the key deliverables developed as part of the prior exercises, is described below.

2. Lean Six Sigma Project Charter

Following are the sections that comprise the project charter, which defines the problem to be investigated. The project charter is shown in Figure 4.2.

Project Name: SHS Discipline Process Improvement

Project Overview: SHS is one of the largest high schools in the Orange County Public School, system with more than 3400 students and 340 faculty members. The student

Project Name: High School Discipline Process Improvement.

Project Overview: The Sunshine High School Discipline process lacks standardization, and has delays in processing students through the Dean's discipline process.

Problem Statement: The discipline process lacks consistency between offenses and actions, as well as across classrooms. There are also significant delays in processing students through the discipline deans' offices.

Customer/Stakeholders: (Internal/external) students, parents, faculty, administration, school board, security/law enforcement, society.

What is important to these customers – CTS: Minimize classroom disruptions, minimize school discipline issues, level of knowledge in code of conduct, consistency of offenses, and actions.

Goal of the Project: Improve discipline process, by reducing time by XX%, and improving consistency by XX%.

Scope Statement: Begins with student misconduct either in halls or classrooms, includes classroom discipline, Dean's discipline, Attendance contract, in-school and out-of school discipline processes.

Financial and Other Benefit(s): Reduction in number of offenses and repeat offenses.

Potential Risks: Data not available, resistance to change from students and faculty, time constraints.

Milestones: 2/2 to 4/28.

Project Resources: David Christiansen, Kevin Cochie, Marcela Bernardinez, Khalid Buradha, Jose Saenz, Dr. Sandy Furterer.

FIGURE 4.2 Project charter template.

population is very diverse, made up of many nationalities. The campus is divided into an East Campus that consists exclusively of freshman students, and a West Campus that consists of sophomores through seniors. The leadership team consists of a principal, three assistant principals, and nine deans. The discipline program is charged with the responsibility of providing a safe and effective learning environment. The discipline system is overseen by one assistant principal and three discipline deans. This program is affected by many factors, including student attendance, student adherence to code of conduct, and classroom management and discipline. The discipline program is a system of sub-processes that work together to achieve an environment conducive to quality learning.

Problem Statement: The discipline process at SHS is inefficient and inconsistent. The students can wait from one to several class periods in the administration office waiting for the paperwork to get processed by the referring teacher, or to be seen by the discipline dean. Additionally, the classroom discipline process varies, as well as the discipline consequences given for various discipline infractions.

The SHS Lean Six Sigma team will work on improving the discipline program at SHS. The SHS discipline program consists of multiple subprocesses that affect one another. These subprocesses are complex and are each affected by multiple factors, including student background, academic standing, and other variables that will be analyzed for correlation.

Ultimately, the customer of the discipline program is the parent or guardian of the student. Their desires as a customer will be captured then matched to process technical steps through QFD.

Improvements to one or more of the subprocesses will be made by this Lean Six Sigma team, as well as control plans to assist in implementation and control of the improvements. The project team has neither control over selected implementations nor ultimate control of the changes.

This project will follow the Lean Six Sigma DMAIC process and will generate a formal written report and presentation. Selected Lean Six Sigma tools will be used throughout the project with the intent of providing tutorial explanations to the SHS process owners and champion.

In the Define phase, the team began to understand the problem and process to be improved. They developed a detailed project description, or project charter, that describes the problem statement, project goals, and scope statement. A stakeholder analysis was also performed to identify the critical customers and stakeholders that are impacted by the process to be improved. The concerns and how the stakeholders are affected are also defined in this phase. A SIPOC was developed to provide a high-level view of the processes to be improved. A detailed work plan was generated to provide guidance to the team for how they would successfully complete the activities of the DMAIC problem-solving methodology for this project.

Customers/Stakeholders: Faculty, students, parents, school administrators, school board, and security/police officers.

What Is Important to These Customers – CTS: Reduction in the number of discipline referrals, reduction in the number of classroom disruptions, consistency in application of the discipline consequences, and knowledge of the code of conduct.

Goal of the Project: Complete a comprehensive DMAIC analysis of this process/system using Six Sigma tools and methodology. The end state of the project will yield recommendations to the SHS administrative staff for improvements of the process.

Scope Statement: This project will analyze the discipline system of SHS. This analysis will map the subprocesses of the discipline system to include:

- Classroom Discipline process: defined as in-class discipline by a faculty member to include initiation of a dean's office referral.
- Discipline Action process: defined as the processes that occur once a student and/or referral arrives to the dean's office through the discipline action with data input into the student database followed by feedback to the initiating faculty member.
- Attendance Contract process: defined as the process a student undergoes to receive and adhere to an attendance contract.

The project will map the subprocesses that are executed within the discipline system and the perceptions/satisfaction levels of these processes from the viewpoint of the system customers (administrators, faculty, students, and parents).

Upon completion of surveying the discipline system customers, the project team will focus on improving one or several of the discipline system's subprocesses with the intent of providing recommendations for improvement of the processes that will positively impact the overall discipline system and academic environment.

Principal Project Deliverables/Outputs:

Define:

- Project charter
- Stakeholder analysis
- Define report

Measure:

- SIPOC diagram (high-level process map)
- Process flow diagram (detailed process map)
- CTS: Key outputs of the process from the customers' view
- Key metrics: Key inputs of the process
- Pareto charts: Graphical depiction of target improvement areas
- Measure report

Analyze:

- Cause and effect diagrams
- Summary of data
- Summary of improvement areas (recommended)
- Cost of quality analysis
- Analyze report
- Benchmarking

Improve:

- Recommended improvement plans
- QFD diagram: matches CTSs to process steps or improvement areas
- Revised process flow and information flow diagrams (recommended)
- Improve/Control report

Control:

- Recommended control plan
- Improve/Control report

Projected Financial and Other Benefits: Potential benefits to improving the discipline process includes an enhanced academic environment to facilitate student learning, fewer discipline issues of repeat offenders, and decreased probability of potential liability issues regarding faculty usage of the disciplinary process.

Risk Management Matrix: The risk management matrix is shown in Figure 4.3.

Project Resources:
Project Leader: Kevin Cochie
Division/Department: SHS administration
Process Owner: Discipline deans

Potential risks	Probability of risk (H/M/L)	Impact of risk (H/M/L)	Risk mitigation strategy
Data not available	H	H	Identify issues early to the principal Collect manual data
Resistance to change from students	L	M	Change strategy
Resistance from faculty	M	H	Change strategy
Time constraints	M	H	Good project planning

FIGURE 4.3 Risk management matrix.

Process Champion: Principal
Project Sponsor: Discipline dean
CI Mentor/MBB: Sandra Furterer
Finance: To be determined
Project Team Members: Marcela Bernardinez, Khalid Buradha, Kevin Cochie, and Jose Saenz
Estimated milestones are shown in Figure 4.4.

Critical Success Factors:

- Partnership with SHS administration: The success of this project hinges on close partnership between the Six Sigma team and the SHS administrators and process owners.
- Complete process mapping: This process is detailed and complex. Successful data gathering and analysis is heavily dependent on the Six Sigma team becoming well versed with the procedures within the discipline process.
- CTS identification: Customer CTS variables must be identified from the standpoint of the customers identified in the stakeholder analysis. This may include more than one primary customer base.

Management Approach

Scope Management Approach: This project will be managed by the project leader, but responsibility of the success hinges upon a collective effort from all team members. Communication between the team members shall flow cross functionally. Electronic mail will be a prime source of communication outside of class and group meetings, therefore it is imperative that when communicating with other team members, the Master Black Belt, or a process owner from SHS, all other team members shall be copied on the communication.

Issues Management Approach: All issues will be documented through weekly team meetings by the team secretary. Issues for resolution at the team level shall be settled by the team collectively. Issues that rise above the team level will be settled by the Master Black Belt/Professor and/or the project champion.

Milestones	
Phase	**Estimated completion date**
Define	January 11
Measure	February 3
Analyze	March 2
Improve	April 6
Control	April 27

FIGURE 4.4 Milestones.

3. Stakeholder Analysis

A critical part of the Define phase is to perform a stakeholder analysis to understand the people impacted by the project. There are primary stakeholders, which are usually the main internal and external customers of the process being improved. The secondary stakeholders are affected by the project, but not in as a direct manner. Figures 4.5 and 4.6 show the primary and secondary stakeholders for the discipline process, and their major concerns. Note that + represents a positive impact or potential improvement, whereas – represents a potential negative impact to the project.

Figure 4.7 shows the commitment level of each major stakeholder group at the beginning of the project (Sholtes, Joiner, and Striebel 2003).

4. Team Ground Rules and Roles

The team brainstormed the ground rules related to their attitudes and the processes or behaviors they would adhere to while working with each other, as shown below.

	Stakeholders	Who are they?	Potential impact or concerns	+/–
PRIMARY	SHS faculty	Customer: This includes all SHS permanent faculty and substitute teachers. They are customers of the discipline system. Their input into the system is referrals into the system with the expected output of a disciplinary action.	▪ Standardized processes ▪ Reduction of errors and rework ▪ Continuity of infraction enforcement ▪ Resistance to enforcing codes	+ + + –
	SHS students	Customer: This includes more than 3500 students that attend SHS. They are customers of the discipline system as they are the inputs to the system and the expected outcome is a fair and consistent reaction to infractions.	▪ Reduction of repeat offenses ▪ Increase of academic performance ▪ Resistance to imposition of strict policies	+ + –
	SHS parents/ guardians	Customer: This includes all parents or guardians of the students of SHS. Their children are the inputs to the discipline system. The expected output is a safe environment conducive to a positive learning environment for their children.	▪ Increase of knowledge of code of conduct ▪ Reduction of communication gaps ▪ Resistance to change current procedures	+ + –

FIGURE 4.5 Primary stakeholder analysis definition.

	Stakeholders	Who are they?	Potential impact or concerns	+/−
SECONDARY	SHS administration	Stakeholder: The assistant principals and deans are charged with a tremendous responsibility of educating young adults to include quality academic programs and a safe learning environment free of classroom disruption. Oversight of the discipline system and it's subprocesses.	▪ Reduce instances of classroom disruption ▪ Resistance to change of discipline procedures that impact administrative focus areas	+ −
	SHS security and law enforcement	Stakeholder: SHS security and law enforcement is responsible for the oversight of campus security. They require swift and consistent enforcement by the process owners of the discipline system to assist in maintaining good order and discipline on the school campus.	▪ Reduction of campus related security issues ▪ Resistance to change of discipline procedures that impact campus security	+ −
	County public school system	Stakeholder: The school district is financially liable for the security and safety of all students within the entire school system. The public school system requires good order and discipline on all campuses and within all classrooms.	▪ Reduce instances of classroom and campus disruptions from discipline infractions ▪ Potential OCPS restrictions on recommended improvements (bureaucracy)	+ −

FIGURE 4.6 Secondary stakeholder analysis definition.

Attitudes

- Be as open as possible, but honor the right of privacy
- Information discussed in the team will remain confidential. With regards to people's opinions, what is said here stays here
- Everyone is responsible for the success of the meeting
- Be a team player. Respect each other's ideas. Question and participate
- Respect differences
- Be supportive rather than judgmental
- Practice self-respect and mutual respect
- Criticize only ideas, not people
- Be open to new concepts and to concepts presented in new ways. Keep an open mind. Appreciate the points of view of others
- Be willing to make mistakes or have a different opinion

Commitment level	Administration	Faculty	Students	Parents	OCPS	Security/Courtesy officers
Enthusiastic	★					
Helpful	★	★		★		★
Hesitant		★				
Indifferent			★		★	
Uncooperative			★			
Opposed						
Hostile						

FIGURE 4.7 Stakeholder resistance to change.

- Share your knowledge, experience, time, and talents
- Relax. Be yourself. Be honest

Processes

- Use time wisely, starting on time, returning from breaks and ending meetings promptly
- Publish agenda and outcomes
- Ask for what we need from our facilitator and other group members
- Attend all meetings. Be on time
- Absenteeism permitted if scheduled in advance with the leader
- When members miss a meeting, we will share the responsibility for bringing them up to date
- Maintain 100% focus and attention while meeting
- Stay focused on the task and the person of the moment
- Communicate before, during, and after the meeting to ensure that action items are properly documented, resolved, and assigned to a responsible individual and given a due date
- Phones or pagers on "stun" (vibrate, instead of ring or beep) during the meetings
- One person talks at a time
- Participate enthusiastically
- Do not interrupt a person's speech
- Keep up-to-date

5. Project Plan and Responsibilities Matrix

The detailed project plan is shown in Figure 4.8, with tasks to be completed, due date, deliverables, and resources. It includes the person (or people) responsible for each activity.

6. SIPOC

The SIPOC identifies the processes that are part of the scope of the improvement effort, as well as the suppliers, customers, inputs and outputs of these processes. There are three processes that are part of the scope of this project: classroom discipline process, Dean's office discipline process, and the attendance or behavioral contract

ID	Task name	Duration	Start	Finish
1	UHS six sigma project start	0 days	Mon 1/10/05	Mon 1/10/05
2	Define phase	18 days	Mon 1/10/05	Wed 2/2/05
3	Define the problem	2 days	Mon 1/10/05	Tue 1/11/05
4	Define the scope of the project	2 days	Wed 1/12/05	Tue 1/13/05
5	Define the potential benefits of the project	2 days	Fri 1/14/05	Mon 1/17/05
6	Prepare project charter	2 days	Tue 1/18/05	Wed 1/19/05
7	Prepare initial process flow charts and sar	2 days	Tue 1/20/05	Fri 1/21/05
8	Customer / stakeholder analysis	2 days	Mon 1/24/05	Tue 1/25/05
9	Make work plan	1 day	Wed 1/26/05	Wed 1/26/05
10	Responsibilities matrix	1 day	Thu 1/27/05	Thu 1/27/05
11	Determine next steps	1 day	Thu 1/27/05	Thu 1/27/05
12	Prepare team member participation log	1 day	Thu 1/27/05	Thu 1/27/05
13	Prepare define phase report and presentat	4 days	Fri 1/28/05	Wed 2/2/05
14	Measure phase	20 days	Thu 2/3/05	Wed 3/2/05
15	Meet with UHS dean to finalize process flo'	1 day	Thu 2/3/05	Thu 2/3/05
16	Determine student, faculty and parent surve	1 day	Fri 2/4/05	Fri 2/4/05
17	Conduct student, faculty and parent surve	1 day	Mon 2/7/05	Mon 2/7/05
18	Collect survey results	1 day	Thu 2/8/05	Thu 2/8/05
19	Analyze survey results	12 days	Wed 2/9/05	Thu 2/24/05
20	Prepare SIPOC	7 days	Wed 2/9/05	Thu 2/17/05
21	Draw process flow charts	10 days	Wed 2/9/05	Thu 2/22/05
22	Make pareto charts using survey results	10 days	Wed 2/9/05	Thu 2/22/05
23	Identify critical quality characteristics	8 days	Wed 2/9/05	Fri 2/18/05
24	Key metrics	2 days	Wed 2/9/05	Thu 2/10/05
25	Define items for resolution	1 day	Wed 2/9/05	Wed 2/9/05
26	Process capability index	2 days	Wed 2/9/05	Thu 2/10/05
27	Determine next steps	1 day	Wed 2/9/05	Wed 2/9/05
28	Prepare team member participation log	1 day	Wed 2/9/05	Wed 2/9/05
29	Prepare measure phase report and presen	14 days	Fri 2/11/05	Wed 3/2/05
30	Analyze phase	25 days	Thu 3/3/05	Wed 4/6/05
31	Determine causes and effects of the problem	6 days	Thu 3/3/05	Thu 3/3/05
32	Define items for resolution	1 day	Thu 3/3/05	Thu 3/3/05
33	Make improvement plans	3 days	Thu 3/3/05	Mon 3/7/05

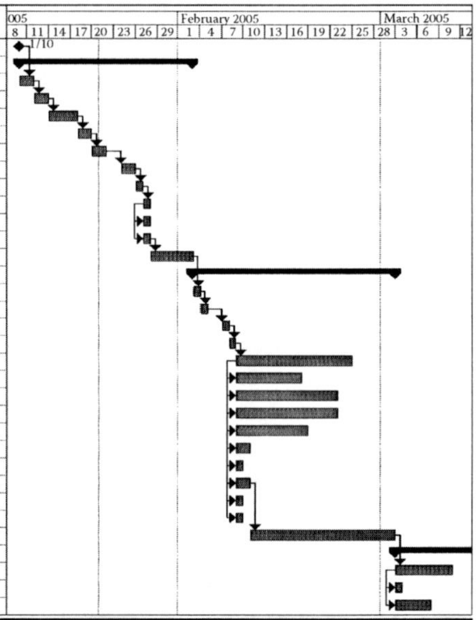

ID	Task name	Duration	Start	Finish
34	Prepare recommendations for improvement			
35	Cost Benefit analysis			
36	Prepare team member participation log			
37	Prepare measure phase report			
38	Prepare measure phase presentation			
39	Improve phase			
40	Quantification of improvement plans			
41	Define items for resolution			
42	Prepare team member participation log			
43	Control phase			
44	Proposed control mechanisms			
45	Verification of improvements / results			
46	Prepare team member paticipation log			
47	Final report			
48	Team assessment			
49	Project assessment			
50	Document lessons learned			
51	Prepare team member participation log			
52	Prepare final project report			
53	Prepare final project presentation			
54	UHS six sigma project end			

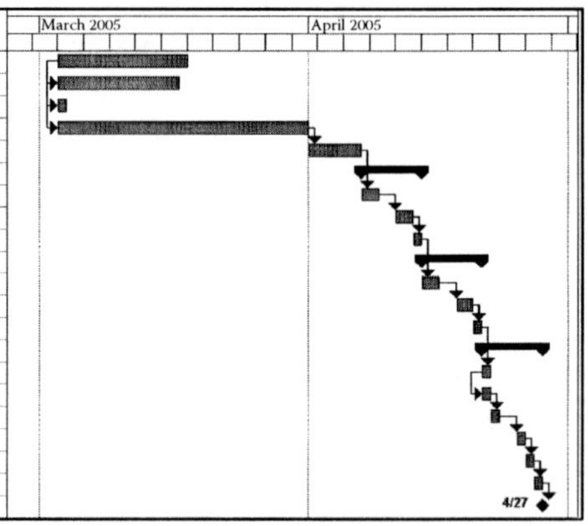

FIGURE 4.8 Project plan.

process. The input to the classroom discipline process is the misconduct performed by the students. The faculty and administration identify the students' misconducts. The output of the classroom discipline process is a student discipline consequence that the student must serve, of which the parent and faculty are notified.

The input to the dean's office discipline process is also student misconduct. The student performs the misconduct, and the faculty completes the referral form which is also an input to the dean's office discipline process. The output of the dean's office discipline process is a discipline consequence that is performed by the student, and notification is provided to the faculty and parent. The input to the attendance or behavior contract process is the attendance or behavior problem of the student identified by the faculty or attendance office. The output of the attendance or behavioral contract process is the attendance or behavior contract, that the student must get signed, with appropriate notifications to the faculty and parent. The SIPOC is shown in Figure 4.9.

7. Team Member Bios

Team members created a short bio describing their background and skills for the project.

Marcela Bernardinez
Marcela Bernardinez was born in 1980 in San Miguel de Tucuman, Argentina but raised in Venezuela because her parents decided to move. After she finished high school in Venezuela, she decided to have a new experience, meet new people, find new opportunities, and discover a new world, so she came to the U.S. to study industrial engineering. She has been in the U.S. for six years, and it has been a challenge to arrive at where she is now. Marcela has a bachelor's degree from the University of Central Florida in industrial engineering, and is pursuing her master's degree in industrial engineering at the same university. In addition, she is a member of the Institute of Industrial Engineers and Society of Hispanic Professional Engineers.

Supplier	Input	Process	Output	Customers
• Students • Faculty • Admin	• Student misconduct	Perform classroom discipline	• Discipline consequence	• Students • Faculty • Parent
• Students • Faculty • Dean's office	• Student misconduct • Referral form	Perform dean's office discipline process	• Discipline consequence	• Students • Faculty • Parent
• Students • Faculty	• Attendance problem • Behavior problem	Manage attendance and behavioral	• Attendance contract • Behavior contract	• Students • Faculty • Parent

FIGURE 4.9 SIPOC.

Khalid A. Buradha

Khalid Buradha is a field supervisor in the Inspection Department of Saudi Aramco Company. His contribution was focused on monitoring and managing the quality system of the final product (Projects > $2 million). He holds a bachelor's degree in electrical engineering from Tulsa University in Oklahoma. He was a maintenance engineer for steam and natural gas plants. Part of his main duties was to put the down equipment back in service, and to enhance the performance and reliability of plant equipment, as well as developing several databases for the department.

Kevin S. Cochie

Captain Kevin S. Cochie is an active duty officer in the U.S. Army. His specialty is Special Operations Aviation. He pilots the MH-47E Chinook and MH-6 Littlebird, and served combat time in Afghanistan and Iraq. He holds a bachelor's degree in design engineering technology from the University of Central Florida and a master's degree in management from Troy State University.

Jose G. Saenz

Jose is native to Panama. He has a bachelor's degree in industrial engineering management and is pursuing a master's degree in quality engineering at the University of Central Florida. He worked as a cost analyst for Towerbank, a local bank from Panama, where he contributed to the implementation of an activity based costing system. Jose is an active member of ASQ, and the treasurer of the ASQ Student Chapter of the Orlando section.

Dr. Sandy Furterer

Dr. Furterer is the course instructor for the Total Quality Improvement course at the University of Central Florida, as well as the Master Black Belt for the project. She has extensive consulting experience in process improvement, Lean Six Sigma, and information systems improvement. Dr. Furterer has a bachelor's degree and master of science degree in industrial and systems engineering from Ohio State University, an MBA from Xavier University, and a PhD in industrial engineering from the University of Central Florida.

8. Define Phase Presentation

The Define phase presentation summarizing the written Define phase presentation is included in the downloadable instructor materials.

DEFINE PHASE CASE DISCUSSION

1. **Define Phase Written Report**

   1.1 How did your team ensure the quality of the written report? How did you assign the work to your team members? Did you face challenges of

team members not completing their assigned tasks in a timely manner and, if so, how did you deal with it?

1.2 Did your team face difficult challenges in the Define phase? How did your team deal with conflict in your team?

1.3 Did your instructor and/or Black Belt or Master Black Belt mentor help your team better learn how to apply the Lean Six Sigma tools, and how?

1.4 Did your Define phase report provide a clear vision of the project, why or why not?

1.5 How could you improve your Define phase report based on the Define phase report given in the book? How could you improve the Define phase report in the book?

2. Lean Six Sigma Project Charter

Review the project charter presented in the Define phase report.

2.1 A problem statement should include a view of what is going on in the business, and when it is occurring. The problem statement should provide data to quantify the problem. Does the problem statement in the Define phase case study example written report provide a clear picture of the business problem? Rewrite the problem statement to improve it.

2.2 The goal statement should describe the project team's objective and be quantifiable (if possible). Rewrite the Define phase case study example's goal statement to improve it.

2.3 Did your project charter's scope differ from the example provided? How did you assess what was a reasonable scope for your project?

3. Stakeholder Analysis

Review the stakeholder analysis in the Define phase report.

3.1 Is it necessary to identify the many stakeholders as in the example case study?

3.2 Is it helpful to group the stakeholders into primary and secondary stakeholders? Describe the difference between the primary and secondary stakeholder groups.

4. Team Ground Rules and Roles

4.1 Discuss how your team developed your team's ground rules. How did you reach consensus on the team's ground rules?

5. Project Plan and Responsibilities Matrix

5.1 Discuss how your team developed their project plan and how they assigned resources to the tasks. How did the team determine estimated durations for the work activities?

6. SIPOC

6.1 How did your team develop the SIPOC? Was it difficult to start at a high-level, or did the team start at a detailed level and move up to a high-level SIPOC?

7. Team Member Bios

7.1 What was the value in developing the bios, and summarizing your unique skills related to the project? Who receives value from this exercise?

8. Define Phase Presentation

8.1 How did your team decide how many slides/pages to include in your presentation?
8.2 How did your team decide upon the level of detail to include in your presentation?

SHS DISCIPLINE PROCESS IMPROVEMENT LEAN SIX SIGMA PROJECT MEASURE PHASE

THE PROCESS

The discipline program is a system of subprocesses that work together to achieve an environment conducive to quality learning. The discipline system has been divided into the following subprocesses:

Classroom discipline and referral initiation: This subprocess consists of the actions that occur in the classroom when a faculty member observes a student infraction of the code of conduct. This subprocess is initiated with a student infraction and ends with classroom discipline imposed or the initiation of a referral to the dean's office. Once a student misbehaves within the classroom, the teacher will decide, based upon the severity of the infraction, whether they will give an in-class discipline consequence (typically an additional assignment), separate them from the class, or call their parent. If the teacher decides to call the parent, they will access the student's information card for the phone number. If they are able to contact the parent, they will talk to them and discuss the student's behavior. Sometimes the contact phone number is not correct, so the teacher will investigate to get the correct phone number (email the discipline dean, or ask the student for the correct phone number). The teacher will decide if they want to meet with the parent and will meet with them as appropriate. The teacher will then give the student their discipline consequence. If the infraction is severe (level 2 or above), the faculty member will send the student to the discipline dean's office and complete a discipline referral form, and then send this to the discipline dean.

Dean's office discipline actions: This process begins once the dean's office receives a student from a faculty referral, includes the processing of that student, data entry of the completed referral, and feedback to the faculty member who initiated the referral. There are severity levels of misconducts (refer to safety/discipline referral form on p. 76). If the student committed a misconduct of level 2, 3, or 4, or one that the teacher deemed necessary to send to the discipline dean, they will receive a discipline referral. The teacher will complete the form, but may need to wait until the end of the class period. The student may wait at the discipline dean's office until the form is sent by the teacher, and the discipline dean is free. This wait can range from 20 minutes to 1.5 hours. The dean will access the student's file and history of offenses. If the misconduct is level 2, 3, or 4, the dean will contact the parent. If the misconduct is level 4 or higher, the dean opens a police investigation. The dean will complete the form, and give the student the discipline consequence. The student should complete the discipline action. The dean will follow-up to ensure that the student completed the discipline action. If the student does not complete the action, then the dean will assign another discipline action. If the misconduct is habitual, or the student refuses to complete the action, the dean will institute a behavior contract. This is a contract with the student, the dean and the parent that provides for consequences and an agreement of improved behavior. The dean will complete the referral form and give the form to the office assistant, who will enter the information into the student database. A copy of the form is put into the student's file and given to the faculty member.

Attendance contract: This process is initiated when a student is put onto an attendance contract and includes the process to track the contract and subsequent discipline penalties if the contract is violated. When the student has more than five unexcused or more than ten excused absences, the attendance dean will initiate a student attendance contract with the student. An initial meeting is held with the student, the attendance dean and the parents/guardians. The student must get signatures from each teacher when they attend each class for the length of the contract (i.e., one month). The attendance dean will verify that the contract is completed each month. If the student violates the attendance contract, the student will receive consequences and a discipline referral and consequence will be completed. Once the student completes the attendance contract, the contract is stored in the student's file.

In-school or out-of-school suspension: A critical subprocess is determination of whether the student is put into in-school or out-of-school suspension. It is important to determine if the students who spend time suspended out of school have lower academic performance than those with in-school suspension. When the student receives a suspension, their teachers are emailed. The teacher provides the material that the student will miss during their suspension. The in-school suspension teacher will verify that the teachers send the material, if they do not they will follow-up with the teachers and discipline dean. The latter will then follow-up with the teachers to ensure that the students receive the missing work. In general, the teachers provide the needed materials. The student will attend the in-school suspension and the In-school Suspension Teacher will log their attendance. The student who received the out-of-school suspension will return to school. Their teachers will ensure that the work was made up.

Faculty and student focus group to determine critical to satisfaction criteria: The Lean Six Sigma team held a focus group with a representative sample of faculty and another with a group of students. The purpose of the focus group was to identify what is important to the faculty and the students regarding the discipline process. There were four main CTS criteria derived from the focus groups:

1. Minimize classroom disruptions
2. Minimize school discipline referrals
3. Level of knowledge of the student code of conduct
4. Consistency of offenses and actions

Student database information: It is important to administration that the Lean Six Sigma project provides a statistically based analysis of student discipline data in the student information system to understand which students have the highest percentage of discipline referrals by class, race/ethnicity, and socio-economic class, as well as the percentages by type of discipline referral. The downloadable instructor materials includes the student database information, "SHS Case Study Data.xls;" and the time to process the discipline referrals "Disc_time.xls."

Voice of customer (VOC) surveys: SHS process owners and administrators will assist in deployment of faculty, and student surveys, as well as interviews of key stakeholders within the administrative staff and leadership team. The downloadable instructor materials includes the faculty and student survey data: "faculty survey data.xls;" "student survey data.xls."

MEASURE PHASE EXERCISES

1. **Measure Report**
   Create a measure phase report, including your findings, results and conclusions of the Measure phase.

2. **Process Maps**
   Create level-1 and level-2 process maps for each of the following processes:
   - Classroom discipline and referral initiation
   - Dean's office discipline actions
   - Attendance contract
   - In-school or out-of-school suspension

3. **Operational Definitions**
   Develop an operational definition for each of the four identified CTS criteria:
   - Minimize classroom disruptions
   - Minimize school discipline referrals
   - Level of knowledge of the student code of conduct
   - Consistency of offenses and actions

4. **Data Collection Plan**
   Use the data collection plan format in Figure 4.10 to develop a data collection plan that will collect VOC and voice of process (VOP) data during the Measure phase.

5. **VOC Surveys**
   Create a VOC survey to better understand the faculty's requirements related to the discipline process elements:

Critical to satisfaction	Metric	Data collection mechanisms (surveys, interviews, focus groups, etc.)	Analysis mechanisms (statistics, statistical tests, etc.)	Sampling plan (sample size, sample frequency)	Sampling instructions (who, where, when, how)
Minimize classroom disruptions					
Minimize school discipline referrals					
Level of knowledge of the student code of conduct					
Consistency of offenses and actions					

FIGURE 4.10 Data collection plan.

a. Validate the initial critical to satisfaction characteristics identified above
b. Level of teacher/parent contact
c. Level of teacher/parent/counselor contact
d. Teacher classroom control
e. Consistency of classroom discipline
f. Consistency of discipline dean actions
g. Faculty knowledge of code of conduct
h. Student knowledge of code of conduct
i. Amount of training of students on the code of conduct
j. Time spent on training the students per semester on the code of conduct
k. Whether the faculty feel it is important to review the code of conduct with the students
l. Whether the faculty feel it is important to provide course information to students who are suspended
m. The length of time that the faculty take to complete the discipline referral form after sending the student to the Discipline Dean
n. The level of training by the faculty on the discipline referral process
o. Whether the faculty attempt to contact the parents when students have chronic discipline problems
p. The most effective in-class discipline action
q. Whether the faculty count "tardies" in their classes
r. Whether the faculty feel that it is important to count tardies
s. Whether the faculty feel it's important to log attendance
t. Whether the faculty have classroom rules that they enforce

u. Whether the faculty is satisfied with classroom discipline
v. Whether the faculty is satisfied with the discipline dean's discipline process
w. Whether the faculty is satisfied with the overall discipline process at the high school

Create a VOC Survey to better understand the students' requirements related to the discipline process.

a. Level of discipline imposed by classroom teachers
b. Level of discipline imposed by discipline and/or attendance deans
c. Number of teachers you observed calling students' parents
d. Level of knowledge of student with student code of conduct
e. How was the student trained on the student code of conduct?
f. How much time did the teacher spend on the student code of conduct training?
g. How many of your teachers log tardies?
h. If you received a discipline referral, what was your parent's reaction?
i. If you received a discipline referral, what was your parent's reaction to the discipline consequence?
j. If you received a discipline referral, how long did you wait to see the discipline dean?
k. Which discipline action is the most effective when disciplining students?
    - Attendance/behavior contract
    - Special work assignment
    - Lunch/after school suspension
    - Saturday School
    - Out-of-school suspension
    - In-school suspension

6. **Pareto Chart**
   Create a Pareto chart using the data in Figure 4.11 related to discipline offenses by type.

7. **VOP Matrix**
   Create a VOP matrix using the VOP matrix template detailed below to identify how the CTS, process factors, operational definitions, metrics, and targets relate to each other (Figure 4.12).

8. **Benchmarking**
   Perform a benchmarking of local high schools or school districts to understand how they handle their discipline processes.

9. **Statistical Analyses**
   Review the student database, "SHS Data for Case Study.xls" and perform the following statistical analyses:

Offense type	Frequency of occurrence
Dress code	274
Failure to report for detention	97
Unauthorized absences	255
Fighting	57
Other-level 1	188
Tardiness	247
Profane/obscene language	91
Class disruption	188
Insubordination	120
Repeat miscellaneous	111
Disrespect for others	64
Insubordination level 2	91
Repeated miscellaneous other	83
Other	395
Gross insubordination	73

FIGURE 4.11 Pareto chart data.

CTS	Process factors	Operational definition	Metric	Target

FIGURE 4.12 VOP matrix template.

(A) Create a histogram for the number of discipline referrals.
(B) Calculate the mean and standard deviation for the number of discipline referrals.
(C) Does the data follow a normal distribution?

10. Validate the Measurement System

Develop a plan for performing an attribute gage R&R study to assess the consistency across the discipline deans for the consequences that they give for several of the commonest consequences. Refer to offenses.xls data.

11. COPQ

Brainstorm potential COPQ for the case study for the following categories:

- Prevention
- Appraisal
- Internal failure
- External failure

12. Measure Phase Presentation

Prepare a presentation (PowerPoint) from the case study exercises that provides a short (10–15 minutes) oral presentation of the Measure Phase deliverables and findings.

MEASURE PHASE

1. Measure Report

A report of the Measure phase for the SHS discipline process improvement project, including the key deliverables developed as part of the prior exercises, is detailed below. The Measure phase of the DMAIC process is designed to gain information on the process performance and develop problem and/or process improvement statements. The objectives of this phase in relation to the SHS discipline improvement project are as follows:

- Map the current process
- Gather initial data and determine current process performance
- Confirm key customer requirements of the process
- Organize and stratify all data collected

To define the current state of this process, face-to-face interviews and meetings with the deans were conducted. Consequently, a process map of the entire discipline system was developed. This process map or workflow chart is a schematic picture of the process being studied. It also shows the major steps in the process. In the Analyze phase of the DMAIC cycle, each step in this process will be examined with respect to time, value, and cost.

In general, the discipline system has three major components or subprocesses:

1. **Classroom Discipline:** This subprocess starts when student misconduct occurs in the classroom. Based on the severity of the misconduct, an immediate discipline action is imposed or a referral to the dean's office is initiated.
2. **Dean's Office Discipline:** This subprocess begins when a misconduct is observed on the school premises or when a referral is received (an output of the classroom discipline subprocess). Depending upon the offense type, appropriate actions/sequences against the offenders will be established. These actions are tracked to ensure that the student completes his/her discipline consequence. Failure to do so will lead to new subsequent discipline actions.
3. **Attendance/Behavioral Contract:** This process is initiated when a student is placed on an attendance contract because of his/her misconduct (tardiness or class skipping). The student will obtain signatures in the attendance log/calendar from their teachers and parents daily. The attendance log/calendar has to be submitted to the attendance office on monthly a basis. Violators to the contract rules will face new subsequent discipline penalties.

2. Process Maps

The process maps for each subprocess of the discipline process are shown in Figures 4.13, 4.14, and 4.15.

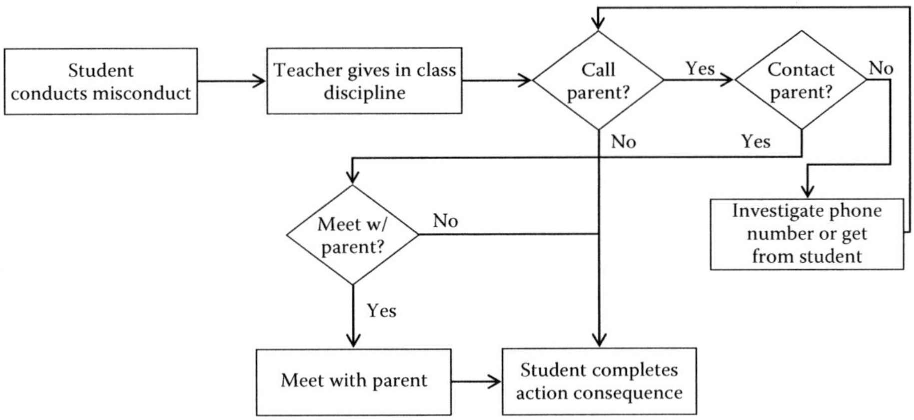

FIGURE 4.13 Classroom discipline process map.

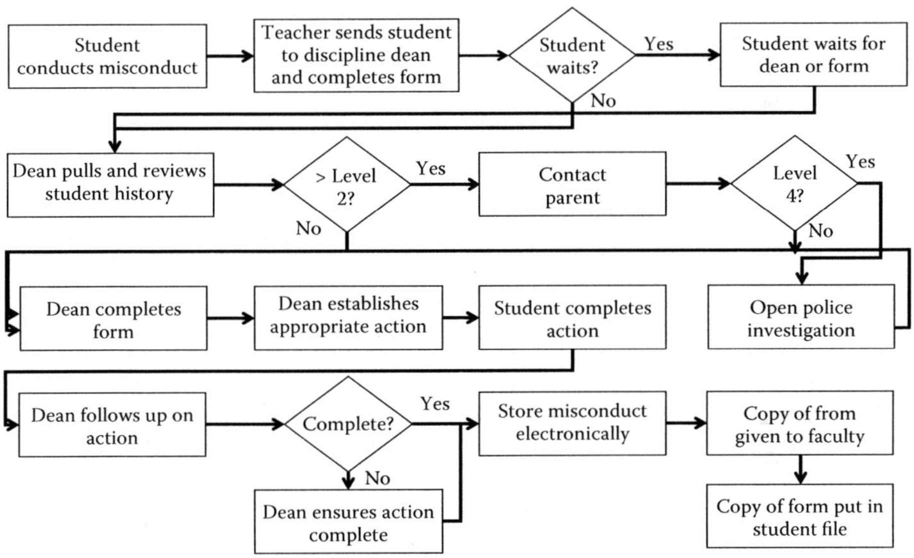

FIGURE 4.14 Dean's office discipline process map.

3. Operational Definitions

CTS: Minimize classroom disruptions.
Defining the Measure: Number of classroom disruptions per week.
Purpose: To determine the number of discipline disruptions in the classroom.
Clear way to measure the process: Determine a statistically valid sample of each type of classroom, by subject and class grade (9, 10, 11, 12) and level (i.e., Advanced Placement, International Baccalaureate, Honors, General, Basic).

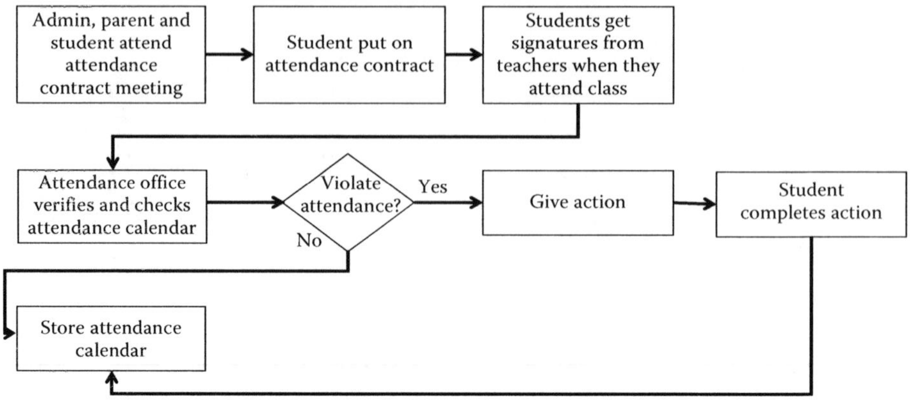

FIGURE 4.15 Attendance/behavioral contract process map.

CTS: Minimize school discipline referrals.
Defining the Measure: Number discipline referrals by week.
Purpose: To determine the number of discipline referrals by period.
Clear way to measure the process: Extracted from student database, average and standard deviation for timeframe sampled.

CTS: Level of knowledge of student code of conduct.
Defining the Measure: Number discipline referrals by type.
Purpose: To determine the root causes of discipline referrals and most frequently occurring types, to infer the level of knowledge of the code of conduct.
Clear way to measure the process: Extracted from student database, average and standard deviation for discipline referrals by type.

CTS: Consistency of offenses and actions.
Defining the Measure: Number of actions to offense types.
Purpose: To determine the consistency of offenses to actions given.
Clear way to measure the process: Extracted from the student database, listing actions to offenses in the sample. Also gain VOC from faculty and student surveys to understand perceived consistency of discipline process.

4. Data Collection Plan

The data collection plan for the project is shown in Figure 4.16.

In the Measure phase of a DMAIC cycle, the Lean Six Sigma team collected the necessary data through the SHS student database to gain insight into the current discipline process performance and to identify areas of improvement.

In general, the data collection revealed the following facts:

- Most offenders are freshmen and represent 42% of the offender population.
- As repeat offenders (more than one referral, more than two referrals, etc.) are analyzed, the percentage of freshmen accounting for them increases.

Critical to satisfaction (CTS)	Metric	Data collection mechanism	Analysis mechanism	Sampling plan	Sampling instructions
Minimize classroom disruptions	Number of discipline referrals per week	Surveys, focus groups	Histograms, pareto charts, basic statistics, hypothesis tests	Faculty (goal 50), subset of students (goal 400)	Faculty: email request sent by principal requesting participation. Students: select classes take survey in computer lab
Minimize school discipline referrals	Number of discipline referrals per week	School database	Basic statistics, hypothesis tests	All students	Extracted from database for year to date
Level of knowledge of code of conduct	Number of referrals by type	School database	Pareto charts, basic statistics, hypothesis tests	All students	Extracted from database for year to date
Discipline offense and action consistency	Number of consistent offenses to actions	School database	Pareto charts, basic statistics, hypothesis tests	All students	Extracted from database year to date

FIGURE 4.16 Data collection plan.

- The academic performance averages (GPAs) for students with no discipline issues (2.99 GPA) are higher than the offenders (2.19 GPA).
- Students with discipline issues have more absent days (4 more on average) than students who have no discipline issues.
- Among all of the categories, repeat offenders who score in the lower 30% or the reading Florida Comprehensive Assessment Test (FCAT) have the lowest academic performance average (1.53).
- Students who are given out-of-school suspension have statistically lower GPAs (1.86 GPA) than students who receive punishment, but not out-of-school suspension (2.26 GPA).
- Approximately 20% of the violation codes account for 80% of the referrals issued to students.

5. VOC Faculty and Student Surveys

Survey Methodology

Surveys are powerful tools used to measure perceptions or importance of stakeholders of process characteristics. Perceptions are sometimes skewed to individual personalities, but they are reality to those who work within the process. Process owners must be attentive to these perceptions because they influence the overall operation and efficiency of the system. Perceptions are vital to workforce buy-in when collective efforts are required to maximize the efficiency of the process. This is particularly true for the discipline process because so many individuals influence the process. The students, faculty, and administration comprise many personalities, social backgrounds, and philosophies. For this system to operate efficiently and effectively, these individuals must buy-in to working collectively for a common goal of the entire organization. This includes consistency across classroom discipline, consistent backing of faculty discipline actions, and consistent enforcement of code of conduct violations.

The Six Sigma team developed two surveys (downloadable instructor materials) to collect data from the faculty and students. The goal of these surveys was to capture the student and faculty perceptions of the discipline program at SHS.

The development of the survey was a very detailed process. The SHS Six Sigma project team first conducted a brainstorming session to decide what data needed to be collected from the surveys. The session was conducted by constructing two affinity diagrams. The purpose of the affinity diagram is to organize brainstormed data into categories. The first diagram was labeled "What data do we want to collect?" The second affinity diagram was then made by constructing questions that supported the "ideas" from the first affinity diagram. The end result was two draft surveys.

Draft surveys were reviewed by the team's Master Black Belt. Once modifications were made to the question construct, format, and order, the surveys were put in front of a group of administrators from SHS for review. The discipline deans and attendance deans reviewed the surveys and provided input for modification. This included adding several questions and modifying several answer responses for other questions.

The next step was to meet with a focus group of faculty members to ensure that the survey would collect appropriate key information pertaining to the discipline process. This session was very productive and was conducted by applying a formal brainstorming session. The affinity diagram that resulted yielded several areas they thought were important to classroom discipline. The major issue this group noted was that the faculty tended not to contact parents on discipline-related issues, a lack of parent involvement, and a lack of parent/teacher/counselor integration. This input was then incorporated into the surveys and final drafts were produced. Once the final surveys were approved by the university's Internal Review Board (IRB) and the SHS principal, the team established the logistics for administering the surveys.

It was determined that the entire faculty population of 173 teachers would be sampled, as well as 500 students. The principal selected six teachers to solicit volunteers from their students to take the student survey. All six teachers were eager to participate and the end result was that more than 540 students responded to the survey. The team performed a statistical analysis on the sample to ensure that the demographics of the sample mirrored the student population demographic breakdown.

Both surveys were converted to web-based forms and posted on the web to streamline administration of the surveys and data collection time. The faculty was notified by multiple emails soliciting their participation. The students were sampled over a two-day period by having their teachers bring them to a computer laboratory on the SHS campus.

6. Pareto Chart

A Pareto analysis may be used to categorize and prioritize attribute data (qualitative). It helps prioritize improvement efforts and identify the most frequent problems. Moreover, it states that approximately 20% of root causes of problems account for approximately 80% of all problems encountered. Consequently, to significantly reduce problems in the organization, you can focus on a few key problem areas, i.e., the 20%.

Most frequent offenses:

Based on the safety/discipline referral form (p. 76), the team found that there are 15 possible level-1 offenses, 13 possible level-2 offenses, 17 possible level-3 offenses, and 20 possible level-4 offenses, resulting in 65 possible offenses.

The team reviewed 2204 offenses from the database with the purpose of identifying the most frequent offenses. The dates of the offenses range from the beginning of the school year through February 23. A Pareto chart was developed, shown in Figure 4.17.

The data revealed that 14 out of 65 offense types (22%) account for 82% of all discipline issues in the school. The other category on the Pareto chart was large, but comprised many discipline consequences with very small percentage occurrences, so the team focused on the largest number of higher-percentage discipline occurrences. Figure 4.18 shows the students' response when asked to rate the most effective discipline consequence. Attendance/behavioral contract, lunch and after-school detention, in-school suspension, Saturday school, out-of-school suspension, and special work assignments were the highest-rated consequences.

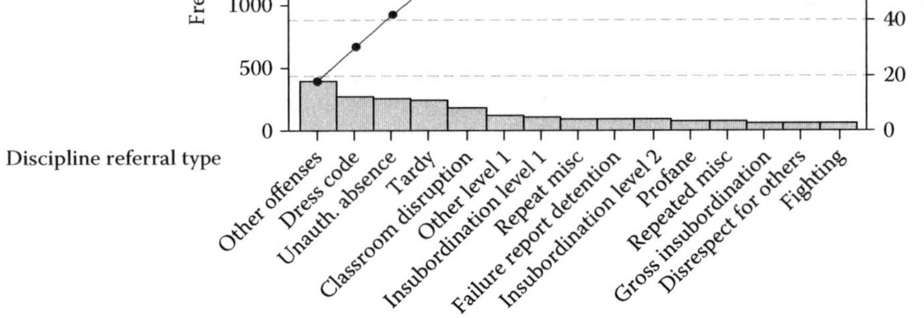

Frequency	395	274	255	247	188	120	111	97	91	91	83	73	64	58	57
Percent	18	12	12	11	9	5	5	4	4	4	4	3	3	3	3
Cum %	18	30	42	53	62	67	72	77	81	85	89	92	95	97	100

FIGURE 4.17 Pareto chart of offenses.

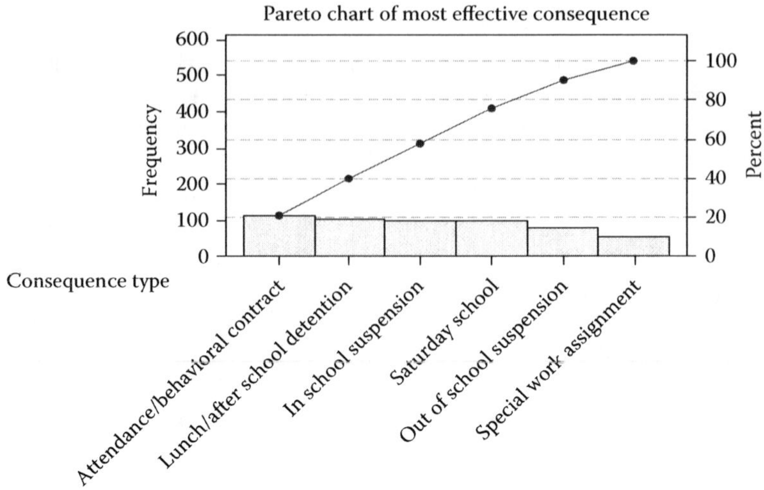

Frequency	112	102	100	98	76	52
Percent	20.7	18.9	18.5	18.1	14.1	9.6
Cum %	20.7	39.6	58.1	76.3	90.4	100.0

FIGURE 4.18 Pareto chart of most effective consequences.

7. VOP Matrix

The VOP matrix demonstrates the integration of CTS criteria to the factors that may impact the CTSs, as well as the operational definition of the CTS along with the metrics and target goals for the metrics. Because the main focus of the Six Sigma project is to meet the customers' requirements or CTSs, this helps to ensure that we are measuring the process to meet the needs of the customers. The VOP matrix is shown in Figure 4.19.

8. Benchmarking

The Lean Six Sigma team benchmarked other high schools, specifically in the disciplinary area. The Lean Six Sigma team used benchmarking to measure the SHS disciplinary system performance against similar high schools.

Benchmarking was performed on other high schools (including private and public high schools) in Florida. More than 20 public and private high schools were involved in this benchmarking process. After the information was gathered from other high school's performance in the disciplinary area, the team created a benchmarking matrix revealing some of the facts found in this analysis; these facts are summarized here.

This benchmarking analysis reveals many interesting approaches implemented by other high schools that could be considered in the process of improving the SHS disciplinary system. The team noticed there are some high schools that developed their own disciplinary procedures and policies, such as the attendance

CTS	Factors	Operational definition	Metric	Target
Minimize classroom discipline issues	Freshmen training on code Clear guidelines	Training exists and is performed Clear guidelines exist	Number disruptions	Reduce number disruptions by 50%
Classroom discipline consistency	Guidelines teacher training	Clear guidelines exist Teacher training each year	Guidelines number of faculty trained	100% of faculty are trained within 3 months of hire or Jan. 1
Teacher/parent/counselor integration	Apathy of parents data missing	Engaged parents assessed by parent survey	Percentage of responses on survey for identified questions	Increase % of ratings in high categories by 10%
Adherence to code of conduct	Training expectations	All students will be trained in code of conduct for 2 hours per semester Clear expectations conveyed	Number of students trained	100% of students are trained within first month of school or transfer
Classroom control	Teacher training	All teachers trained in classroom management mentors for new teachers	Number of teachers trained Number of teachers with mentors Rating of mentoring program	100% of teachers trained 100% of new teachers have a mentor
Teacher/ parent contact	Apathy of parents Data missing	Engaged parents assessed by parent survey	Percentage of responses on survey for identified questions	Increase percentage of ratings in high categories by 10%
Reduction of referrals	Freshmen No emphasis on common offenses	Guidelines	Freshmen and transfers trained in code of conduct and guidelines	100% of freshmen and transfers trained in code of conduct and guidelines

FIGURE 4.19 VOP matrix.

and tardies policies implemented by Cocoa Beach and Bishop Moore High Schools. These schools discretely state the consequences of being tardy to class and school. Frequent absences and tardies lead to poor academic work, lack of social development, and possible academic failure. Regular attendance has a significant role with respect to school interest, social adjustment, and scholastic achievement.

The Timber Creek High School and Vero Beach High School are currently implementing positive behavior support systems. This system involves proactive strategies for defining, teaching, and supporting appropriate student behaviors to create positive school environments. This application is giving these schools the opportunity to improve lifestyle results (personal, health, social, family, work, recreation) for all students by making problem behavior less prevalent by creating a positive school environment where desired behavior is taught to the student through proactive strategies.

In addition to the positive actions discussed above, other interesting findings include well-defined parent/teacher communication guidelines, progress reports, student's hallway passes, and parent involvement contracts.

9. Statistical Analysis

The average number of discipline referrals in the database for the six months of data for the entire student population was 1.536, with a standard deviation of 5.057. The histogram of the data is shown in Figure 4.20.

Minitab® was used to perform a test for normality. The p-value is less than .005, so the null hypothesis that the data is normal must be rejected, so the data is not normal.

10. Validate the Measurement System

Note: The results of the following gage R&R study are hypothetical, and not based upon actual results.

An attribute gage R&R study was performed to assess the consistency of the application of discipline consequences to offense types. When students receive a discipline referral, there are, in general, several offenses that are simultaneously breached in the code of conduct. The team reviewed the offense database to select several combinations of offenses that occurred in the student population over the

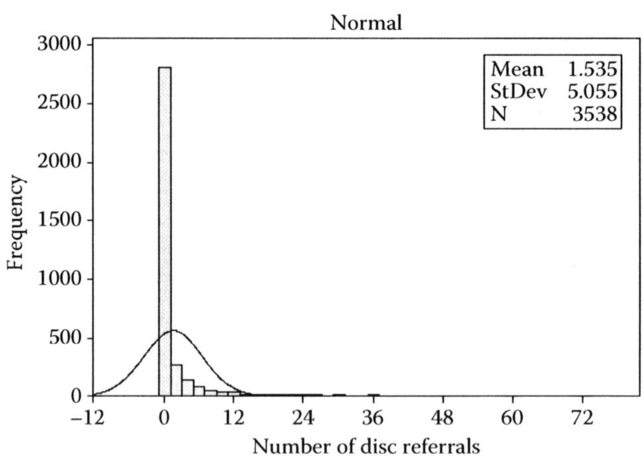

FIGURE 4.20 Histogram of number of discipline referrals.

current academic year. Three discipline deans reviewed the referrals, and assigned discipline consequences. The initial design of the gage R&R is given in Figure 4.21. The table includes the sample number, the offense types reviewed, the actions from an expert panel identifying the "best" discipline action for the offenses, and each of the dean's recommendations for the discipline actions. Figure 4.22 shows the results of the samples for the study. Figure 4.23 shows the summarized results of the study.

The Appraiser score percentage differed from 50 to 90% accuracy across the three discipline deans. This demonstrates the repeatability, or whether the appraiser agrees with their rating on each trial. The core versus attribute scores ranged from 30 to 60%, identifying whether the appraiser agreed with a known standard for the two trials. So even though one of the appraisers was more consistent with himself/herself, h/she did not agree with the standard discipline action. The screen percent effective score was 10%, which indicated that the discipline deans did not agree with each other. The screen percent effective score versus attribute was also 10%, which indicated that the discipline deans did not agree with each other or with the standard action. For a consistent measurement system, these scores should be ≥80%. The gage R&R study shows that the system lacks consistency in assigning discipline actions for the offenses. This could be due to the nature of dealing with people, and the large combination of offense types and discipline actions available.

Known	Offense types	Actions	Dean 1		Dean 2		Dean 3	
Sample		Actual	Try 1	Try 2	Try 1	Try 2	Try 1	Try 2
1	3F, 3Q	A, B, C, R	A, B, C, R	A, C	A, B, R	A, C	A, B, C, R	A, C
2	1L	C	C	C	C	C	A, B, C, G	C
3	1J, 1L	B, C, G	B, C, G	B, C, G	A, B, C, D	A, B, C, D	A, B, H, I	B, C, G
4	3D, 3H, 3Q	A, B, C, R	A, B, C, R	A, B, C, R	B,C, D	B, C, D	A, B, C, R	A, B, C, R
5	2B, 2F, 2I, 2M	B, C, P	B, C, P	B, C	A, B, P	A, B, P	B, C, P	B, C, P
6	3D	A, B, C, R	A, B, R	A, B, R	A, B, C, R	A, B, C	B, R	B, R
7	4B	A, S	A, S	A, S	A, S	A, S	A, S	A, S
8	4F	A, B, F, S	A, B, C, S	A, B, C	A, B, F, S	A, B, F, S	A, B, S	A, B, S
9	2B, 2M	B, C, K	B, C, K	B, C, K	B, C, H	B, C, H	B, C, K	B, C
10	1F, 1L, 2I	B, C, G, I	B, C, G, I	B, C, G, I	B, C	B, C	B, C, G, I	B, C, G

FIGURE 4.21 Attribute gage R&R design and results.

Known	Offense types	Actions	Dean 1		Dean 2		Dean 3	
Sample		Actual	Try 1	Try 2	Try 1	Try 2	Try 1	Try 2
1	3F, 3Q	A, B, C, R	Pass	Fail	Fail	Fail	Pass	Fail
2	1L	C	Pass	Pass	Pass	Pass	Fail	Pass
3	1J, 1L	B, C, G	Pass	Pass	Fail	Fail	Fail	Pass
4	3D, 3H, 3Q	A, B, C, R	Pass	Pass	Fail	Fail	Pass	Pass
5	2B, 2F, 2I, 2M	B, C, P	Pass	Fail	Fail	Fail	Pass	Pass
6	3D	A, B, C, R	Fail	Fail	Pass	Fail	Fail	Fail
7	4B	A, S	Pass	Pass	Pass	Pass	Pass	Pass
8	4F	A, B, F, S	Pass	Fail	Pass	Pass	Fail	Fail
9	2B, 2M	B, C, K	Pass	Pass	Fail	Fail	Pass	Fail
10	1F, 1L, 2I	B, C, G, I	Pass	Pass	Fail	Fail	Pass	Fail

FIGURE 4.22 Gage R&R results.

11. COPQ

Based on the COPQ classifications, the SHS may have the costs described below.

1. **Internal failure cost:**
   - (a) *Cost of processing unnecessary referrals:* it is a reduced productivity cost because the dean and his or her staff have to incur labor-hour costs and resources in dealing with unnecessary referrals.
   - (b) *Cost of over processing due to the referral routing procedure:* it is a reduced productivity cost because there are more personnel (i.e., dean's secretary) involved in processing the referral.
   - (c) *Cost of dealing with repeat offenses:* it is a rework cost because the administration has to process a referral again due to a repeat offense. It also impacts productivity because the dean and staff incur labor-hour costs and resources "reworking" offenders.

2. **Appraisal cost:**
   - (a) *Cost of attendance contract inspection:* it is an inspection cost because the administration performs random inspections of attendance contract files.
   - (b) *Cost of reviewing student discipline issues:* it is a control cost because the school may incur labor-hour cost reviewing the students' discipline records, and retrieving information from the database.

SCORING REPORT

DATE: __________

NAME: HS Process Improvement

PRODUCT: Offenses and Action Types

BUSINESS: 0

Attribute Legend

1 pass

2 fail

Known population		Operator #1		Operator #2		Operator #3		Y/N	Y/N
Sample #	Attribute	Try #1	Try #2	Try #1	Try #2	Try #1	Try #2	Agree	Agree
1	Pass	Pass	Fail	Fail	Fail	Pass	Fail	N	N
2	Pass	Pass	Pass	Pass	Pass	Fail	Pass	N	N
3	Pass	Pass	Pass	Fail	Fail	Fail	Pass	N	N
4	Pass	Pass	Pass	Fail	Fail	Pass	Pass	N	N
5	Pass	Pass	Fail	Fail	Fail	Pass	Pass	N	N
6	Pass	Fail	Fail	Pass	Fail	Fail	Fail	N	N
7	Pass	Pass	Pass	Pass	Pass	Pass	Pass	Y	Y
8	Pass	Pass	Fail	Pass	Pass	Fail	Fail	N	N
9	Pass	Pass	Pass	Fail	Fail	Pass	Fail	N	N
10	Pass	Pass	Pass	Fail	Fail	Pass	Fail	N	N
APPRAISER SCORE[1]			70.00%		90.00%		50.00%		
CORE VS. ATTRIBUTE			60.00%		30.00%		30.00%		

SCREEN % EFFECTIVE SCORE[3] -> 10.00 %

SCREEN % EFFECTIVE SCORE VS. ATTRIBUTE[4] -> 10.00 %

FIGURE 4.23 Attribute gage R&R summary.

3. **Preventive cost:**
   (a) *Cost of training of the code of conduct:* it is a training cost because the administration spends resources in training its faculty regarding the code of conduct and the discipline policies of the school. On the other hand, the faculty incurs labor-hour cost in teaching the code of conduct to students.
   (b) *Cost of setting policies and procedures:* it is a policies and procedures cost because the school may spend resources and labor-hour cost in setting the school policies regarding discipline and student behavior.

12. Measure Phase Presentation

The Measure phase presentation summarizing the written Measure phase presentation is included in the downloadable instructor materials.

MEASURE PHASE CASE DISCUSSION

1. **Measure Report**
   1.1 Review Measure report and brainstorm some areas for improving the report.
   1.2 How did your team ensure the quality of the written report? How did you assign the work to your team members? Did you face any challenges of team members not completing their assigned tasks in a timely manner and, if so, how did you deal with it?
   1.3 Did your team face difficult challenges in the Measure phase? How did your team deal with conflict in your team?
   1.4 Did your instructor and/or Black Belt or Master Black Belt mentor help your team better learn how to apply the Lean Six Sigma tools in the Measure phase, and how?
   1.5 Did your Measure phase report provide a clear understanding of the VOC and the VOP, why or why not?

2. **Process Maps**
   2.1 While developing the process maps, how did your team decide how much detail to provide on the level-2 process maps?
   2.2 Was it difficult to develop a level-2 from the level-1 process maps? What were the challenges?

3. **Operational Definitions**
   3.1 Review the operational definitions from the Measure phase report, define an operational definition that provides a better metric for assessing the level of knowledge and training of the student code of conduct.
   3.2 Discuss why it may be important for the faculty and students to be familiar with the student code of conduct.

4. Data Collection Plan

4.1 Incorporate the enhanced operational definition developed in number 3 above into the data collection plan from the Measure phase report.

5. Voice of Customer Surveys

5.1 How did your team develop the questions for the faculty and/or student survey? Did you review them with other students to assess whether the questions met your needs?
5.2 Create an affinity diagram for the main categories on the faculty or student survey, grouping the questions into the higher-level "affinities." Was this an easier way to approach and organize the questions of the surveys?

6. Pareto Chart

6.1 Discuss how the Pareto chart provides the priority for investigating root causes and variables that impact the most frequent discipline offenses.

7. VOP Matrix

7.1 How does the VOP matrix help to tie the CTSs, the operational definitions and the metrics together?

8. Benchmarking

8.1 Was it difficult to find benchmarking information specific to discipline types and processes?

9. Statistical Analyses

9.1 Statistical analyses showed that the number of discipline referrals by student is not a normal distribution. What ramifications does this have for the statistical analysis that should be performed?

10. Validate the Measurement System

10.1 Describe the approach that you took to develop the attribute gage R&R. How did you select the actions to be included in the study. How did you envision developing the "experts" operational definition for the actions to be given for the discipline offenses?

11. COPQ

11.1 Would it be easy to quantify and collect data on the costs of quality that you identified for the case study exercise?

12. Measure Phase Presentation

12.1 How did your team decide how many slides/pages to include in your presentation?
12.2 How did your team decide upon the level of detail to include in your presentation?

ANALYZE PHASE EXERCISES

1. **Analyze Report**
   Create an Analyze phase report, including your findings, results, and conclusions of the Analyze phase.

2. **Cause and Effect Diagram**
   Create cause and effect diagrams for the following effects:
   - Why does 42% of the offender population account for freshmen?
   - Why do 22% of the offense codes (14) account for 80% of the infractions committed?
   - Why do repeat offenders continue to commit code of conduct infractions?

3. **Cause and Effect Matrix**
   Create a cause and effect matrix for the following effects:
   - Why do 42% of the offender population account for freshmen?
   - Why do 22% of the offense codes (14) account for 80% of the infractions committed?
   - Why do repeat offenders continue to commit code of conduct infractions?

4. **Why-Why Diagram**
   Create a Why-Why diagram for why students must wait to see the discipline dean when getting a discipline referral.

5. **Process Analysis**
   Prepare a process analysis for the following processes:
   - Classroom discipline process
   - Dean's office discipline process
   - Attendance/behavioral contract process

6. **Histogram, Graphical, and Data Analysis**
   (a) Perform a histogram and graphical analysis for the following variables from the Student Discipline Database:
      - Number of discipline referrals
      - Student GPA
      - Number of excused and unexcused absences

   (b) Perform data analysis on the student database, "SHS Case Study Data.xls"
      - Number of referrals by grade (freshmen, sophomore, junior, senior)
      - Number of repeat offenders by grade (freshmen, sophomore, junior, senior)

7. **Waste Analysis**

   Perform a waste analysis for the following high school discipline processes:

   - Classroom discipline process
   - Dean's office discipline process
   - Attendance/behavioral contract process

8. **Correlation Analysis**

   Perform a correlation analysis for the following variables:

   - Number of discipline referrals correlated to GPA
   - Gender related to GPA
   - Race related to GPA
   - Other variables of interest in the student database, "SHS Case Study Data.xls"

9. **Regression Analysis**

   Perform a regression analysis to try to predict the number of discipline referrals, based on grade, gender, GPA, number of excused absences, number of unexcused absences, age, and number of days suspended.

10. **Basic Statistics**

    Calculate the mean and standard deviation for the following variables:

    - GPA
    - Unexcused absences
    - Excused absences
    - Number of discipline referrals across all students
    - Number of discipline referrals across students with discipline referrals

11. **Confidence Intervals**

    Calculate a confidence interval about the mean and the variance for the following variables:

    - GPA
    - Unexcused absences
    - Excused absences
    - Number of discipline referrals across all students
    - Number of discipline referrals across students with discipline referrals

12. **Hypothesis Testing**

    Perform the following hypothesis tests:

    - Is GPA different for students with discipline issues and for those without?
    - Is GPA different for students suspended versus not suspended?
    - Is average number of discipline referrals greater by gender?

We want to determine if there is a statistically significant difference for the following variables for students with at least one discipline referral:

- Is the GPA different for students suspended versus not suspended?
- Is average number repeat discipline referrals greater by gender?

13. Analysis of Variance (ANOVA)

Perform an ANOVA to determine the following for all students:

- Is GPA different by race?
- Is the average number of discipline referrals different by GPA?
- Is the average number of discipline referrals different by race?

Perform an ANOVA to analyze the following hypotheses for students with discipline issues:

- Is GPA different by race?
- Is the average number discipline referrals different by grade average?
- Is the number discipline referrals different by race?

14. Survey Analysis

- Perform survey analysis for the faculty survey data "Faculty Survey Data.xls." Include Pareto charts for each question, and chi-square analysis.
- Perform survey analysis for the student survey data "Student Survey Data.xls."

Include Pareto charts for each question, and chi-square analysis.

15. DPPM/DPMO

- Calculate the DPMO and related sigma level for the discipline process, assuming a 1.5 sigma shift, for the following data:

Opportunities for failure:

- Faculty member fails to complete the discipline referral form.
- There is a long wait at the discipline dean's office.
- The student fails to complete his/her discipline consequence.

Defects:

- Number of defects where faculty member fails to complete the discipline referral in a month = 5.
- Number of times a student waits at the discipline dean's office in a month = 20 units.
- Number of discipline referrals per month = 120.

16. Process Capability

Calculate the process capability for the discipline time "disc_time.xls" with the following specifications:

- Lower specification limit: 10 minutes.
- Upper specification limit: 30 minutes.

17. Analyze Phase Presentation

Prepare a presentation (PowerPoint) from the case study exercises that provides a short (10–15minutes) oral presentation of the Analyze phase deliverables and findings.

ANALYZE PHASE

1. Analyze Report

Following is a report of the Analyze phase for the SHS Discipline Process Improvement project, including the key deliverables developed as part of the prior exercises.

The Analyze phase of the DMAIC process is designed to gain insight into the root causes of the problems, as well as understand the process variables. The objectives of this phase in relation to the SHS Discipline Improvement Project are as follows:

- Understand the root causes
- Understand the capability of the processes
- Develop relationships between variables
- Analyze the process for value-added and nonvalue-added activities
- Identify and eliminate process waste
- Understand the defects per million opportunities and the sigma levels

2. Cause and Effect Diagram

Root cause analysis is a very important activity in a Lean Six Sigma project. It is where the data are analyzed in detail and tools are used to determine the root causes of problems and inefficiencies. Too often, data are collected and project team members, champions, or knowledge workers jump to conclusions based on raw data. Lean Six Sigma and DMAIC prevent this from happening. Several tools are very useful in determining root causes.

In this project, three areas of primary improvement were chosen to determine the root causes of inefficiencies. The team brainstormed to determine root causes for the three areas. The areas of interest were based on data derived from the analysis of the school data base data and the data collected from the faculty and student surveys. The three areas were:

1. Why does 42% of the offender population account for freshmen?
2. Why do 22% of the offense codes (14) account for 80% of the infractions committed?
3. Why do repeat offenders continue to commit code of conduct infractions?

The Six Sigma team first constructed fishbone diagrams (cause and effect diagrams) to get to the root causes of these three areas. As seen in Figure 4.24, the fishbone diagram for the effect labeled 42%, four branches were constructed as potential areas where causes exist. The branches were related to students, training given to students, campus layout, administration, and faculty. From there, the Six Sigma team brainstormed potential causes to the effect. Several root causes were identified as the

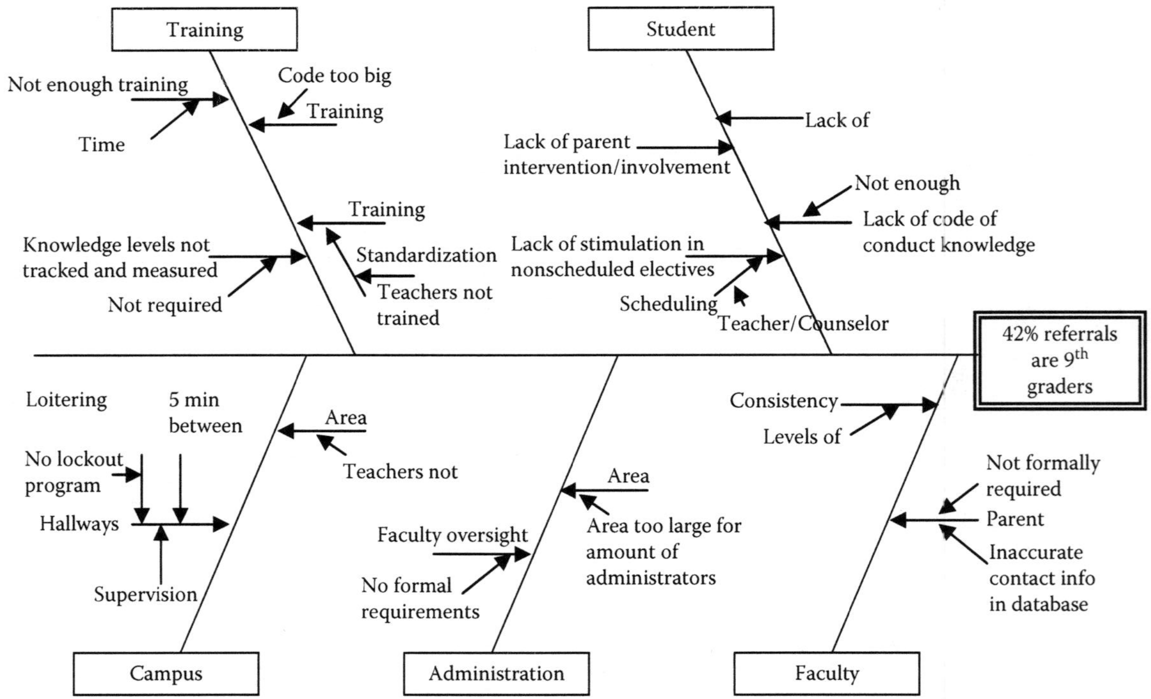

FIGURE 4.24 Cause and effect diagram: Why are most offenders freshmen?

primary reasons freshmen account for so many of the offenders. First, by nature of the training, the teachers are not formally trained in the discipline program and the amount of training the freshmen receive is inadequate. This is also evident by the data collected from faculty and student surveys. On average, the amount of training the freshmen receive per semester is < 50 minutes. The freshmen are new to the high school and come from multiple different middle schools. The fact that over 40% of the offenders being freshmen could be a result of not understanding the code of conduct, the discipline policies of SHS, and lack of knowledge of appropriate behavior. Additionally, further analysis of this 42% population revealed that most of their infractions are level-1 and level-2 dress code and attendance violations. When looking for the root cause of the attendance violations, it was determined that several factors contribute to the effect. First, the students systemically loiter in the hallways between classes. There is little sense of urgency to move from one class to another. This, coupled with the absence of faculty members in the hallways in-between class and their nonenforcement of logging tardies into the system, contribute to the problem of level-1 attendance infractions. A joint lock-out program between administration and faculty from the discipline/attendance system that could potentially significantly reduce the amount of attendance violators is absent. Reducing attendance violations would have second- and third-order effects because the amount of constructive classroom time would increase while the number of students receiving attendance contracts would decrease.

When looking at the next area of potential improvement in Figure 4.25, common infractions, many of the same root causes contribute to 22% of the offense codes making up 80% of the number of offenses committed. The root causes for common offenses among the entire school lie with the training of students in their behavioral practices. Research proves that adolescents can be taught appropriate social behavior (Metzler, Biglan, Rusby, & Sprague, 2001). Of the 14 offense codes that account for the preponderance of offenses, most are level-1 and level-2 offenses. Attendance violations and dress code violations account for 40% of these offenses. When looking at the fishbone diagram for this effect, the root cause is drawn to the basic fact that there is no formal training program for faculty or students on the code of conduct. The administration of SHS must give consideration to the possibility that the county code of conduct guide is not sufficient for this size of school with the demographics involved. The county code of conduct broadly covers unacceptable behavior for the entire county school system. SHS does not possess a SHS-specific guide for behavior within the high school. Absent with this document is a specific training program for faculty or students that emphasizes instruction on the common types of infractions that account for most of the offenses that are committed.

The last area of potential improvement the Lean Six Sigma Team focused on was the area of repeat offenders (Figure 4.26). Repeat offenders are defined as students who commit two or more violations of the student code of conduct. A cause and effect diagram for this area was constructed to determine the source of causes that contribute to the repeated noncompliance of the rules.

Of the repeat offenders, the number of freshmen violators increased from 42% to over 60% when looking at students with five or more violations. Again, this root cause returned us to the freshmen class and their lack of adherence to the code of

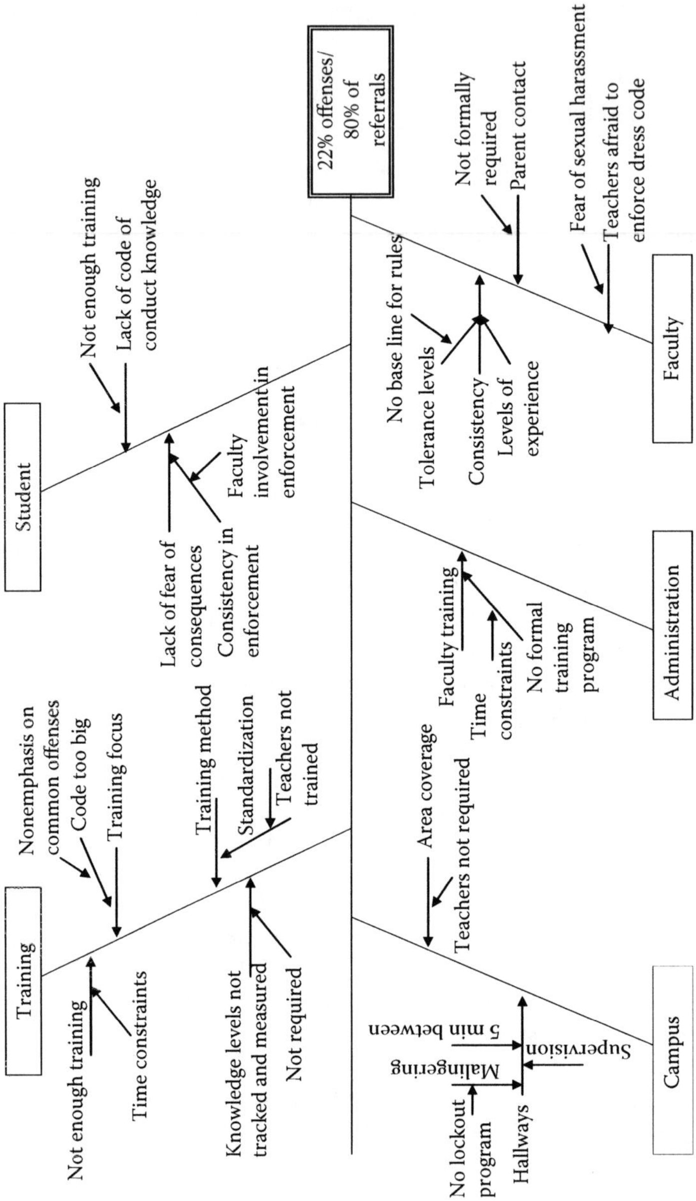

FIGURE 4.25 Cause and effect diagram: Why do 14 offense codes account for 80% of referrals?

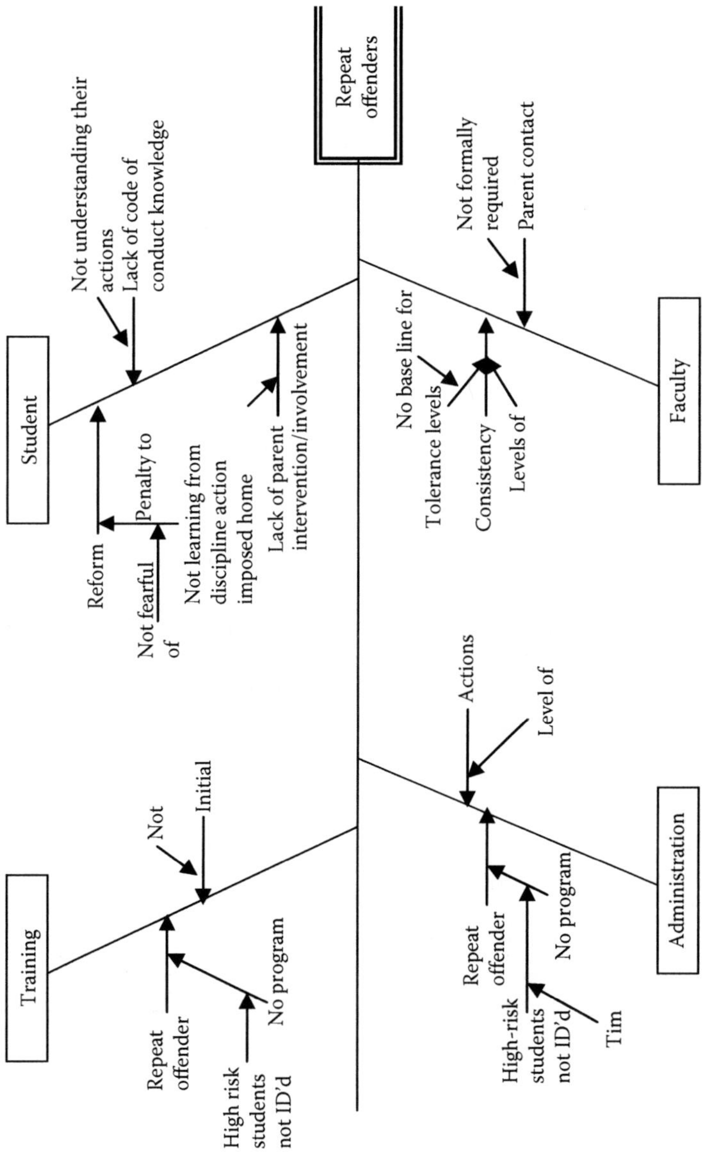

FIGURE 4.26 Cause and effect diagram: Why do students continue to commit offenses?

conduct due partially to inadequate training. Additionally, the Six Sigma team determined that many students commit repeat offenses because of the lack of positive reinforcement in their home environments as well as no special programs developed for high-risk students and repeat offenders. Aside from developing specialized programs to reform the repeat offenders, the team also looked at the sociological aspect of the student. Lack of parental involvement is a significant factor in the reinforcement of appropriate behavior in any student. Students in today's society lack much of the parental involvement that students once had. Thirty percent of SHS students who responded to the student survey said their parents have no opinion of them getting into trouble at school, nor do they have an opinion on the type or amount of punishment imposed. These students are disciplined the same way other students are disciplined, which is a reaction to negative behavior. This prompts the question of whether or not a positive behavior support system would be appropriate for these students, as well as all students at SHS. This cause and effect diagram depicts that students do not reform because they do not understand that their actions/behavior are inappropriate. Additionally, no positive behavioral support system is in place that *teaches* the students appropriate behavior. Such a system would be proactive in lieu of a reactive negative support system.

3. Cause and Effect Matrix

A cause and effect matrix was used to understand if the same root causes contribute to multiple effects. It establishes the relationship $Y = F(X)$, where Y equals the output variables, and X represents the input/process variables or root causes. The cause and effect matrix for the discipline process is shown in Figure 4.27. The total score can be used to understand where process improvement recommendations should be focused in the Improve phase. The consistency of the process; lack of training on the student code of conduct and the discipline process; and student lack of maturity

	Effects				
	42% of offenders are freshmen	22% of offense codes=80% of infractions	Why repeat offenders?	Total	Relative weighting
Causes/importance:	10	3	5		
No training	9	3	9	144	2
Lack of parental involvement	3		9	75	5
Faculty experience	1	1	3	28	8
Lack of resources	9	3		99	4
Lack of consistency	9	9	9	162	1
No id of high-risk students			9	45	6
Students not understanding appropriate behavior			9	45	6
Student lack of maturity	9		9	135	3

FIGURE 4.27 Cause and effect matrix.

are the top three causes that should be focused on in the Improve phase to identify improvement areas that can eliminate these root causes first.

4. Why-Why Diagram

An additional tool utilized to help determine root causes was the "5 Whys." The 5 Whys is a tool that causes a team to continue to ask the question "Why?" 3–5 times to drive the team deeper than a first-order cause to a deeper root cause. Again, in this project we chose the three major areas of improvement discovered during the database analysis to derive root causes. The 5 Whys were used in conjunction with the cause and effect diagrams to determine the root causes.

The first 5 Whys analysis shown in Figure 4.28, and correlates to the first cause and effect diagram. Why do 9th graders account for the bulk of the offenders in the high school? When looking at the Pareto diagram of the offenses they commit, they are mostly level-1 and level-2 offenses that largely consist of dress code violations, attendance violations, and disrespect. Why is this? They are new to the high school and to the discipline policies and procedures at this high school. Why? Because their level of knowledge is lower than the upper classmen who have institutional knowledge from years prior, yet these freshmen get the same amount of training as the upper classmen receive. Why do they receive the same amount of training? The root answer to this is because there is no special program in place for the freshmen to educate them on the appropriate behavior at this school.

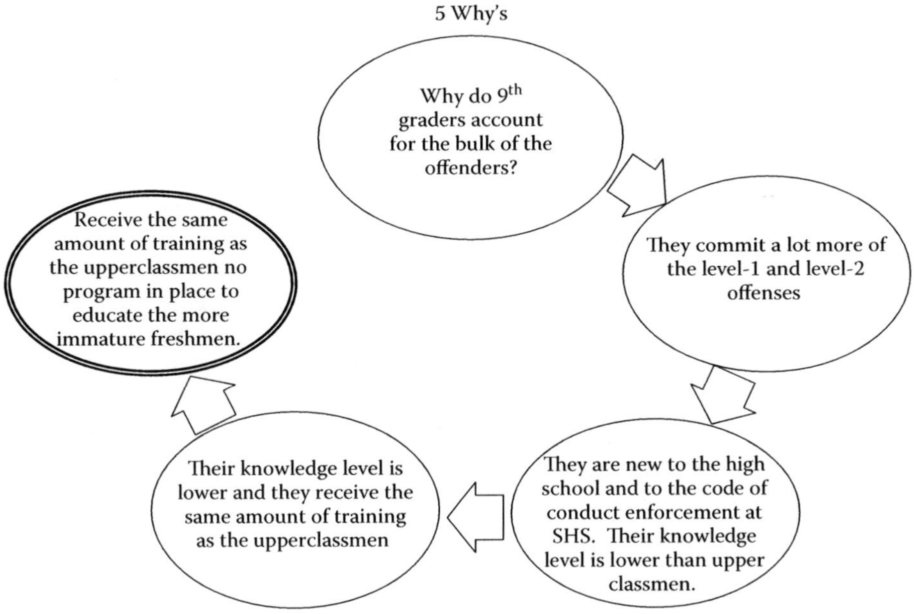

FIGURE 4.28 Why-Why diagram: Why do 9th graders account for the bulk of the offenders?

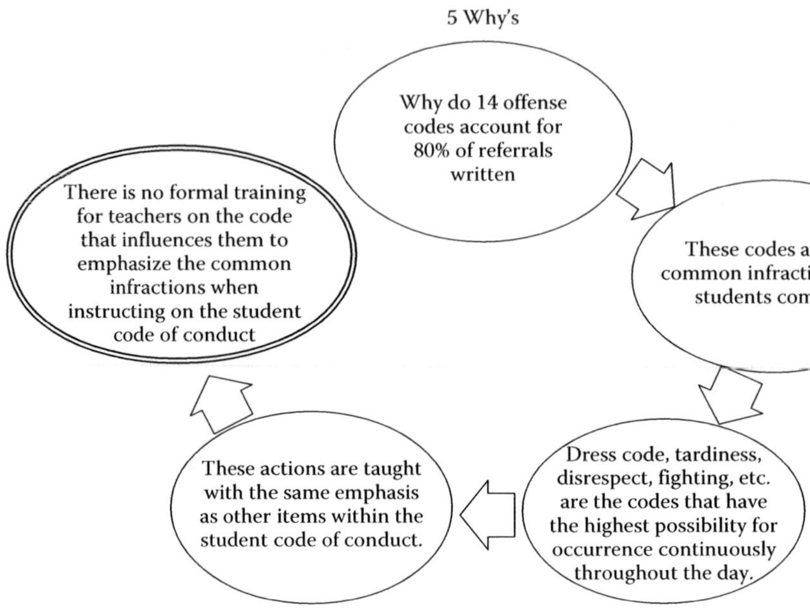

FIGURE 4.29 Why-Why diagram: Why do 14 offense codes account for 80% of referrals written?

The second 5 Whys analysis (Figure 4.29) correlates to the second cause and effect diagram and looks at why 14 of the 64 offense codes account for 80% of the offenses that are committed. The Six Sigma team went through only four iterations of this model to derive the root cause. Among other reasons noted in the cause and effect diagram, the team determined that the reason why these common offenses are so frequently committed is because these are the offense codes that have the highest opportunity for occurrence. For example, at every moment of the day, a student can be out of dress code or have the opportunity to skip class, whereas other offense codes are not as opportunistic. However, these offense codes are given the same amount of focus when training the students on the code of conduct. This is because there is no special policy in place that first trains the faculty on the code of conduct or the SHS policy, and there is no program in place that trains the students in a consistent manner. The absence of such a policy neglects the potential of focusing on the common infractions when giving instructions on appropriate behavior. For example, if the school had a specific written policy on discipline at SHS, this policy might give the faculty a framework for instruction that emphasizes the common offenses. This framework is ideally written under the umbrella of the county's code of conduct.

Lastly, a 5 Whys analysis was conducted on why repeat offenders do not reform (Figure 4.30). This analysis took the team beyond five iterations of why to arrive at a basic root cause. Many factors influence why or why not a student is reformed after an act of misbehavior. SHS has multiple students that are repeat offenders as defined by students who commit two or more offenses. Factors that range from not

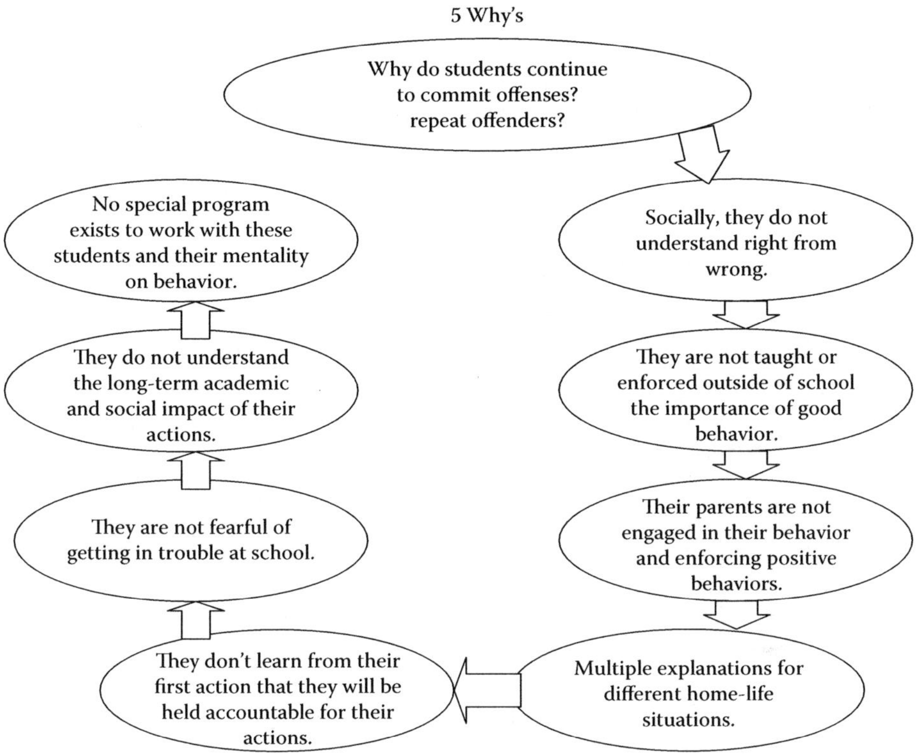

FIGURE 4.30 Why do students continue to commit offenses?

understanding their actions, to inadequate training contribute to repeat offenses by the same student. This analysis took the team to the root cause that no program is in place at SHS that identifies high-risk behavioral students before they become repeat offenders, or once a student becomes a repeat offender, no monitoring program is in place to continuously monitor these students to track their behavior and help them reform. A program, through design, could identify students by race, gender, GPA, and middle-school referral history to identify potential behavioral problems prior to them committing multiple offenses during the school year. Not in place, but possible, a high school could complete an analysis of the incoming freshmen class to identify "high-risk" students for behavioral problems. By identifying this high-risk group, faculty members could be notified in an effort to focus behavioral training and policy to this subgroup.

5. Process Analysis

A process analysis was performed to identify the nonvalue-added activities in the classroom discipline process, the dean's discipline process and the attendance/behavioral contract process. This analysis can be used to focus improvement activities in the Improve phase. Although one could argue that the entire discipline process is nonvalue-added, and that students should just behave, we can still differentiate between value-added and nonvalue-added activities. A value-added activity could be that a student receiving and serving their discipline consequence can provide value if he/she corrects his/her behavior. Another value-added activity might be meeting with parents to resolve the discipline issue. Figure 4.31 shows the process analysis.

6. Histogram, Graphical, and Data Analysis

Data Analysis

Most offenders are freshmen, representing 42% of the offender population. To comprehend these offenders' behaviors, further analysis was performed on this sample utilizing the school database inputs.

Unique characteristics of this sample are noted below:

- The average number of absent days is nine.
- Fourteen percent of offenders received out-of-school suspension, and the average time is three days.
- When looking to the number of offenses committed by an offender in this sample, the average is two offenses.
- Attendance issues account for 22% of the offenses.
- Dress code violations account for 21% of the offenses.

Figure 4.32 shows the number of offenses by offense type.

Freshmen Offenders (Last Year)

The following analysis was performed on data from August 1 through February 23 to compare with the sample pulled from the same date range for this study. A list of offenders from the database was generated on this period, but the graduated seniors who committed offenses during this timeframe were not found in the school database.

A total of 918 offenses, plus senior offenses occurred during that timeframe last year. The data revealed that most offenses are committed by freshmen, representing 44% of the offender population. Although the number of offenses has been reduced by 156 compared with current data, the freshmen offenders' percentage did not change significantly. Even if you added senior offenders (approximately 100 offenders) the percentage of freshmen offenders would only drop to 42%, the same proportion of offenders for this study.

In summary, it is a positive observation that the total number of referrals has decreased, but the high percentage of freshmen offenders did not decrease.

7. Waste Analysis

To identify possible areas for improvement, the SHS Six Sigma Team used a systematic approach for identifying and eliminating waste through continuous improvement.

Process step	Nonvalue-added	Value-added
Classroom discipline process		
Student performs misconduct	X	
Teacher gives in-class discipline consequence		X
Call parent	X	
Contact parent		X
Look for phone number	X	
Meet with parent		X
Student completes consequence		X
Dean's discipline process		
Student performs misconduct	X	
Teacher sends student to discipline dean		X
Teacher fills-out form	X	
Student waits for dean	X	
Dean pulls up student info		X
Contact parent (level 2 or greater		X
Open police investigation (level 4)		
Dean completes discipline form	X	
Dean assigns consequence		X
Student completes consequence		X
Dean checks that student completes consequence	X	
Store discipline form electronically		X
Give copy to faculty	X	
Put copy in student folder	X	
Attendance/Behavioral contract process		
Attend meeting		X
Student put on contract	X	
Student gets signatures	X	
Attendance officer verifies	X	
Student violates attendance	X	
Give action		X
Student completes action		X
Store attendance calendar	X	

FIGURE 4.31 Process analysis for discipline process.

The Lean Six Sigma team identified the following types of wastes in relation to the SHS disciplinary process. Efforts to reduce these wastes will Lean the discipline process. The waste analysis is shown in Figure 4.33.

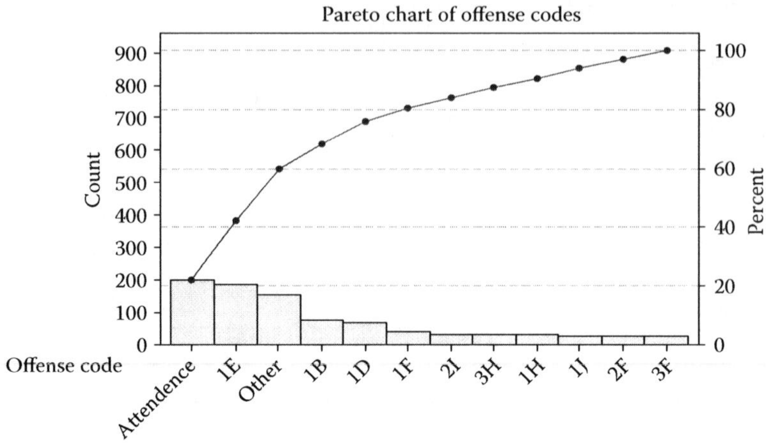

	Attendence	1E	Other	1B	1D	1F	2I	3H	1H	1J	2F	3F
Count	199	186	157	76	70	40	33	31	30	29	29	27
Percent	21.9	20.5	17.3	8.4	7.7	4.4	3.6	3.4	3.3	3.2	3.2	3.0
Cum %	21.9	42.4	59.8	68.1	75.9	80.3	83.9	87.3	90.6	93.8	97.0	100.0

FIGURE 4.32 Pareto chart of the number of offenses by offense type.

8. Correlation Analysis

Correlation analysis was performed to assess the relationship between variables in the student database. The following hypotheses were tested:

Null hypotheses:

- Number of discipline referrals correlated to GPA
- Number of unexcused absences correlated to GPA
- Number of excused absences correlated to GPA
- Gender related to GPA
- Race related to GPA
- Number of days suspended related to GPA
- Number of days suspended related to number of discipline referrals
- Number of discipline referrals and grade
- Number of repeat discipline referrals and grade
- Number of repeat discipline referrals and age

Minitab was used to perform a correlation analysis for the above hypotheses. The only correlation that was significant (showing a relationship between two variables) was between grade and the number of discipline referrals.

9. Regression Analysis

We performed a regression analysis to determine if there was a linear model that could help predict the number of discipline referrals, based on the following variables:

- Grade, gender, GPA, number of excused absences, number of unexcused absences, age, and number of days suspended.

Description	Type of waste	Explanation
Unnecessary referrals	Over production	Unnecessary referrals are offenses that could have been handled in-class by the faculty member, such as sending a student to the Dean's office right away instead of imposing lunch detention on the student or calling the student's parent.
Data entering errors	Defects	Since the information from the student referrals is uploaded by a person, errors may be present. There might be errors in the input of data into the database, resulting in misleading information.
Repeat offenders	Correction	Students are constantly being corrected for repeat violations of the code of conduct. This is a result of not sufficient instruction on appropriate behavior. Repeat offenses occur on common types of level-1 and -2 infractions.
Filling referral forms	Inventory	This type of waste might be present when hard copies such as student calendars are stored in files. This is considered to be an inventory waste because this information would be pulled from the school database.
Student signing off calendars	Motion	This process requires a great deal of motion to sign off the attendance calendar. By enforcing the classroom management program system, students will be logged during class periods reducing the number of students walking in the hallways.
Referral routing procedure	Processing	By creating an electronic referral system, the dean's secretary can be eliminated from the processing procedure. Instead of the dean writing the offense action on the referral form and giving it to the Dean's secretary for input, he/she could simply input the offense action into the electronic referral system themselves.
Student signing-off calendars	Processing	When students are placed on attendance/behavioral contracts, they are required to obtain signatures from their faculties and parents. This task is time consuming for the students, teachers and parents.
Referral routing procedure	Transportation	The referral system is antiquated considering the school and OCPS has a fairly modern and extensive IT network.
Data input in the database	Waiting	When the referral is filled-out by the dean, it is handed to his secretary for data input. The referral has to wait until the secretary is not busy working on another task to upload it in the database.
Classroom disruption	Waiting, and People	When the class time has been interrupted by misconduct, students have to wait for the professor who is writing a referral to continue the lecture.
Students loitering in the hallways	Waiting	When students fail to get to class on time and faculty members do not strictly enforce the tardy system, teachers and students end-up waiting for the entire class to get seated and ready to conduct class.

FIGURE 4.33 Waste analysis.

The R-square (adjusted) was 26.3%. Because we want an R-squared value > 64%, we did not find a good model that predicted the number of discipline referrals.

10. Basic Statistics

From the student population, we identified 743 students who have one or more offenses in their record:

- 21% of the students have 1 or more discipline referrals
- 42% of offenders are freshmen (Figure 4.34)
- 49% of offenders are of Hispanic race (Figure 4.34)
- 16% of the students are repeat offenders with more than one discipline referrals, 9th graders represent the highest percent of the repeat offenders (Figure 4.35)

The mean and standard deviation for the following variables are:

- GPA: mean = 2.82, standard deviation = 0.97
- Unexcused absences: mean = 5.42, standard deviation = 7.84
- Excused absences: mean = 1.48, standard deviation = 2.58
- Number of discipline referrals across all students: mean = 1.53, standard deviation = 5.06
- Number of discipline referrals across students with discipline referrals: mean = 7.37, standard deviation = 8.93

Grade	% Students by grade		Race	% Students by race	% Students in student population
9	43%		Black	13%	8%
10	25%		Hispanic	49%	39%
11	20%		Caucasian	34%	45%
12	12%		Other	4%	8%

FIGURE 4.34 Percentage of students with discipline referrals by grade and race.

Grade	% Students by grade		Race	% Students by race	% Students in student population
9	38%		Black	12%	8%
10	28%		Hispanic	50%	39%
11	21%		Caucasian	34%	45%
12	13%		Other	3%	8%

FIGURE 4.35 Percentage of students with discipline referrals that are repeat offenders (> referrals) by grade and race.

11. Confidence Intervals

The team calculated confidence intervals about the mean for the following variables:

- GPA: (2.79, 2.85)
- Unexcused absences: (5.16, 5.68)
- Excused absences: (1.39, 1.56)
- Number of discipline referrals across all students: (1.37, 1.70)
- Number of discipline referrals across students with discipline referrals: (7.07, 7.66)

12. Hypothesis Testing

We analyzed the following hypotheses using hypothesis tests:

- Is the GPA different for students with discipline issues and for those without?
- Is the GPA different for students suspended versus not suspended?
- Is the average number of discipline referrals greater by gender?

We found that there is a significant difference in GPA for students with discipline issues (2.17) and those without (2.99). The GPAs for students with and without discipline issues is shown in Figure 4.36, with the boxplot of GPA in Figure 4.37.

	GPA
No discipline referrals	2.99
Discipline referrals	2.17

FIGURE 4.36 GPA for students with discipline issues and those without.

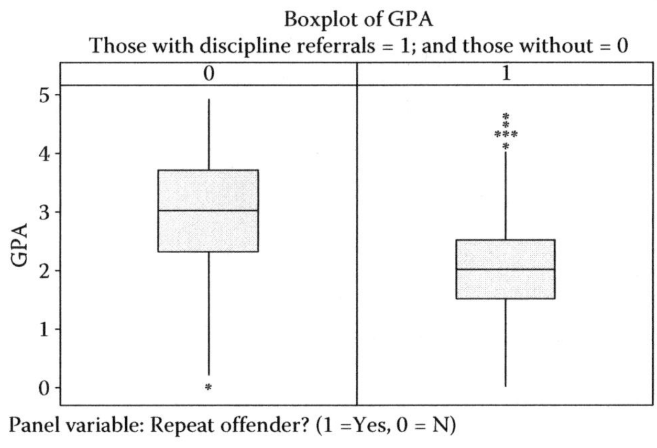

FIGURE 4.37 Boxplot of GPA for students with discipline issues and those without.

We found that GPA is different for students suspended (1.90) versus those not suspended (2.86), as shown in Figures 4.38 and 4.39, respectively.

13. ANOVA

We analyzed the following hypotheses using ANOVA:

- GPA is the same by race
- Average number of discipline referrals is the same by grade
- Average number of discipline referrals is the same by race

We found that the GPA is significantly different by race. See Figure 4.40 for the average GPAs by race, and Figure 4.41 for the boxplot of GPA by race.

	GPA
Not suspended	2.86
Suspended	1.90

FIGURE 4.38 Average GPA for students suspended and not suspended.

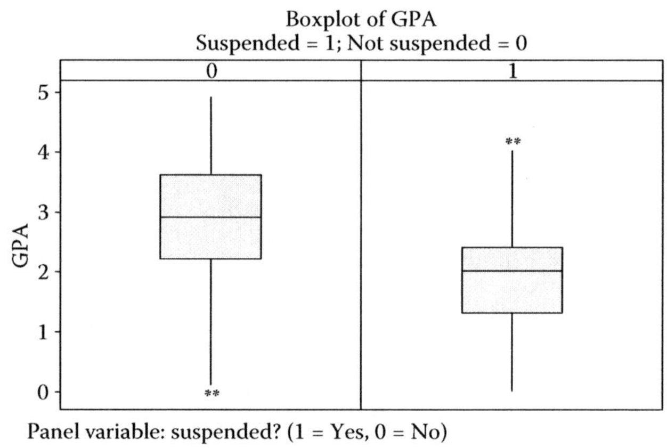

FIGURE 4.39 Boxplot of average GPA for students suspended and not suspended.

Race	GPA
Asian	3.47
Black	2.54
Hispanic	2.63
Indian	2.66
Mixed	2.67
White	2.97

FIGURE 4.40 GPA by race.

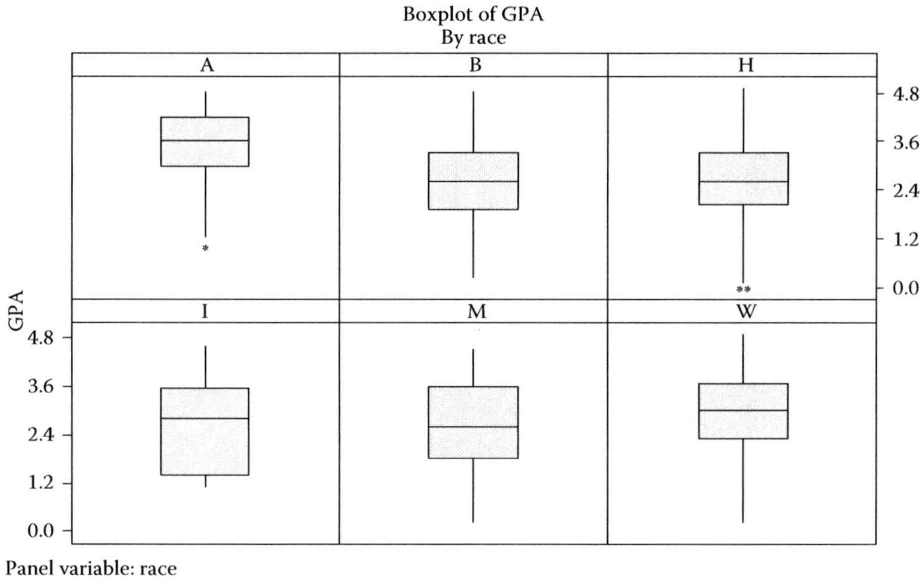

FIGURE 4.41 Boxplot GPA vs. race.

Grade	Average number discipline referrals
9	6.37
10	8.20
11	7.52
12	8.78

FIGURE 4.42 Average number of discipline referrals by grade.

The average number of discipline referrals is different by grade. See Figure 4.42 for the average number of discipline referrals by grade, and Figure 4.43 for the boxplot of discipline referrals by grade.

The average number of discipline referrals is different by race. See Figure 4.44 for the average number of discipline referrals by race, and Figure 4.45 for the boxplot of discipline referrals by race.

14. Survey Analysis

Faculty Survey Summary of Data

The faculty survey that was used to gather the faculty VOC is included in the instructor's material.

The faculty survey revealed interesting perceptions on the level of knowledge of the code of conduct, training of the code of conduct, and other points of interest.

Seventy-five percent of the respondents noted that the most important thing to them regarding the discipline program is the minimization of classroom discipline

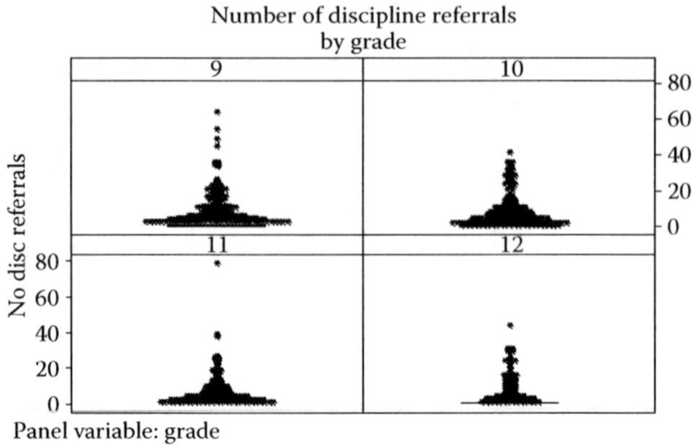

FIGURE 4.43 Boxplot of discipline referrals by grade.

Race	Number of discipline referrals
Asian	5.76
Black	7.04
Hispanic	7.30
Indian	6.50
Mixed	5.50
White	7.66

FIGURE 4.44 Average number of discipline referrals by race.

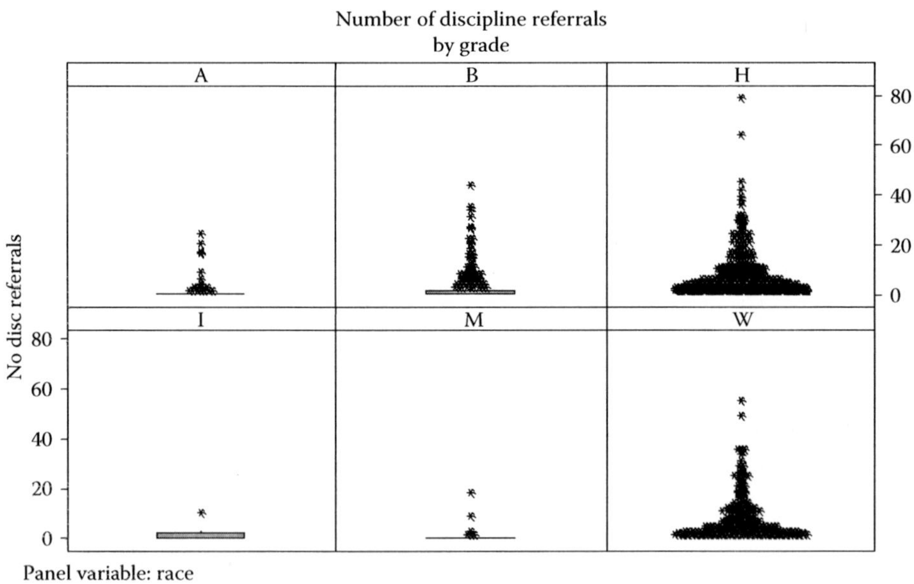

FIGURE 4.45 Boxplot of discipline referrals by race.

issues. This means they want a program that minimizes classroom disruption, disrespect, tardies, and other discipline issues. When regarding discipline in the classroom, the number-one weakness that the faculty identified was a lack of consistency across the board when it comes to classroom discipline policies and administration of policies. The faculty believes that some faculty members tend to let students "slide" when it comes to code of conduct infractions, thus making it harder to enforce these infractions in their classrooms. Additionally, the faculty believes that there is a lack of parent/teacher/counselor integration. They believe that these three areas are not linked in a fashion that enables the student to be highly successful.

The respondents generally ranked their personal level of knowledge of the code of conduct high, their peers' level of knowledge moderate, and the student level of knowledge low. At the same time, they felt that the level of training of the code of conduct is "not enough" and they admit that they spend "less than a period" instructing the students on this topic.

Several areas of concern were identified by responses from the faculty survey. First, 31% responded that it is "not very important" to prepare school work for students who are in-school-suspended. That would equate to almost one-third of the faculty who do not understand the academic importance to those students who are serving in in-school-suspension.

Second, 36% of the respondents sometimes or never log tardies into the attendance system. This is a clear indicator of the problem of so many students remaining in the hallways when the bell rings after the end of the period. However, 87% of the respondents think that it is important to log tardies into the system. This data is backed-up by the student survey in that 57% of the respondents state that three or fewer of their teachers count them tardy when late to class.

Several areas show favorable perceptions of the faculty. Seventy-four percent of the respondents are satisfied or very satisfied with classroom discipline at SHS. Seventy-six percent of the respondents are satisfied or very satisfied with the Dean's office discipline at SHS. Sixty-nine percent of the respondents are satisfied/very satisfied with the overall discipline at SHS. Complimentary of these data is the 69% response rate that the administration will back them when it comes to discipline-related issues.

Also positive to note is the favorable response for potentially implementing a positive behavioral system (PBS) at SHS. The goal of the PBS is to create a baseline standard for classroom discipline, as well as rewarding students positively or negatively for overall behavior. Seventy-three percent of the respondents stated they would support a PBS at SHS.

Open-ended questions offered some constructive recommendations from faculty members. A few of the constructive comments provided by faculty members are shown below:

Remove chronic disrupters into an 'at-risk' program.

More parent involvement and motivation is necessary! Most of the problems come from teachers who let their students talk while they are teaching, and then those students think they can do so in another teacher's class. Teachers need to learn to be the boss!

Review Code of Conduct with all incoming freshman and parents. A required night/program they must attend before they enroll their child. This could be video that middle of the year transfers must watch with parents.

Maybe semester trips for students without any form of discipline referrals. My school did it when I was in high school and it worked for us.

Consistency would be the key ingredient that's necessary to improve the discipline process at SHS.

Consistency... too many teachers allow students to wander hallways without passes, allow food/drink in class, let classes go early, don't count tardies, etc.

Demographics concluded that the respondents covered a wide spectrum of faculty members. The age range, length of teaching experience, and curriculum were fairly evenly distributed while type of classes massed on general education and combination. This indicated that the population of students that the respondents are exposed to is diverse.

Student Survey Summary of Data

The student survey that was used to gather the student VOC is included in the instructor's material. The student survey revealed interesting points of interest. Four freshmen teachers' classes and two upper-class teachers' classes participated in the survey. The result was 543 students who participated, of which 71% were freshmen and 29% were upperclassmen. The response of the student survey skews toward the freshmen response, but coincidentally, the database data points toward the freshmen campus as the source of most of the discipline issues.

Sixty-nine percent of the student respondents feel the discipline imposed by their classroom teachers is "just right" and 52% feel the punishment imposed by the discipline deans is "just right."

Eighty-nine percent of the students perceived to have viewed only three or fewer of their teachers contacting parents when it comes to classroom-related discipline issues. This is an interesting contradiction to 31% of the faculty respondents stating that the most effective form of in-class discipline is calling the student's parent! This issue is further complicated by faculty comments on issues with inaccurate parent contact information in the school database. Furthermore, it was estimated by one dean that the success rate for contacting parents is <20%.

Of the students who took the survey and had a referral, the students believed that 34% of their parents had "no opinion" of them getting in trouble. Furthermore, the students believed that 31% had no opinion of the punishment their child received at school. This possibly infers that up to one-third of the parents of SHS students have no opinion or are disengaged from their children getting into trouble at SHS.

Another interesting note is that, when asked what discipline actions the students dislike the most, attendance contract, detention, and in-school detention were the most disliked. These actions cause an inconvenience to the students and/or take time away from them. These actions should prove to be the most effective form of punishment to deter repeat offenders.

15. DPPM/DPMO

The DPPM for the discipline time (see Process Capability in the next section) is 6,71,990, which equates to a little over 1.0 sigma.

16. Process Capability

We performed a process capability analysis for the average discipline time. The lower specification limit was identified as ten minutes, and the upper specification limit was 30 minutes. The process is not capable related to the discipline time at the dean's office (Figure 4.46).

17. Analyze Phase Presentation

The Analyze phase presentation summarizing the written Analyze phase presentation is included in the downloadable instructor materials.

ANALYZE PHASE CASE DISCUSSION

1. **Analyze Report**
   1.1 Review the Analyze report and brainstorm some areas for improving it.
   1.2 How did your team ensure the quality of the written report? How did you assign work to your team members? Did you face challenges of team members not completing their assigned tasks in a timely manner and, if so, how did you deal with it?

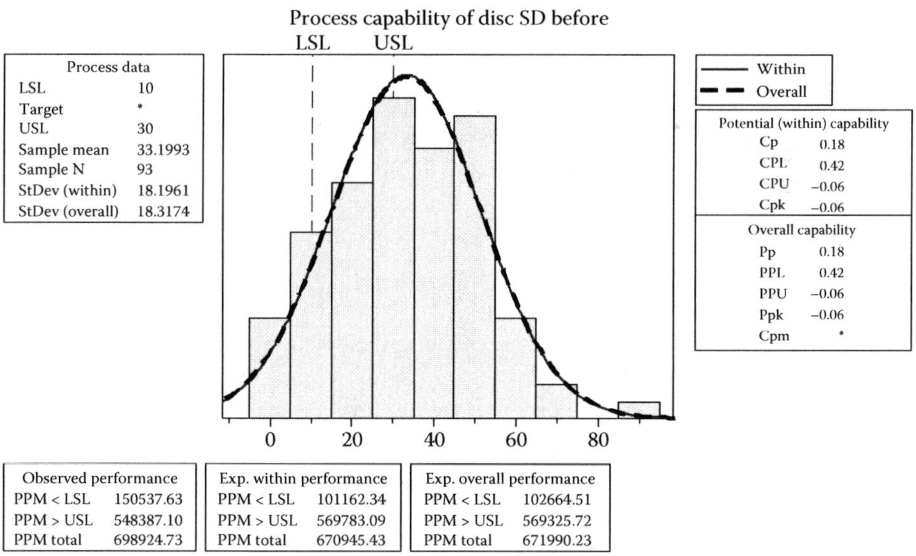

FIGURE 4.46 Process capability of discipline time before improvement. *Note:* Not actual data, for illustrative purposes only.

1.3 Did your team face difficult challenges in the Analyze phase? How did your team deal with conflict in your team?
1.4 Did your instructor and/or Black Belt or Master Black Belt mentor help your team better learn how to apply the Lean Six Sigma tools in the Analyze phase, and how?
1.5 Did your Analyze phase report provide a clear understanding of the root causes of the discipline process, why or why not?

2. Cause and Effect Diagram

2.1 How did your team determine the root causes, and how did they validate the root causes?

3. Cause and Effect Matrix

3.1 Did many of the causes apply to many of the effects?

4. Why-Why Diagram

4.1 Was it easier to create the cause and effect diagram, the cause and effect matrix, or the Why-Why diagram? Which of the tools was more valuable for getting to the root causes?

5. Process Analysis

5.1 Discuss how your team defined whether the activities were value-added or nonvalue-added? Was the percentage of value-added activities or value-added-time what you would expect for this type of process and, if so, why?

6. Histogram, Graphical, and Data Analysis

6.1 What type of distribution does your data appear to be from a graphical analysis?
6.2 Can you test your distribution statistically and determine a likely distribution, what is it?
6.3 Did you have outliers in your data?

7. Waste Analysis

7.1 What types of waste were prevalent in the discipline process and why?

8. Correlation Analysis

8.1 Were there significant variables that were correlated? Do they appear to have a cause and effect relationship, and why?

9. Regression Analysis

9.1 Were you able to identify a model that can predict GPA? Why or why not?

10. Basic Statistics

10.1 What conclusions can you draw from the basic statistics?

11. Confidence Intervals

11.1 What are your conclusions from the confidence intervals that you calculated?

12. Hypothesis Testing

12.1 What were your key findings for your hypothesis tests?
12.2 What conclusions can you make from a practical perspective?
12.3 How might you use these findings in the Improve phase?

13. ANOVA

13.1 What were your key conclusions in your ANOVA?

14. Survey Analysis

14.1 What were the significant findings in the faculty survey?
14.2 What were the significant findings in the student survey?
14.3 Did your survey assess customer satisfaction with the discipline process?
14.4 Was there consistency in the responses between the faculty and the students?

15. DPPM/DPMO

15.1 What is your DPPM/DPMO and sigma level. Is there room for improvement, and how did you determine that there is room for improvement?

16. Process Capability

16.1 What conclusions can you draw from the process capability study? Is your process capable? Is your process stable and in control? Can you have a process that is in control, but not capable, and how?

17. Analyze Phase Presentation

17.1 How did your team decide how many slides/pages to include in your presentation?
17.2 How did your team decide upon the level of detail to include in your presentation?

IMPROVE PHASE EXERCISES

1. Improve Report

Create an Improve phase report, including your findings, results and conclusions of the Improve phase.

2. **Recommendations for Improvement**
   Brainstorm the Recommendations for Improvement.

3. **Revised QFD**
   Revise your QFD from the Measure phase to map the Improvement Recommendations to the Critical to Satisfaction Characteristics.

4. **Action Plan**
   Create an action plan for demonstrating how you would implement the improvement recommendations.

5. **Future State Process Map**
   Create a future state process map for the following processes:
   - Classroom Discipline process
   - Dean's Office Discipline process
   - Attendance/Behavioral Contract process

6. **Revised VOP Matrix**
   - Revise your VOP matrix from the Measure phase with updated targets

7. **Training Plans, Procedures**
   Create a training plan, and a detailed procedure for one of the discipline processes.

8. **Improve Phase Presentation**
   Prepare a presentation (PowerPoint) from the case study exercises that provides a short (10–15 minutes) oral presentation of the Improve phase deliverables and findings.

IMPROVE PHASE

1. Improve Report

A report of the Improve phase for the SHS discipline process improvement project, including the key deliverables developed as part of the prior exercises, is described below.

The Improve phase of the DMAIC process is designed to identify improvement recommendations, implement them and then assess the improvement. The objectives of this phase in relation to the SHS discipline improvement project are as follows:

- Identify the improvement recommendations
- Develop action plans for implementation
- Pilot the improvement recommendations
- Assess the improvement

2. Recommendations for Improvement

In the Improve phase, the analysis of the data measured is used to make improvements to the process or system. The Lean Six Sigma team put together multiple recommendations that would improve the overall discipline system and subsequently the academic environment as a whole. With the recommendations, suggested implementation plans are included as a guide for the SHS leadership team to follow when ultimately designing and implementing changes to the system. The recommendations that follow are based on data collected from the student database, teacher and faculty surveys, interviews with leadership team staff, interviews with faculty members, and benchmarking of best practices in similar educational systems.

Recommendation #1: Create a unique and tailored SHS discipline program.
This is a written publication that outlines the specific SHS policy on discipline and behavior. The publication would cover all aspects of discipline to include training, faculty responsibilities, and student responsibilities. The guide would outline specific punishments that would result from specific offenses by the student, i.e., clarifying the rules. Simply running the discipline program under the umbrella of the county code of conduct is not sufficient. Many schools have taken their county conduct policies one step further and published their own guide to discipline. This creates a baseline for classroom discipline which will help to create consistency among classrooms.

Recommendation #2: Create discipline dashboard for the SHS principal.
Develop control charts and post on a "dashboard" weekly discrete data that will be used to make decisions on the discipline/attendance programs. The dashboard is developed by the principal with consultation from the discipline and attendance deans. The dashboard is given to the principal at the end of each week and is a "visual snapshot" of the current state of both programs. Control charts are statistical tools used to monitor processes and to ensure that they remain "in control" or stable. Moreover, it helps to distinguish between process variation resulting from common causes and variation resulting from special causes.

Data Basis: A tool/mean is needed to assist the SHS administration to closely monitor the discipline process. A discipline dashboard is a very effective tool to communicate and decide on what type of action is needed if an out of control pattern is recognized.

Recommendation #3: Create a behavioral program specifically designed for the 9th grade campus.

Consider implementing a PBS for the 9th grade campus. The PBS is a research proven system that creates consistency in classroom discipline (i.e., creating baselines for acceptable behavior). A PBS is data-driven and can be individually tailored for the specific school it is implemented in.

The implementation of a PBS is not simple, but the rewards for the work put forth are invaluable to the teaching environment as a whole. This team recommends a PBS team be formed as soon as possible to create the tailored PBS for the freshmen campus. This team should be formed from the Discipline Deans' staff, teachers with strong discipline skills, counselors or teachers with child development experience. The team should be trained in the specifics and history of a PBS before they convene to create the SHS PBS. The team should then concentrate their efforts to create a system by the next school year for implementation with next year's freshmen class.

Data Basis: Forty-two percent of the offender population at SHS is freshmen. Although the total number of referrals at SHS has dropped with the inception of the 9th grade campus, the percentage of freshmen offenders remained the same. Additionally, incoming freshmen receive the same amount of training on the code of conduct as upperclassmen who have institutional knowledge and experience at the high school. Most teacher respondents say they spend a period or less time per semester reviewing the content of the code of conduct, and they also say that the amount of time spent on this training is not enough. The preponderance of the freshmen's first week at SHS should consist of orientations to the school, programs, and the discipline guide. Sometimes school leadership may be hesitant to take that much time away from classroom time but, in the long run, establishing solid discipline expectations will result in better classroom environments and more class time due to reduction in disruptive situations. Research has shown that punitive school and classroom environments, unclear rules and expectations, and inconsistent application of consequences contribute to increased levels of student antisocial behavior and truancy (Metzler, Biglan, Rusby, and Sprague 2001). This is the current situation within the freshmen ranks. Discipline is not reinforced in the parental home environments in today's society as it was years ago. Students are not as fearful of the consequences they face for inappropriate behavior. Because this team or SHS can not impose positive parental support for the parents of 3500 students, the issue of teaching appropriate behavior must be approached from a different angle. A PBS allows the school to create a positive environment to teach and reinforce positive behavior. This system is perfect for the freshmen campus at SHS because the maturity level is very close to that at a middle-school level where PBS have proved successful in multiple studies.

Recommendation #4: Identify high-risk freshmen prior to the school year, and monitor their status in the first quarter of the school year.
This program should be done the month prior to the start of the school year when SHS has a fairly solid roster of the incoming freshmen class. A query can then be generated using FileMaker Pro, then sorted to identify freshmen who, in middle school, have multiple referrals, a low GPA, attendance problems, and low FCAT scores. This population is then considered the high-risk population for the incoming freshmen class. These will be the majority of students who fall within the 42% population of offenders. With this population generated, they can now be put into classrooms with teachers known for strong classroom management skills. This entire recommendation can be accomplished by discipline deans identifying high-risk students and having the guidance counselors ensure they are put into the appropriate classrooms where they are set-up for success. For example, a high-risk student taking

biology should be put in Valenza's class. A high-risk student taking English should be put into Burley's class.

Data Basis: Students with low GPAs and low FCAT scores statistically have discipline issues. There is no special program in place that identifies these students before they commit offenses. With minimal effort, this can become a proactive system in lieu of the current reactive system.

Recommendation #5: Emphasize common offenses when training students on acceptable behavior at SHS.

Concentrate training on the 14 common offenses that account for 80% of the offenses committed. When developing the SHS Discipline Guide, publish specific consequences for the common types of offenses. For example, publish a table that mirrors specific consequences for five or more tardies, ten or more, etc. When publishing this policy for the common offenses, all teachers will have a baseline of consequence to train the students.

Data Basis: Twenty-two percent of the offenses on the referral form (14 offenses) account for 80% of the offenses committed. The number-one weakest area of classroom discipline answered by the faculty survey was a lack of consistency among the teachers. Additionally, open-ended comments for this survey revealed the common theme that they want better consistency in discipline. The published policy with emphasis on the common offenses will provide this baseline for which consistency will result.

Recommendation #6: Impose on the faculty members the importance of logging tardies and their responsibility to monitor the hallways in between classes.

Instead of administrative staff patrolling the hallways and pushing students to class, administrators should focus their efforts on managing the faculty to push the flow of students through their hallways. As part of teacher orientation, train the faculty on the importance of accurate tardy reporting and their responsibilities.

Data Basis: By contract, teachers are required to be in the hallways greeting their students at every class. This is not happening at SHS and the preponderance of teachers do not get out in the hallways. This allows for students loitering on their way to class, and increases the chances for hallway disruptions and probability of a high tardiness rate.

Recommendation #7: Establish an alternate consequence schedule for students who are in the lower 30% FCAT population.

When a student is sent to the dean's office for a referral, one of the first actions by the dean is to check to see if the student is in the lower 30% FCAT population. If that student is in that population, then special consideration is given for the punishment of that student. Unless absolutely necessary, this student should not be given out-of-school-suspension.

Data Basis: Students who rank in the lower 30% statistically have lower GPAs. Students who are given out-of-school suspension have statistically lower GPAs then those who do not serve this punishment. By giving a student who is in the lower

30% FCAT out-of-school suspension is a detriment to the student's chance of improving. If possible, give students in the lower 30% in-school suspension where they can be controlled and given additional reading assignments.

Recommendation #8: Create a faculty reward system for active discipline and classroom management skills.

Implementation: Create a small committee who can monthly select a faculty member who demonstrates outstanding classroom management skills and practices.

Data Basis: Currently, there is no positive environment award system in place to influence faculty members to actively promote good discipline at SHS. The level of classroom management skills varies across the faculty. If good teachers are rewarded for good practices, others will notice and learn from those teachers' tacit knowledge.

Recommendation #9: Create a parental involvement contract for repeat offenders.

This is a written contract that will apply only to those students that have repeat offenses. This contract could be part of the code of conduct contract in the student enrollment process to acknowledge not only the parents of those repeat offenders, but also to the whole population. Parents of repeat offenders will acknowledge their written commitment to this contract and this can be performed at the beginning of each school year or semester. This contract can include a minimum number of community service hours or school involvement if their adolescent becomes a repeat offender. The primary goal of creating this parental involvement contract for those repeat offenders is to increase parental reaction when students misbehave at school.

This requirement could be initiated at the beginning of the school year and be part of the written acknowledgement of the SHS discipline policy guide and the county's code of conduct. The requirement of this parental contract is included in the SHS Discipline guide and would outline the consequences the parents must meet should their adolescent become a repeat offender at the school. Whether the stipulations are service at the school or mandatory parental counseling, the contract is designed to influence more parental involvement for the repeat offenders.

Data Basis: When looking at the repeat offender population survey respondents, 30% state their parents have no opinion at all when their son/daughter misbehaves in school. Evidence shows that when schools work together with families to support learning, children tend to succeed not just in school, but also throughout life. Recent research has shown that, particularly for students who have reached high school, the type of parental involvement that has the most impact on student performance requires their direct participation in school activities.

Recommendation #10: Create a knowledge-sharing program for classroom management best practices.

This will include the involvement of teachers to discover best practices of effective classroom management. By creating this knowledge-sharing program, teachers will have the opportunity to share their strategies with respect to class room management with other teachers. Teachers will learn from their colleagues how to work with students who have many types of special needs and apply various management techniques to help students become self-regulated learners. This program could help the entire faculty population in learning how to increase student motivation, build student–teacher relationships, and increase home–school communication. The main purpose of creating this knowledge-sharing program is to enable teachers from the SHS School to learn from the experiences, methodologies and achievements of colleagues.

Various information and communication technologies may be used by teachers to communicate and share their ideas and inputs on the topic. A knowledge management system for this area could be as simple as a best practices committee that publishes a bi-semester newsletter, to a more complicated information technology design that stores best practices in a database.

Data Basis: The underlying theme to the respondents in the faculty survey is that they desire consistency in discipline and among their peers in classroom discipline. The demographics among the teaching staff are very broad and the discipline management ability equally broad. Many of the teachers who have weaker classroom management/discipline abilities could leverage from the experience that lies with many of the more experienced teachers. Currently, there is no knowledge-sharing system in place that gives teachers the opportunity to share their experiences, techniques, and methodology with respect to classroom management. By creating this system, teachers and students will benefit from such a positive improvement recommendation.

Recommendation #11: When a referral is issued and is necessary to impose a disciplinary action, utilize lunch/after school detention (action code G) and in-school suspension (action code P) as the primary actions for the most common offenses found. If a student repeats the offense, consider imposing an attendance/behavioral contract (action code I) to the student because it is the most undesirable action according to the student survey.

Data Basis: When looking at the most frequent actions taken during the study period, verbal reprimand (action code C), counseling and direction (code B) and parental contact (action code A) were the most frequent discipline consequences. However, the Student Survey revealed that the students consider attendance/behavioral contract, lunch/after school detention, and in-school suspension as the three most effective (undesirable) disciplinary actions, respectively. Therefore, if those actions are applied more often, the likelihood that the student will repeat the offense may decrease.

Recommendation #12: Consider imposing in-school suspension rather than out-of-school suspension unless absolutely necessary.

Data Basis: In-school suspension ranked as the third most undesirable action imposed by the school. Moreover, it was statistically proven that students who are suspended tend to have a lower GPA in contrast with those who are not suspended. Consequently, out-of-school suspension is one of the factors that affect the academic performance of a student, which may be mitigated by applying in-school suspension.

Recommendation #13: Automate the referral process by developing a reliable program. This program should replace the current referral process.

Seek funding to support a project for developing a program that can automate and manage classroom referrals. To minimize the use of referral forms (paper based), a replica of the form should be electronically developed. Limited access has to be given to faculty members. Teachers could enter the necessary data while the student walks to the dean office. A referral will be generated and queued under the new referral list. As soon as the student walks into the dean's office, the dean can pull that referral along with the student history just by one simple click. After deciding which consequences are assigned, the deans can enter the required information into the program. Once this step is done, an automated email goes to the teachers and to the parents explaining the nature of the misconduct and consequences.

To have a successful program, essential requirements of the automated system are:

- It should be capable of handling every task in the referral process.
- Deans'/faculty inputs should be incorporated during the development cycle.
- The interface should be user-friendly.
- It should have the ability to retrieve/pull-up historical data from different databases or systems. The system should be integrated with the existing programs.
- The system should be reliable and available.
- Some of the fields in the program have to be mandatory to be filled to prevent any type of error.

Data Basis: The referral system is antiquated considering the school and county have a fairly modern and extensive IT network.

3. Revised QFD

The revised QFD maps the Improvement Recommendations to the CTS criteria (Figure 4.47).

4. Action Plan

A Pareto chart (Figure 4.48) shows the prioritized list of recommendations to identify which improvement recommendations should be implemented first.

Figure 4.49 is a summary of the action plan with the recommended improvements and the time frame to implement them.

5. Future State Process Map

A revised process map incorporating the improvement recommendations is shown in Figure 4.50.

6. Revised VOP Matrix

The revised VOP matrix is included in Figure 4.51, with the most recent targets.

Project name: HS discipline process

Customer requirements \ Technical requirements	Importance	(1) Tailored discipline process	(2) Discipline dashboard	(3) 9th grade positive behavioral system	(4) High risk freshmen students	(5) Common offenses	(6) Logging tardiness	(7) Faculty reward system for discipline	(8) Parental involvement for repeat offend	(9) Best discipline process practices	(10) Use focused discipline consequences	(11) Impose in-school suspension vs. OSS	(12) Automate discipline process
Reduction of number of discipline referrals	38	9	9	9	9	9		5		9	5	5	5
Classroom discipline consistency	19	5	3	5	3	3	3	9			9	9	9
Teacher/parent/counselor integration	15								9				
Adherence to code of conduct	10	5	5	3	9	1	3	5					
Classroom control	8	9	5	5	9	9	1	3		9	9	9	
Teacher/parent contact	7								9				
Minimize classroom discipline issues	3	9	5	9	9	5	5	5		9	5	5	5
Absolute weight		586	504	534	588	496	110	450	198	441	448	448	376
Relative weight		2	4	3	1	5	12	6	11	9	7	7	10

FIGURE 4.47 Revised QFD.

7. Training Plans and Procedures

Train incoming freshmen and parents during summer orientation on the new standardized discipline process. Train current students in assemblies. Train faculty during faculty meetings on the new standardized discipline process.

8. Improve Phase Presentation

The Improve phase presentation can be found in the downloadable instructor materials.

IMPROVE PHASE CASE DISCUSSION

1. **Improve Report**
   1.1 Review the Improve report and brainstorm some areas for improving it.
   1.2 How did your team ensure the quality of the written report? How did you assign the work to your team members? Did you face challenges of team members not completing their assigned tasks in a timely manner and, if so, how did you deal with it?

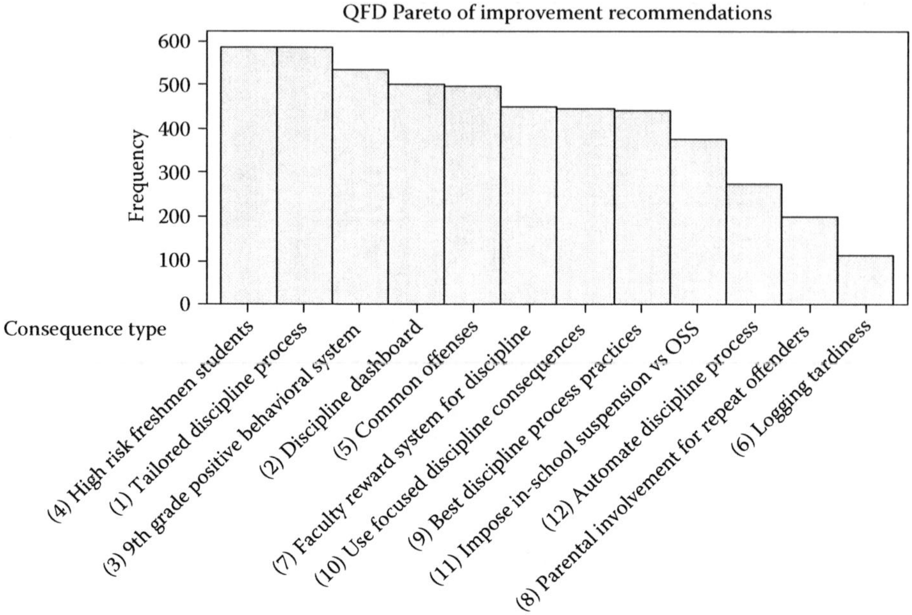

Frequency	588	586	534	504	496	450	448	441	376	276	198	110
Percent	11.7	11.7	10.7	10.1	9.9	9.0	8.9	8.8	7.5	5.5	4.0	2.2
Cum %	11.7	23.4	34.1	44.2	54.1	63.1	72.0	80.8	88.3	93.8	97.8	100.0

FIGURE 4.48 Pareto chart of improvement recommendations.

Recommendations	Priority	Time frame
(4) High-risk freshmen students	588	Month 1
(1) Tailored discipline process	586	Month 3
(3) Ninth grade (positive) behavioral system	534	Next fall
(2) Discipline dashboard	504	Month 1
(5) Common offenses	496	Month 3
(7) Faculty reward system for discipline	450	Months 3–6
(10) Use focused discipline consequences	448	Month 3
(9) Best discipline process practices	441	Month 3
(11) Impose in-school suspension vs out-of-school suspension	376	Month 3
(12) Automate discipline process	276	Long term
(8) Parental involvement for repeat offenders	198	Month 12
(6) Logging tardiness	110	Month 12

FIGURE 4.49 Action plan.

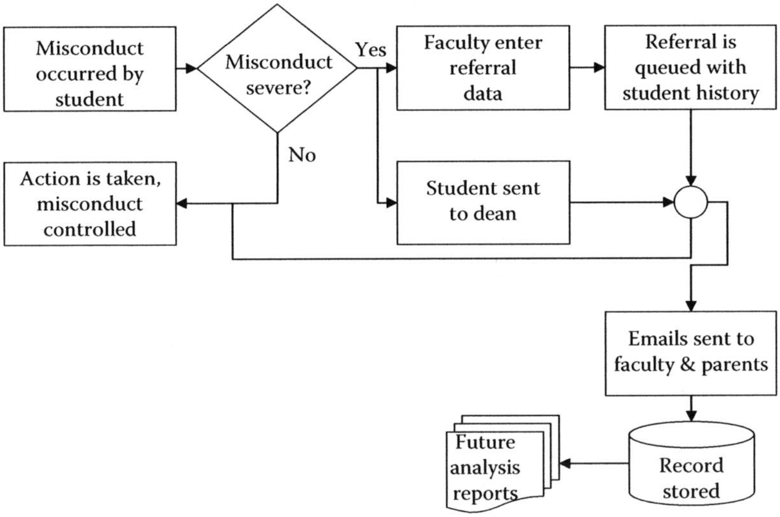

FIGURE 4.50 Future process map.

1.3 Did your team face difficult challenges in the Improve phase? How did your team deal with conflict on your team?
1.4 Did your instructor and/or Black Belt or Master Black Belt mentor help your team better learn how to apply the Lean Six Sigma tools in the Improve phase, and how?
1.5 Did your Improve phase report provide a clear understanding of the root causes of the discipline process, why or why not?
1.6 Compare your improve report to the improve report in the book, what are the major differences between your report and the author's report?
1.7 How would you improve your report?

2. Recommendations for Improvement

2.1 How did your team generate ideas for improvement?
2.2 What tools and previous data did you use to extract information for the improvement recommendations?
3.3 How do your recommendations differ from the one's in the book?

3. Revised QFD

3.1 Does the QFD support the alignment with the CTS characteristics?
3.2 How will you assess customer satisfaction?

4. Action Plan

4.1 How did your Lean Six Sigma team identify the timings for when to implement your recommendations?

CTS	Factors	Operational definition	Metric	Target
Minimize classroom discipline issues	Freshmen training on code clear guidelines	Training exists and is performed clear guidelines exist	Number disruptions	Reduce number disruptions by 50%
Classroom discipline consistency	Guidelines teacher training	Clear guidelines, exist Teacher training each year	Guidelines Number of faculty trained	100% of faculty are trained within 3 months of hire or Jan. 1
Teacher/ parent/ counselor integration	Apathy of parents data missing	Engaged parents assessed by parent survey	% of responses on survey for identified questions	Increase % of ratings in high categories by 10%
Adherence to code of conduct	Training expectations	All students will be trained in code of conduct for 2 hours per semester Clear expectations conveyed	Number students trained	100% of students are trained within first month of school or transfer
Classroom control	Teacher training	All teachers trained in classroom management Mentors for new teachers	Number teachers trained Number teachers with mentors Rating of mentoring program	100% of teachers trained 100% of new teachers have a mentor
Teacher / parent contact	Apathy of parents data missing	Engaged parents assessed by parent survey	% of responses on survey for identified questions	Increase % of ratings in high categories by 10%
Reduction of referrals	Freshmen no emphasis on common offenses	Guidelines	Freshmen and transfers trained in code of conduct and guidelines	100% of freshmen and transfers trained in code of conduct and guidelines

FIGURE 4.51 Revised VOP Matrix.

5. **Future State Process Map**
    5.1 Compare your future state process map to the one in the book. How does it differ? Is yours better, worse, the same?

6. **Revised VOP Matrix**
    6.1 Does the VOP matrix provide alignment between the CTSs, the recommendations, metrics and target?

7. **Training Plans, Procedures**
    7.1 How did you determine which procedures should be developed?
    7.2 How did you decide what type of training should be done?

8. **Improve Phase Presentation**
   8.1 How did your team decide how many slides/pages to include in your presentation?
   8.2 How did your team decide upon the level of detail to include in your presentation?

CONTROL PHASE EXERCISES

1. **Control Report**
   Create a Control phase report, including your findings, results and conclusions of the Control phase.

2. **Hypothesis Tests, Design of Experiments (DOE)**
   Note: The data provided incorporate approximated values based on summarized data, used for instructional purposes.
   Compare the number of discipline referrals for the entire student population before and after improvements.

3. **Mistake Proofing**
   Create a mistake proofing plan to prevent errors from occurring in the discipline process.

4. **Control Plan**
   Develop a Control plan for each improvement recommendation from the Improve phase report.

5. **Process Capability, DPMO**
   Calculate the process capability for the revised time to process students through the dean's office discipline process. Note: This data is hypothetical and for illustrative purposes only.

6. **Control Charts**
   Create an idea for applying control charts to control the discipline process.

7. **Replication Opportunities**
   Identify some potential replication opportunities within the high school, and within the school district.

8. **Standard Work, Kaizen**
   Create a plan for standardizing the work.

9. **Dashboards/Scorecards**
   Create a dashboard or scorecard for tracking and controlling the discipline process.

10. Control Phase Presentation

Prepare a presentation (PowerPoint) from the case study exercises that provides a short (10–15 minutes) oral presentation of the Control phase deliverables and findings.

CONTROL PHASE

1. Control Report

A report of the Control phase for the SHS Discipline Process Improvement project, including the key deliverables developed as part of the prior exercises, is described below.

The purpose of the Control phase of the DMAIC process is to design, develop and incorporate controls into the improved processes. The objectives of this phase are to:

- Assess the gains that were realized by implementing the improvement recommendations in the Improve phase
- Develop the control plan to maintain the gains
- Standardize the process
- Develop future plans for improvement

2. Hypothesis Tests, DOE

We performed a two-proportion test on the total number of discipline referrals before the improvements (5430) and after improvement (4698). The *p*-value was 0, so we reject the null hypothesis that there is no difference in the number of discipline referrals before and after the improvements. Therefore, we conclude that the total number of discipline referrals was significantly reduced after the improvement recommendations were implemented.

3. Mistake Proofing

Ideas for mistake proofing are:

- Automate the discipline process so that the teacher can create the discipline referral on-line, on-time, or scan the document and send via email to the office.
- Implement a process to verify student and parent contact information. Audit data by calling parents on a sampling basis once a month.

4. Control Plan

The control plan for the recommendations is show in Figure 4.52.

5. Process Capability, DPMO

The average discipline time before the improvements were implemented was 33.2 minutes and after was 29.50 minutes. We first tested for normality, and concluded

Recommendations	Control plan
High risk freshmen students	This recommendation can be controlled by evaluating at the mid term and end of semester marks how many of these students encounter discipline actions.
Tailored discipline process	This recommendation would be controlled and evaluated after the fall semester when discipline data is measured against the prior year's data. Additionally, teacher feedback for the policies created within the guide should be solicited with revisions planned for future editions.
9th grade (positive) behavioral system	The use of the dashboard would assess whether the positive behavioral system would have a positive impact on reducing the number of discipline referrals school-wide.
Discipline dashboard	Implement control charts to monitor number of discipline referrals. If the process appeared to be out of control, the deans will know immediately as they are indicated on the control charts. The principal will know weekly the status of the process as reported on his dashboard.
Common offenses	Provide a weekly report on the principal's dashboard that shows the number of offenses by type for the previous week. The principal can then react to that data to provide his guidance to the teachers to impart on their students.
Faculty reward system for discipline	Create a small committee who can monthly select a faculty member who demonstrates outstanding classroom management skills and practices.
Use focused discipline consequences	Monitor the number of referrals issued per month and compare the results against previous years to determine if the referrals have been reduced and to make sure the gain attained is sustained.
Best discipline process practices	Monitoring and evaluating the teacher's participation on this recommended system would be managed through the school established procedures of teacher reviews. This program can be performed at the end of each school year.
Impose in-school suspension vs out of school suspension	After the fall semester, compare the students who were out-school suspended and in-school suspended, and determine if those who received in-school suspension have less repeat offenses and better academic performance than those who received out-school suspension.
Automate discipline process	On a monthly/weekly basis, generate a report to highlight how many offenses were committed. Identify the time to receive referrals at the Dean's office after implementing the automated system. After analyzing that report, a proper action has to be taken to mitigate the situation if needed.
Parental involvement for repeat offenders	This recommendation would be controlled by the number of repeat offender parents that actually get involved.
Logging tardiness	Measure the number of teachers seen at their doors every period. This is always a reflection of the number of tardies reported and one of the items that is reported weekly to the principal on his "dashboard."

FIGURE 4.52 Control plan. *Note:* Not actual data, for illustrative purposes only.

that the discipline time appears to be a normal distribution. We checked for equal variances, found a p-value of 0, and concluded that the variances are not equal. We then did a t-test for unequal variances, where the p-value was 0.085. We failed to reject the null hypothesis, and concluded that there is not a significant difference in the discipline time before and after the improvements were implemented. There is still room for improving the process to reduce the discipline waiting time.

6. Control Charts

The team proposed implementing an NP control chart, identifying the number of discipline referrals as the quality characteristic. The initial control chart, prior to the process being in control, looked as in Figure 4.53.

7. Replication Opportunities

The discipline process improvements could be replicated in other high schools in the school district, especially those with similar student demographics.

8. Standard Work, Kaizen

The discipline process procedures were documented and standardized across the discipline deans. The best practice classroom management techniques were shared across the faculty.

9. Dashboards/Scorecards

The dashboard provides a systems view of the entire process and the critical metrics. Following is the proposed principal's dashboard, which includes tracking number of discipline referrals by week; the numbers of discipline referrals by severity level; the

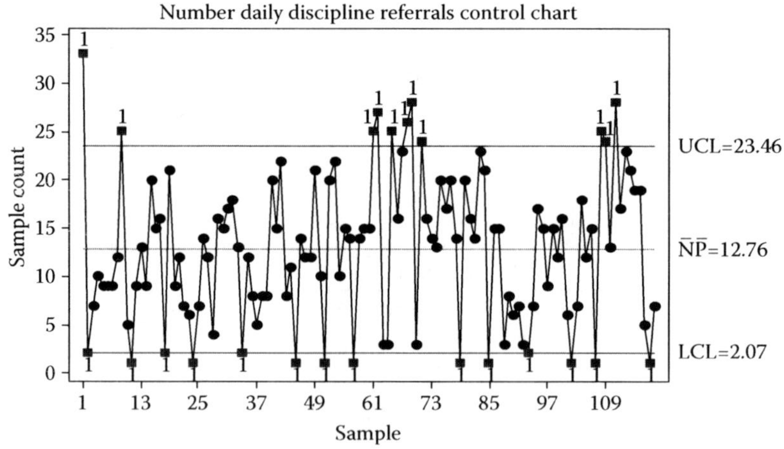

FIGURE 4.53 NP control chart. *Note:* Not actual data, for illustrative purposes only.

Discipline scorecard	
Week ending:	8/12/2007
Weekly number of discipline referrals	42
Level 4	0
Level 3	0
Level 2	0
Level 1	42
9th Graders	20
10th Graders	13
Number behavior contracts:	35
Number attendance contracts:	56
Number days missed excused:	219
Number days missed unexcused:	1096

FIGURE 4.54 Discipline scorecard.

number of discipline referrals by ninth and tenth graders; the number of students on behavior and attendance contracts; the numbers of days missed that were excused and unexcused across the student body. Target levels were identified that would identify the area to be green (OK), yellow (warning carefully watch), and red (investigate the problem). A sample dashboard is presented in Figure 4.54.

10. Control Phase Presentation

The Control phase presentation can be found in the downloadable instructor materials.

CONTROL PHASE CASE DISCUSSION

1. **Control Report**
    1.1 Review the Control report and brainstorm some areas for improving the report.
    1.2 How did your team ensure the quality of the written report? How did you assign the work to your team members? Did you face challenges of team members not completing their assigned tasks in a timely manner and, if so, how did you deal with it?
    1.3 Did your team face difficult challenges in the Control phase? How did your team deal with conflict on your team?

1.4 Did your instructor and/or Black Belt or Master Black Belt mentor help your team better learn how to apply the Lean Six Sigma tools in the Control phase, and how?
1.5 Compare your Control report to the Control report in the book, what are the major differences between your report and the author's report?
1.6 How would you improve your report?

2. Hypothesis Tests, Design of Experiments

2.1 How did you assess the improvement for the CTS?

3. Mistake Proofing

3.1 How well did your team assess the mistake proofing ideas to prevent errors?

4. Control Plan

4.1 How well will your control plan ensure that the improved process will continue to be used by the process owner?
4.2 Are their additional control charts that could be used to ensure process control?

5. Process Capability, DPMO

5.1 Did you validate that your process was in control before calculating the process capability?
5.2 Why is this important?

6. Control Charts

6.1 For this project, did you find attribute or variable control charts to be more applicable for controlling this process.

7. Replication Opportunities

7.1 How did your team identify additional replication opportunities for the discipline process within the high school, and within the school district?

8. Standard Work, Kaizen

8.1 How might you use a kaizen event to have identified process improvement areas, or ways to standardize the process?
8.2 How would you recommend ensuring that the process owners follow the standardized work procedures?

9. Dashboards/Scorecards

9.1 How would your dashboard differ if it was going to be used to present the results of the discipline process to the school board, or be used across several schools?

10. Control Phase Presentation

10.1 How did your team decide how many slides/pages to include in your presentation?

10.2 How did your team decide upon the level of detail to include in your presentation?

REFERENCES

Metzler, C., Biglan, A., Rusby, J., Sprague, J, Evaluation of a comprehensive behavior management program to improve school-wide positive behavior support. *Education and Treatment of Children*, 24 (4), 448–79, 2001.

Scholtes, P., Joiner, B., Streibel, B, *The Team Handbook*, 3rd ed., Oriel Incorporated, Madison, WI, 2003.

5 Financial Services Improvement in City Government—A Lean Six Sigma Case Study

Sandra L. Furterer

CONTENTS

FINANCIAL PROCESS OVERVIEW

Lean Six Sigma can improve the efficiency of processes, improve the quality of service to citizens, and reduce the costs of providing these services. The author worked with a local government's financial administration department to implement Lean Six Sigma. The goal of the project was to streamline the processes and subsequently reduce the financial process cycle time. The city is a 7000-citizen municipality in the state of Ohio. It is a city manager form of government where the city manager manages the city employees and implements policy defined by the mayor and city council members. The finance director reports to the city manager, and is responsible for developing and managing the financial budgets, the financial processes, the mayor's court processes, income tax collection, utility billing, and collection processes.

The financial processes include payroll, purchasing and accounts payable, accounts receivable, monthly reconciliation and budgeting. The finance clerk generates paychecks for administrative personnel, the police department, the fire department, the public works department and city council. The International Association of Fire Fighters (IAFF) represents the fire fighters who require union dues to be held from the members' pay once a month to be submitted to the union. The processing also includes pension matching, making pension payments and reporting. The payroll department also processes income tax payments, garnishments, child support and other withholdings to the appropriate agencies. Employees receive paychecks every two weeks. Pension reporting is performed on a monthly basis. The customers of the payroll process are internal city employees and external agencies that receive withholding payments and reports. The financial director realizes that the current processes, with respect to the processes before the Lean Six Sigma program is implemented, are inefficient, error-prone, lengthy, and have an extensive number of nonvalue-added steps. The entire payroll, pension reporting, withholding payment process takes 13–70 employee hours per pay period, depending if information processing problems occur.

The purchasing and accounts payable processes enable city personnel to purchase materials, products, and services to run the city. Purchase requisitions are generated by personnel. The finance clerk generates the purchase order, which is then approved by the city manager, the finance director, and city council, if necessary. Invoices are received by the finance director and processed by the finance clerk, with the appropriate approvals and signatures. Payments to vendors are frequently late. Multiple invoices for the same payment are frequently received and must be reviewed to determine if they have been paid. The up-front purchasing process takes approximately 7–10 days to generate and approve the purchase orders after the approved purchase requisition is received. The purchase orders are filed until the invoices are received. The entire accounts payable process takes approximately two weeks to process a batch from initial invoice receipt to vendor payment.

The finance clerk records revenue receipts and deposits revenue checks into the bank. In the current process there is a lag between when the revenue checks are received in the finance department and when they are entered into the financial system and deposited into the bank due to process inefficiencies and workload capacity issues.

The finance clerk is responsible for reconciling the financial records on a monthly basis. Reconciliation includes comparing the bank statements for the payroll account, a general account, and several investment accounts, to the financial system entries. Due mainly to process inefficiencies or workload capacity issues (or both), monthly reconciliation currently is rarely performed in a timely manner. Sometimes the finance director reconciles the books and other times it is outsourced to an accountant.

The finance director is responsible for managing the budgeting process throughout the city. He receives budget requests from department managers, consolidates them into a city budget, prepares budget reports for state and county agencies, and makes budget journal entries into the financial information system. The finance director is also responsible for ensuring that expenditures are within the approved budgets, as well as providing budget information to city management. There are some training issues with respect to using the financial system for budgeting, as well as duplicate data entry into multiple information systems. The financial information system is also limited with respect to a user-friendly ad-hoc budget reporting system.

DEFINE PHASE EXERCISES

It is recommended that the students work in project teams of 4–6 students throughout the Lean Six Sigma Case Study.

1. **Define Phase Written Report**
   Prepare a written report from the case study exercises that describes the Define phase activities and key findings.

2. **Lean Six Sigma Project Charter**
   Use the information provided in the Financial Process Overview section above, in addition to the project charter format to develop a project charter for the Lean Six Sigma project.

3. **Stakeholder Analysis**
   Use the information provided in the Financial Process Overview section detailed above, in addition to the stakeholder analysis format, to develop a stakeholder analysis, including stakeholder analysis roles and impact definition, and stakeholder resistance to change.

4. **Team Ground Rules and Roles**
   Develop the project team's ground rules and team members' roles.

5. **Project Plan and Responsibilities Matrix**
   Develop your team's project plan for the DMAIC project. Develop a responsibilities matrix to identify the team members who will be responsible for completing each of the project activities.

6. **SIPOC**
   Use the information provided in the Financial Process Overview section detailed above to develop a SIPOC of the high-level process.

7. **Team Member Biographies (Bios)**
   Each team member should create a short bio of themselves so that the key customers, stakeholders, project champion, sponsor, Black Belt and/or Master Black Belt, can get to know them, and understand the skills and achievements that they bring to the project.

8. Define Phase Presentation

Prepare a presentation (PowerPoint) from the case study exercises that provides a short (10–15 minutes) oral presentation of the Define phase deliverables and findings.

DEFINE PHASE

1. Define Phase Written Report

Following is the Define phase report. A successful implementation of the Lean Six Sigma problem-solving approach and Quality and Lean tools will be measured by the reduction of process inefficiencies, the reduction of the time it takes to process the financial transactions, and the assignment of appropriate staffing levels to handle the workload. No quantitative or qualitative measures of process or quality characteristics exist for any of the financial processes.

The DMAIC problem-solving methodology from the Six Sigma approach was used to improve the financial processes. The goal of the Define phase of the DMAIC Six Sigma problem-solving process is to define the need for improving the financial processes, develop the project charter, and perform the stakeholder analysis.

2. Lean Six Sigma Project Charter

The finance director identified the need to streamline the financial processes. The finance clerk complained of needing additional staff and not being able to complete her work. She was responsible for the purchasing, accounts payable, accounts receivable, payroll and monthly reconciliation and closing. The vendor payments were frequently late, resulting in vendors constantly calling the finance department requesting payment. The revenue receipts were frequently held in the finance department for more than one week before processing and depositing. The estimated current payroll processing time was 13–70 hours, with a mean time of 40 hours. Employees frequently complained about payroll paycheck errors. The monthly reconciliations were not performed on a regular basis. Adjustment journal entries were frequently made months after the error should have been discovered.

The Lean Six Sigma Quality facilitator, the process analyst, and the consulting manager interviewed the finance personnel to understand the financial department goals, the project scope and objectives. Figure 5.1 shows the project charter describing the problem, the goals and scope of the project, the customers and stakeholders and what is important to their satisfaction (CTS), financial benefits, and potential project risks. The goal of the project is to streamline the financial processes, reduce cycle time, and improve quality and accuracy of the processes. The scope of the project is the financial processes, including payroll, purchasing and accounts payable, accounts receivable, monthly reconciliation, and budgeting. Potential financial benefits are in the cost avoidance of not having to hire additional resources, and all work being done by one person, instead of 1.5 full-time equivalents, which could result in a fully loaded payroll cost of $66,000.

Project name: Financial Process Improvement

Problem statement: The finance director identified the need to streamline the financial processes. The finance clerk complained of needing additional staff and not being able to complete her work. Vendor payments were frequently late, resulting in vendors constantly calling the finance department requesting payment. The revenue receipts were frequently held in the finance department for over a week before processing and depositing. The estimated current payroll processing time ranged from 13–70 hours, with a mean time of 40 hours. Employees frequently complained about payroll paycheck errors. The monthly reconciliations were not performed on a regular basis. Adjustment journal entries were frequently made months after the error should have been discovered.

Customer/Stakeholders: (Internal/External) financial departments, city departments, external vendors, governmental agencies (tax reporting, county and state, pension)

What is important to these customers – CTS: Accuracy, timeliness.

Goal of the project: To streamline financial processes, reduce cycle time, improve quality and accuracy.

Scope statement: The financial processes include payroll, purchasing and accounts payable, accounts receivable, monthly reconciliation, and budgeting.

Financial and other benefit(s): Cost avoidance, not having to hire additional resources, and all work being done by 1 person, instead of 1.5 FTEs. $66,000.

Potential risks: Stakeholder buy-in; consulting resources not approved by city manager.

FIGURE 5.1 Project charter.

3. Stakeholder Analysis

The Lean Six Sigma team consisted of the finance clerk, who performed the accounts payable, accounts receivable, payroll and pension reporting, and monthly reconciliation processes within the finance department; the finance director, who managed the financial processes, the mayor's court processes, income tax collection, and utility billing and collection, and also performed the budgeting preparation and tracking for the city; a team quality facilitator, who developed the implementation plan and provided technical Quality and Lean principles and tools knowledge; a process analyst, who helped to collect and prepare process documentation; and a consulting manager, who provided business knowledge and direction, and maintained the formal business relationship between the city and the consulting firm. The team quality facilitator, the process analyst and the consulting manager were hired from an external consulting firm. The team profiled the people and cultural state to understand the level of skills and training of the employees, and their resistance or acceptance levels to change. At the start of the project, the finance clerk was very resistant to change. As the project progressed, she became very receptive to the improvement ideas because she saw how it would help her get her work done more quickly and with fewer errors. She also enjoyed getting the attention related to the improvement effort. The finance director was very receptive to change and the improvement effort. He embraced the vision of improved and streamlined financial processes.

The stakeholders are defined in Figure 5.2, and the stakeholder commitment scale is shown in Figure 5.3.

Stakeholders	Who are they?	Potential impact or concerns	+ / –
Finance clerk	City employee who performs the detailed financial processes including processing payroll, accounts payable and accounts receivable.	▪ Standardized processes ▪ Fewer errors ▪ Reduction of time and work ▪ Resistance to change	+ + + –
Finance director	Manager of the finance and administration departments, including finance, Mayor's Court, utility billing and income tax.	▪ Ensure accounting and finance standards and procedures are followed ▪ Citizen and council satisfaction ▪ Avoid hiring additional staff	+ + –
Quality facilitator and process analyst	Provides Black Belt expertise, identifies improvement recommendations, documents process, collects data, performs statistical analyses.	▪ Reduce resistance to change with finance clerk ▪ Complete project on time and within budget ▪ Add value and improve processes	– + +
Consulting manager	Manages client relationship for consulting company.	▪ Client satisfaction ▪ Complete project on-time and within budget	+ +

FIGURE 5.2 Stakeholder analysis definition.

Stakeholders	Strongly against	Moderate against	Neutral	Moderate support	Strongly support
Finance clerk	X				O
Finance director					XO
Quality facilitator and process analyst					XO
Consulting manager					XO
X = At start of project				O = By end of project	

FIGURE 5.3 Stakeholder commitment scale.

4. Team Ground Rules and Roles

The consulting engagement "statement of work letter" described the roles and anticipated involvement of the finance clerk, the finance director, and the consultants. It was clearly identified that the consultants would work with the city to gather and analyze data and provide recommendations based upon their best practice experience to help improve the financial processes. However, it was ultimately the finance department's responsibility to implement the processes and make change happen.

5. Project Plan and Responsibilities Matrix

The quality facilitator created a letter of understanding to document the roles and responsibilities of the team members. The team created a project plan with activities, a timeline, and resources (Figure 5.4). Figure 5.5 identifies the team mission, and team members' roles and responsibilities.

6. SIPOC

The SIPOC describes the scope of the Financial Process Improvement project. The SIPOC is shown in Figure 5.6. The SIPOC provides the stakeholders identified as the suppliers who provide input to the process (timesheets, data, payments, and the customers that receive the outputs from the processes (paychecks, invoice checks, etc.). The SIPOC also identifies the high-level process steps included in the scope of the project including: accounts payable, accounts receivable, payroll, monthly reconciliation, and budgeting.

7. Team Member Bios

The team quality facilitator, Sandy Furterer, is experienced in Six Sigma, quality management, information systems business and systems analysis, and Lean methodologies. She is a Certified Quality Engineer by the American Society for Quality, and holds a bachelor's degree and master of science degree in industrial engineering from The Ohio State University, and an MBA from Xavier University in Cincinnati, Ohio.

The process analyst, Reggie Fitzsimmons, is experienced in process and quality analysis, as well as process improvement methodologies.

The consulting manager and managing partner, Gregg St. John, is experienced in information systems, Lean, and process improvement.

8. Define Phase Presentation

The Define phase presentation can be found in the downloadable instructor materials.

DEFINE PHASE CASE DISCUSSION

1. **Define Phase Written Report**
    - 1.1 How did your team ensure the quality of the written report? How did you assign the work to your team members? Did you face challenges of team members not completing their assigned tasks in a timely manner and, if so, how did you deal with it?
    - 1.2 Did your team face difficult challenges in the Define phase? How did your team deal with conflict on your team?
    - 1.3 Did your instructor and/or Black Belt or Master Black Belt mentor help your team better learn how to apply the Lean Six Sigma tools, and how?
    - 1.4 Did your Define phase report provide a clear vision of the project, why or why not?

Activity number	Phase/Activity	Duration	Predecessor	Resources
1.0	Define			
1.1	Define process improvement need	One day		Quality facilitator, finance director
1.2	Identify goals	Two days	1.1	Quality facilitator
1.3	Form team	Two days	1.2	Finance director, consulting manager
2.0	Measure		1.0	
2.1	Profile current state	14 days		Quality facilitator, process analyst, finance clerk
2.2	Identify problems that contribute to process inefficiencies an errors	Five days	2.1	Quality facilitator, process analyst, finance clerk
2.3	Identify root causes	Five days	2.2	Quality facilitator, process analyst, finance clerk
3.0	Analyze		2.0	
3.1	Analyze gaps from best practice	Five days		Quality facilitator
3.2	Identify improvement opportunities and develop an improvement plan	Five days	3.1	Quality facilitator, process analyst, finance clerk
3.3	Perform cost / Benefit Analysis	Five days	3.2	Quality facilitator
4.0	Improve		3.0	
4.1	Implement improvement solutions	20 days		Finance clerk
4.2	Measure impact of the improvements	Five days	4.1	Quality facilitator
4.3	Document procedures and train employees on the improved procedures	10 days	4.2	Quality facilitator
5.0	Control		4.0	
5.1	Design and implement process performance measures	Five days		
5.2	Implement a continuous process improvement approach	Ongoing	5.2	
5.3	Celebrate success	Half-a-day	5.3	

FIGURE 5.4 Project plan.

Team Mission	
Document the current financial processes to create desk-top procedures and to identify and implement financial process improvements.	
Role	**Responsibility**
Finance clerk as process owner	Provides process knowledge and identifies and implements improvement opportunities.
Finance director as project champion	Establishes team mission and goals. Provides project team resources and support.
Team quality facilitator as Black Belt	Provides team facilitation. Provides technical quality and lean tool knowledge. Provides best practice for financial processes.
Process analyst	Prepares documentation. Collects process data. Identifies improvement opportunities.
Consulting manager	Provides business knowledge and direction. Manages consultants.

FIGURE 5.5 Team mission, roles and responsibilities.

Suppliers	Inputs	Process	Output	Customers
City employees	Time reports	Payroll	Checks, pension reports, taxes paid	City employees, taxing authorities, state, county
Vendors, city employees	Invoices, requests	Accounts payable	POs, checks	Vendors
State, county	Checks, direct deposits	Accounts receivable	Funds available or invested	City departments
City departments	Financial transactions, receipts, checks, invoices, bank statements	Monthly reconciliation	Balanced accounts, adjustments, financial reports	Finance director, council
City departments	Budgeting needs	Budgeting	Budget, appropriations	Council and citizens

FIGURE 5.6 SIPOC.

2. Lean Six Sigma Project Charter

Review the project charter presented in the Define phase report.

2.1 A problem statement should include a view of what is going on in the business and when it is occurring. The problem statement should provide data to quantify the problem. Does the problem statement in the

Define phase report provide a clear picture of the business problem? Rewrite the problem statement to improve it.

2.2 The goal statement should describe the project team's objective, and be quantifiable, if possible. Rewrite the Define phase goal statement to improve it.

2.3 Did your project charter's scope differ from the example provided? How did you assess what was a reasonable scope for your project?

3. Stakeholder Analysis

Review the stakeholder analysis in the Define phase report.

3.1 Should the city council and the city manager (who the finance director reports to) be defined at stakeholders, why or why not?

3.2 Are there any other stakeholders that could have been identified?

4. Team Ground Rules and Roles

4.1 Discuss how your team developed your team's ground rules. How did you reach consensus on the team's ground rules?

5. Project Plan and Responsibilities Matrix

5.1 Discuss how your team developed their project plan and how they assigned resources to the tasks. How did the team determine estimated durations for the work activities?

6. SIPOC

6.1 How did your team develop the SIPOC? Was it difficult to start at a highlevel, or did the team start at a detailed level and move up to a high-level SIPOC?

7. Team Member Bios

7.1 What was the value in developing the bios, and summarizing your unique skills related to the project? Who receives value from this exercise?

8. Define Phase Presentation

8.1 How did your team decide how many slides/pages to include in your presentation?

8.2 How did your team decide upon the level of detail to include in your presentation?

MEASURE PHASE EXERCISES

1. **Measure Report**
   Create a Measure phase report, including your findings, results and conclusions of the Measure phase.

2. **Process Maps**
   Create level-1 and level-2 process maps for each of the following processes:

   - Accounts payable
   - Accounts receivable
   - Payroll
   - Monthly reconciliation

3. **Operational Definitions**
   Develop an operational definition and metric for each of the identified CTS criteria:

   - Cycle time
   - Accuracy of the process
   - Customer satisfaction

4. **Data Collection Plan**
   Use the data collection format to develop a data collection plan that will collect voice of customer (VOC) and voice of process (VOP) data during the Measure phase.

5. **VOC Surveys**
   Create two VOC surveys to better understand the internal customers and the vendors' requirements and CTS characteristics related to the financial process elements.

6. **Pareto Chart**
   Create a Pareto chart using the data in Figure 5.7 related to the number of vendors by year-to-date purchasing activity.

7. **VOP Matrix**
   Create a VOP matrix to identify how the CTSs, process factors, operational definitions, metrics and targets relate to each other.

8. **Statistical Analysis**
   Review the financial process database, "Financial Process.xls" and perform the following statistical analysis:

Year-to-date activity	Number of vendors
< $500	250
$500 to $999	60
$1,000 to $2,999	100
$3,000 to $4,999	25
$5,000 to $9,999	30
$10,000 to $19,999	45
$20,000 to $49,000	10
$50,000 to $100,000	5
Above $100,000	2

FIGURE 5.7 Data for Pareto chart.

A. For the time to process accounts payable batches
- Create a histogram
- Calculate the mean and standard deviation for the time to process payroll
- Do the data follow a normal distribution?

B. For the time to process payroll batches
- Create a histogram
- Calculate the mean and standard deviation for the time to process an AP batch
- Do the data follow a normal distribution?

C. Perform additional analysis based on the financial data provided.

9. COPQ

Brainstorm potential COPQ for the case study for the following categories:
- Prevention
- Appraisal
- Internal failure
- External failure

10. Measure Phase Presentation

Prepare a presentation (PowerPoint) from the case study exercises that provides a short (10–15 minutes) oral presentation of the Measure phase deliverables and findings.

MEASURE PHASE

1. Measure Report

The goal of the Measure phase of the DMAIC Six Sigma problem-solving process is to understand and document the current state of the processes to be improved, identify the process problems that are causing inefficiencies, and errors and their root causes.

2. Process Maps

The team used process flow chart analysis to map the current state processes. These flow charts identified the steps involved in the finance department activities related to budgeting/investments, purchasing/accounts payable, accounts receivable, monthly reconciliation and payroll. Various system functions were identified in the process flows that were used to perform the financial processes. The process flows identified the written (of which few existed) and unwritten policies that governed the processes. The detailed process flow charts are included in Appendix A. It was decided that budgeting would not be included in the scope of the project after the initial process maps were developed, so only the processes performed by the Financial Clerk would be in the project scope.

3. Operational Definitions

No process measures existed for the financial processes prior to the Lean Six Sigma project. The finance clerk estimated the average and range of the processing times based on her experience with the processes. The estimated processing times are displayed in Figure 5.8. The team also profiled the technology to determine if the financial system was meeting their needs. They had implemented the system about six months prior to the project starting, and there were many training issues related to the software. There were also some inefficient information system flows required by the software applications. Ad-hoc financial reporting capability was difficult, time consuming, and required extensive knowledge of data tables and query ability.

The operational definition for measuring the accounts payable process cycle time is defined as the time to process one batch once the batch is organized. It does not include the time waiting for the invoice to be matched once it is received in the mail.

The operational definition for measuring the accounts receivable process cycle time is the time from when the revenue check or receipt is received in the Finance office until it is deposited in the city's bank account.

The operational definition for measuring the payroll process cycle time is the time from when the last timesheet is received from the city departments to when the

Process	Estimated elapsed processing time range	Estimated average elapsed processing time
Payroll and pension reporting	13–70 hours	60 hours
Purchasing/Accounts payable	30–40 hours per batch (only about half of the due invoices are processed every other week)	40 hours
Accounts receivable	40–80 hours (including delay due to workload capacity issues)	60 hours
Monthly reconciliation	40–80 hours (if performed)	60 hours
Budgeting	No estimate available	No estimate available

FIGURE 5.8 Estimated processing times.

payroll is complete and the paychecks or direct deposit information is printed. It does not include the pension reporting processing and printing time.

The operational definition for measuring the monthly reconciliation time is the time it takes to reconcile the books with the bank statements and make any appropriate adjustments and print the appropriate financial reports.

The operational definitions for the defects for each process will be further defined after the defect types are collected using the check sheets discussed in the data collection plan.

4. Data Collection Plan

Because there was no process measurement system in place to assess the CTS criteria related to cycle time, accuracy and customer satisfaction, the data collection plan is a critical tool to help provide a way to measure the CTS. The data collection plan is shown in Figure 5.9. Cycle time and accuracy should be measured at a detailed process level for each subprocess, including accounts payable, accounts receivable, monthly reconciliation, and payroll.

5. VOC Surveys

Note: The data provided is for illustrative purposes only, and was not actually collected during the Lean Six Sigma project.

Two customer surveys were developed, one to assess the VOC requirements for the vendors regarding the accounts payable process and the other for internal customers regarding the payroll process. Following are VOC surveys.

Vendor VOC Survey

The following survey is being used to assess your satisfaction with the city's accounts payable process. Please complete the survey questions, which should take about 5 minutes. We appreciate your time and feedback in helping us improve our financial processes.

1	2	3	4	5
Strongly Disagree	Disagree	Neutral	Agree	Strongly Agree

1. I receive payment for my invoices in a timely manner.
2. I receive accurate payments for my invoices.
3. If I call or see the city for customer service related to my invoice, I receive prompt service.
4. If I call or see the city for customer service related to my invoice, I receive friendly service.
5. If I call or see the city for customer service related to my invoice, my problem gets solved completely the first time.
6. Please provide ideas for how we could improve customer service with the city.

Critical to Satisfaction (CTS)	Metric	Data collection mechanism (survey, interview, focus group, etc.)	Analysis mechanism (statistics, statistical tests, etc.)	Sampling plan (sample size, sample frequency)	Sampling instructions (who, where, when, how)
Cycle time	AP: cycle time – vendor invoice received to paid	Track for four weeks	Mean, standard deviation, control charts	All invoices for one month	Process analyst tracks date received to when paid
	AR: time to deposit funds in bank from when check received	Track for four weeks	Mean, standard deviation, control charts	All revenue receipts for one month	Process analyst tracks date received to when paid
	Recon: time takes to close	Track for two months	Mean, range	Time to close for two months	Process analyst tracks time to close
	Payroll: time to process payroll	Track for two payroll cycles	Mean, range	Time for two payroll cycles	Process analyst tracks time to close
Accuracy of the process	AP: types and number of defects	Check sheet	Pareto chart	Defects for one month	Finance clerk to track on check sheet
	AR: types and number of defects	Check sheet	Pareto chart	Defects for one month	Finance clerk to track on check sheet
	Recon: types and number of defects	Check sheet	Pareto chart	Defects for one month	Finance clerk to track on check sheet
	Payroll: type and number of defects	Check sheet	Pareto chart	Defects for one month	Finance clerk to track on check sheet
Customer satisfaction	Vendors	Survey	Statistical analysis	Survey 20 vendors	Quality facilitator to create survey and collect survey data
	Internal customers	Survey	Statistical analysis	Survey internal city departments: police, fire, streets, admin	Quality facilitator to create survey and collect survey data

FIGURE 5.9 Financial process data collection plan.

Internal VOC Survey

The following survey is being used to assess your satisfaction with the city's payroll process. Please complete the survey questions, which should take about

5 minutes. We appreciate your time and feedback in helping us improve our financial processes.

1	2	3	4	5
Strongly Disagree	Disagree	Neutral	Agree	Strongly Agree

1. I receive my paycheck in a timely manner.
2. I receive an accurate paycheck.
3. If I call or see the finance department for service related to payroll, I receive prompt service.
4. If I call or see the finance department for service related to payroll, I receive friendly service.
5. If I call or see the finance department for service related to payroll, my problem gets solved completely the first time.
6. Please provide ideas for how we could improve customer service with the finance department.

6. Pareto Chart

The quality facilitator and process analyst noticed that there was a large quantity of invoices for a city of this size. The finance clerk was constantly inundated with invoices coming in on a daily basis. As the team further investigated, asking "why" several times, it became evident that there was no centralized purchasing. Although all of the purchase requisitions came into the finance clerk to be approved by the finance director and city manager, each city department decided what they would purchase and who they would purchase it from. Each department ordered their own office supplies from their favorite office store supplier. There was no preferred or certified vendor list for purchases under $10,000. The team decided to analyze the accounts payable data for the year-to-date, and identify the number of vendors by the dollar value that was purchased by each vendor within the first eight months of the year. The resulting Pareto chart (Figure 5.10) shows more than 250 vendors with a total purchase activity for an eight-month period under $500. Each invoice requires

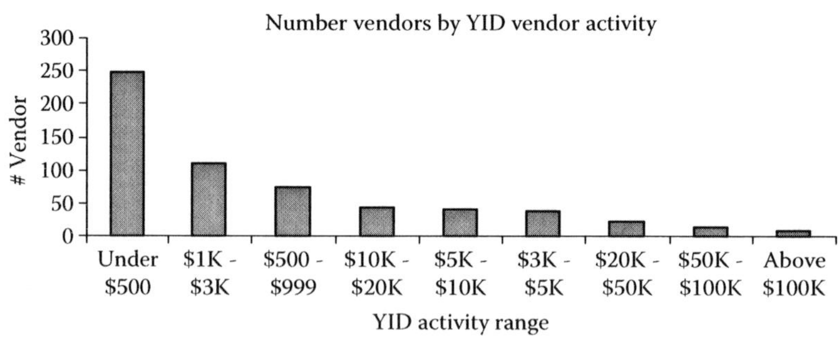

FIGURE 5.10 Pareto chart of year-to-date vendor activity.

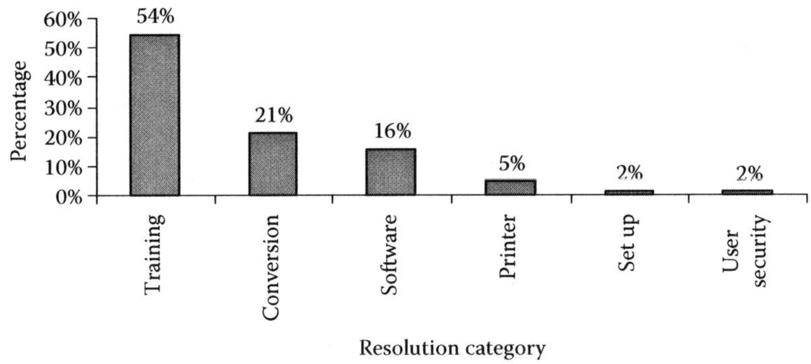

FIGURE 5.11 Information system percentage problem by resolution category.

a purchase requisition to be completed and approved, a purchase order to be created, printed and approved (in duplicate), an invoice to be received and matched with the shipping or receiving paperwork, and the invoice to be entered and processed, a check to be printed and signed (in duplicate), as well as the resultant monthly reconciliation of all of these transactions. An opportunity for consolidating the purchasing activity and eliminating many nonvalue-added activities is identified for analyzing in the Analyze phase.

The finance clerk was constantly overwhelmed when she ran into a problem with the information system. She claimed that the financial system was wrought with problems, and just did not work. She said she constantly had to call into the financial system vendor's information system (IS) help desk to have them fix a problem. The way that the finance clerk dealt with an information system problem was she would call into the vendor's IS help desk, report the problem, and then sit at her desk waiting (not working on another task) for a call back, which could take two hours or more. The team decided to investigate the causes of the IS problems, and understand if the financial system was broke, or whether it was a training issue. The team collected data from the IS vendor's help desk system on the problems reported by the city's finance department, which included problem type, time to resolve, and resolution category. The Pareto chart in Figure 5.11 shows that 54% of the "problems" reported to the IS help desk were related to training (or lack of training) issues, not the "perceived software problems."

7. VOP Matrix

The VOP matrix helped to link the CTS criteria to the metrics, targets and potential process factors that affect the CTS. The VOP matrix was used to summarize the VOP (Figure 5.12). The CTSs were defined as cycle time, accuracy of the process, and customer satisfaction. The factors that potentially impact cycle time were having standard procedures, streamlined processes, training and the volume of the invoices and paychecks. The cycle time for each process was defined to be measured. The accuracy of the processes would be potentially impacted by training in procedures

CTS	Process factors	Operational definition	Metric	Target
Cycle time	▪ Standard procedures exist ▪ Streamlined processes ▪ Training ▪ Volume of invoices and paychecks	Measure each process time	AP: cycle time – vendor invoice received to paid AR: time to deposit funds in bank Recon: time takes to close Payroll: Paid on time per schedule	AP: ten business days AR: two days Recon: ten days Payroll: Paid on time
Accuracy of the process	▪ Training in Procedures and software	Measure each process and defect types	Defects by process and type	95% accuracy
Customer satisfaction	▪ Repeatable process ▪ Collect and assess VOC	Measure customer satisfaction through customer and vendor surveys	% of positive responses for identified survey questions	80% of responses are rated 4 or 5 for identified questions

FIGURE 5.12 VOP matrix.

and the financial software, and would be measured by assessing number and types of defects in each process. Customer satisfaction could be impacted by whether there was a repeatable process and whether the city would collect and measure VOC information. The VOC could be measured through surveys. The proposed target for each of the metrics is also included in the matrix.

8. Statistical Analysis

Note: The data provided are for illustrative purpose only, and were not actually collected during the Lean Six Sigma project.

The average accounts payable batch processing time is four hours, with a standard deviation of 1.6 hours. The average payroll batch processing time is 20.25 hours, with a standard deviation of 5.65 hours.

9. COPQ

The following are potential COPQ for the financial processes related to the following categories:

- Prevention
  - Training on the processes
  - Training on the information system
  - Developing a quality management system

  - Developing a vendor certification program
  - Developing a measurement system
  - Implementing a continuous process improvement program
- Appraisal
  - Certifying vendors
  - Assessing the measurement system
  - Assessing the accuracy and quality of the processes
  - Assessing the customer satisfaction
  - Assessing the process cycle times
- Internal failure
  - Process defects in each of the financial subprocesses found before they reach the internal customers in other departments or external vendors
  - Accounting adjustments during or after monthly reconciliations
- External failure
  - Process defects that reach the vendors or internal customers in other departments
  - Process defects that reach external taxing authorities, or state or county agencies
  - Incorrect or missing garnishments
  - Financial errors or adjustments that city council or the financial auditor discovers
  - Lack of citizen goodwill due to financial errors or adjustments

10. Measure Presentation

The Measure presentation can be found in the downloadable instructor materials.

MEASURE PHASE CASE DISCUSSION

1. **Measure Report**
   1.1 Review the Measure report and brainstorm some areas for improving the report.
   1.2 How did your team ensure the quality of the written report? How did you assign the work to your team members? Did you face challenges of team members not completing their assigned tasks in a timely manner and, if so, how did you deal with it?
   1.3 Did your team face difficult challenges in the Measure phase? How did your team deal with conflict on your team?
   1.4 Did your instructor and/or Black Belt or Master Black Belt mentor help your team better learn how to apply the Lean Six Sigma tools in the Measure phase, and how?
   1.5 Did your Measure phase report provide a clear understanding of the VOC and the VOP, why or why not?

2. **Process Maps**
   2.1 While developing the process maps, how did your team decide how much detail to provide on the level-2 process maps?
   2.2 Was it difficult to develop a level-2 from the level-1 process maps? What were the challenges?

3. **Operational Definitions**
   3.1 Review the operational definitions from the Measure phase report, define an operational definition that provides a better metric for assessing the accounts payable processing time.
   3.2 Discuss how you would recommend improving the operational definitions for process accuracy or defects.

4. **Data Collection Plan**
   4.1 Incorporate the enhanced operational definition developed in number 3 above into the data collection plan from the Measure phase report.

5. **Voice of Customer Surveys**
   5.1 How did your team develop the questions for the internal customers and/or vendor survey? Did you review them with other students to assess whether the questions met your needs?
   5.2 Create an affinity diagram for the main categories on the internal customer or vendor survey, grouping the questions into the higher-level "affinities." Was this an easier way to approach and organize the questions of the surveys?

6. **Pareto Chart**
   6.1 Discuss how the Pareto chart provides an analysis of vendor activity.
   6.2 Discuss how the Pareto chart provides insight into the finance clerk's perceptions of the IS vendor's help desk before and after the Pareto chart was created.

7. **VOP Matrix**
   7.1 How does the VOP matrix help to tie the CTSs, the operational definitions and the metrics together?

8. **Statistical Analysis**
   8.1 How did you determine which statistical analyses were important to perform?
   8.2 What were your important findings from the statistical analyses?

9. Cost of Poor Quality

9.1 Would it be easy to quantify, and collect data on the costs of quality that you identified for the case study exercise?

10. Measure Phase Presentation

10.1 How did your team decide how many slides/pages to include in your presentation?

10.2 How did your team decide upon the level of detail to include in your presentation?

ANALYZE PHASE EXERCISES

1. Analyze Report

Create an Analyze phase report, including your findings, results and conclusions of the Analyze phase.

2. Cause and Effect Diagram

For this project, the cause and effect diagram was created in the Measure phase, but analyzed in the Analyze phase. Create a cause and effect diagram that identify potential causes of process inefficiencies.

3. Cause and Effect Matrix

Create a cause and effect matrix for the following effects:

- Accounts payable defects
- Accounts receivable defects
- Payroll defects
- Monthly reconciliation defects

4. Why-Why Diagram

Create a Why-Why diagram for why payroll processing time takes so long.

5. Process Analysis

Prepare a process analysis for the following processes:

- Accounts payable process
- Accounts receivable process
- Payroll process
- Monthly reconciliation process

6. **Histogram, Graphical, and Data Analysis**
   Perform a histogram and graphical analysis for the following variables:
   - Accounts payable process time
   - Accounts receivable process time
   - Payroll process time
   - Monthly reconciliation process time
   - Monthly reconciliation defect types

7. **Waste Analysis**
   Perform a waste analysis for the following processes:
   - Accounts payable
   - Accounts receivable
   - Payroll
   - Monthly reconciliation

8. **Correlation Analysis**
   Perform a correlation analysis for the following variables:
   - Number of employees per batch related to payroll process time
   - Number of invoices per batch related to accounts payable process time

9. **Regression Analysis**
   Perform a regression analysis to try to predict the time to perform the monthly reconciliation process by the number of defects in each of the processes (accounts payable, accounts receivable and payroll).

10. **Confidence Intervals**
    Calculate a confidence interval about the mean and the variance for the following variables:
    - Accounts payable processing time
    - Accounts receivable processing time
    - Payroll processing time
    - Monthly reconciliation processing time

11. **Hypothesis Testing**
    Perform the following hypothesis tests:
    - Is the processing time for the first 12 payroll cycles significantly different from the last 12 payroll cycles?

12. Survey Analysis

- Perform survey analysis for the vendor VOC survey data "Vendor Survey Data.xls." Include Pareto charts for each question, and chi-square analysis.
- Perform survey analysis for the internal customer survey data "Internal Customer Survey Data.xls." Include Pareto charts for each question, and chi-square analysis.

13. DPPM/DPMO

- Calculate the DPMO and related sigma level for the process, assuming a 1.5 sigma shift, for the following data:

Opportunities for failure:

- Timesheet erroneous data
- Pay rate error
- Payroll processing error

Defects:

- Number of payroll defects per batch: 5

Units:

- Number of paychecks per batch: 100

14. Process Capability

Calculate the process capability for the accounts payable processing time with the following specifications:

- Lower specification limit: two hours
- Upper specification limit: four hours

15. Analyze Phase Presentation

Prepare a presentation (PowerPoint) from the case study exercises that provides a short (10–15 minutes) oral presentation of the Analyze phase deliverables and findings.

ANALYZE PHASE

1. Analyze Report

The goal of the Analyze phase is to analyze the problems and process inefficiencies. Another part of the Analyze phase is to perform a cost and benefit analysis to understand whether the improvements are too costly compared with the estimated benefits to improve productivity and quality.

2. Cause and Effect Diagram

The project team used the process flow charts and several Lean tools (including waste identification and elimination, standardization of operations) to identify and eliminate nonvalue-added activities, and good housekeeping (part of the 5S') to identify process problems such as inefficient sorting and filing of purchase orders and invoices. The team used brainstorming techniques to identify problems.

The team used cause and effect analysis to identify root causes related to people (such as lack of training and skills), methods (lack of standardized procedures), information technology (information system human factors and processing flow was confusing and inefficient), and hardware (broken and inefficient printers). A cause and effect diagram is presented in Figure 5.13.

The team identified gaps comparing the current state processes to best practice financial processes. The team quality facilitator and the process analyst used their understanding of financial processes and the concepts of Lean principles and the process flow charts to identify nonvalued-added activities, especially related to unnecessary work and rework. The team used the concept of implementing improvements that would prevent problems and rework due to printer jams, and inefficient use of the technology to reduce the financial processing time. The team quality facilitator performed an analysis of reported financial information system problems using Pareto analysis and statistical process control charts across the finance and administration department. The purpose was to identify employee training and knowledge gaps with respect to the financial and administrative information system.

3. Cause and Effect Matrix

A cause and effect matrix (Figure 5.14) was created to understand similar causes that produced defects in each of the financial processes. The bureaucratic culture

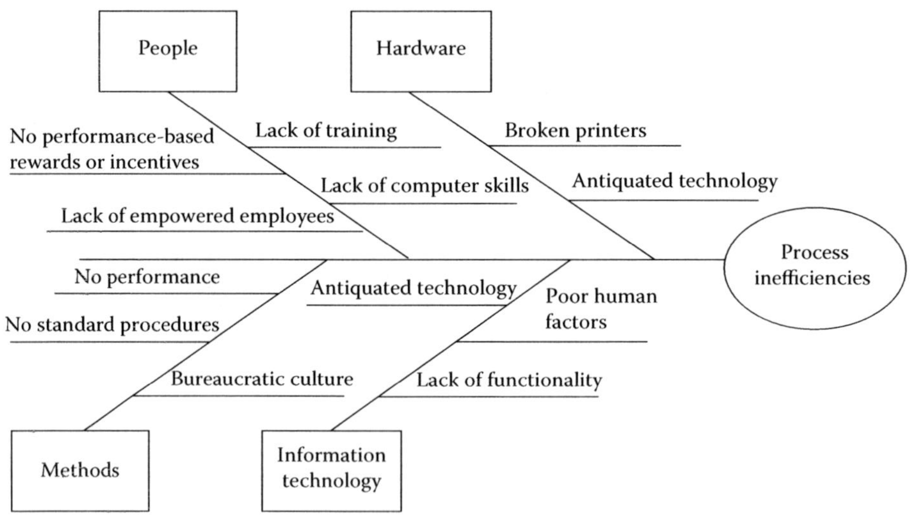

FIGURE 5.13 Cause and effect diagram.

	Effects					
	AP defects	AR defects	Payroll defects	Recon defects	Total	Relative weighting
Causes/Importance	8	4	10	6		
Lack of training	9	9	9	3	216	2
Lack of standard procedures	3	3	3	9	120	4
Antiquated technology	3	3	9	9	180	3
Lack of functionality		9	3	3	84	5
Bureaucratic culture	9	9	9	9	252	1

FIGURE 5.14 Cause and effect matrix.

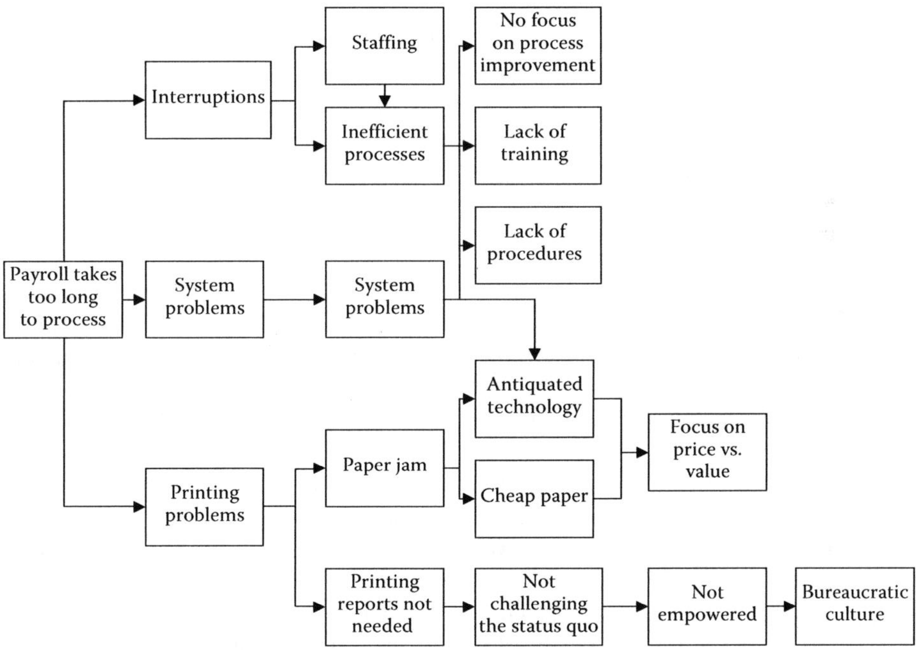

FIGURE 5.15 Why-Why diagram.

contributed most to process defects. Next was lack of training, then antiquated technology. Lack of standard procedures and then lack of functionality were the last two root causes in priority order, contributing to process defects.

4. Why-Why Diagram

A Why-Why diagram (Figure 5.15) was created to identify the root causes for why payroll processing time takes so long. The root causes are similar to what was already identified when generating the cause and effect diagram and matrix. Some of the root causes are lack of training, lack of procedures, no focus on process improvement, bureaucratic culture, focus on price versus value.

5. Process Analysis

The process analysis was performed for the accounts payable, accounts receivable, monthly reconciliation, and accounts receivable processes. The number of value-added versus nonvalue-added activities was compared for each process. The monthly reconciliation process had the highest percentage (93%) of nonvalue-added activities, followed by the accounts receivable process with 86% nonvalue-added activities. The payroll process had 83% of the activities identified as nonvalue-added, while the accounts payable process had 61% nonvalued-added activities. Balancing the books in the monthly reconciliation process is necessary from a financial audit and controls perspective, but the defects and inefficiencies from all of the upfront processes, such as payroll, accounts payable, and accounts receivable flow into the downstream reconciliation process and causes the balancing problems. The focus in the Improve phase should be to improve the upfront processes to reduce reconciliation problems. Figure 5.16 shows the summary of value-added and nonvalue-added percentages in each process. Figures 5.17 and 5.18 show the actual activities that are identified as adding value or not adding value to the processes.

6. Histogram, Graphical, and Data Analysis

Note: The data provided are for illustrative purposes only, and were not actually collected during the Lean Six Sigma project.

Histograms were created for the accounts payable and payroll batch processing times. The histogram of accounts payable batch times appears to follow a normal distribution (Figure 5.19). Additional statistical tests must be run to test for normality.

The payroll histogram distribution does not appear normal from looking at the histogram (Figure 5.20).

Figures 5.21 and 5.22 show the individuals and moving range control charts of the time (in hours) that it took the software vendor to resolve reported information system problems for the city. This showed that problems with the system contributed to out of control conditions and therefore process inefficiencies. The out of control conditions were assigned to a cause related to a computer program archiving process that was extremely difficult to identify because it only happened during a monthly archiving process.

Process	Value added percentage of activities	Nonvalue added percentage of activities
Accounts payable	39%	61%
Accounts receivable	14%	86%
Monthly reconciliation	7%	93%
Payroll	17%	83%

FIGURE 5.16 Financial process value analysis summary table.

Process	Value added activities	Nonvalue added activities	Value added % of activities
Accounts payable	▪ Perform bidding process ▪ Council approves ▪ Enter new vendor ▪ Enter PO in FSS ▪ Approve PO ▪ Fire Dept. calls with amount ▪ Treasurer gives amount needed ▪ Fill out req form ▪ Pay PO	▪ Obtain PO number ▪ Fill out req form ▪ Print PO ▪ Verify premium invoice number ▪ Verify PO exists ▪ Store PO ▪ Send invoice to supervisor ▪ Total invoices on calculator ▪ Print report ▪ Verify total ▪ Fix problems ▪ Print checks ▪ Send checks ▪ File copy	39%
Accounts receivable	▪ Post receipt in system ▪ Deposit at bank	▪ Make copy of check ▪ Total on calculator ▪ Staple deposit slip and copy ▪ Print report ▪ Verify total ▪ Match report to receipts ▪ Fix problems ▪ File receipts ▪ Stamp back of checks ▪ Fill out deposit slip ▪ Staple to report ▪ Store	14%

FIGURE 5.17 Financial process value analysis: accounts payable and accounts receivable.

7. Waste Analysis

The Lean Six Sigma team performed a waste analysis for the following processes: accounts payable, accounts receivable, monthly reconciliation and payroll.

There were multiple instances of each of the eight types of waste across all of the processes (Figure 5.23).

8. Correlation Analysis

Note: The data provided are for illustrative purposes only, and were not actually collected during the Lean Six Sigma project.

The team quality facilitator performed a correlation analysis to determine if there is a relationship between the number of employees and the payroll process time for each batch. There was a negative correlation with a correlation coefficient (r) of –.15, showing little correlation between the batch time and the number of employees.

Process	Value added activities	Nonvalue added activities	Value added % of activities
Monthly reconciliation	▪ Reset month in system	▪ Print reports ▪ Compare report totals to bank statements ▪ Call help desk for help ▪ Fix problem ▪ Reconcile bank statements ▪ List outstanding checks ▪ Compare totals ▪ Verify items ▪ Review check register ▪ Review wire transfers ▪ Make adjustments ▪ Bank make adjustments ▪ File bank statements ▪ File reports	7%
Payroll	▪ Enter time sheets in system ▪ Print checks ▪ Print deduction checks ▪ Print direct deposit vouchers ▪ Perform direct deposit transfer	▪ Verify time sheets ▪ Create manual hours sheet ▪ Print reports ▪ Compare hours totals ▪ Fix hours in system ▪ Print reports ▪ Compare totals ▪ Fix hours ▪ Print payroll reports ▪ Fix printer problems ▪ Redo payroll in system ▪ Void printed checks ▪ Re print checks ▪ Change printer paper ▪ Print successful bank report and send to bank ▪ Fix direct deposit problems ▪ Fix paycheck problems ▪ Write manual check ▪ Bank fixes problem ▪ Fix problem in direct deposit ▪ Bank calls with problem ▪ Write check for general fund ▪ Deposit in bank ▪ File copies of report	17%

FIGURE 5.18 Financial process value analysis: monthly reconciliation and payroll.

The team quality facilitator also performed a correlation analysis between the number of invoices per batch and accounts payable process time. The correlation coefficient was –0.54, which would indicate a slight inverse relationship between the two variables, as number of invoices increases, processing time decreases. One should always assess whether the statistical results make sense and, in this case, it

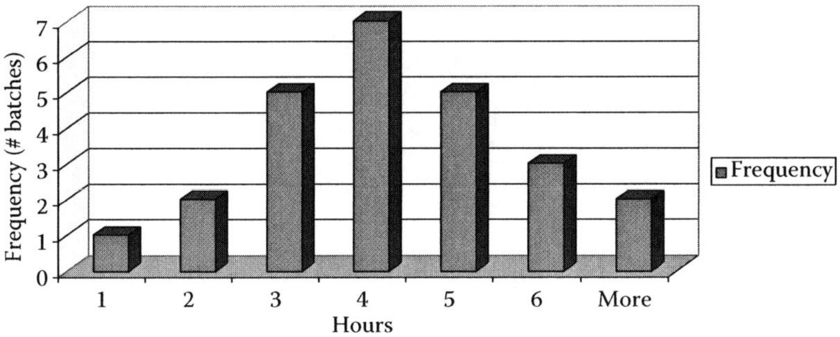

FIGURE 5.19 Histogram of AP batch time.

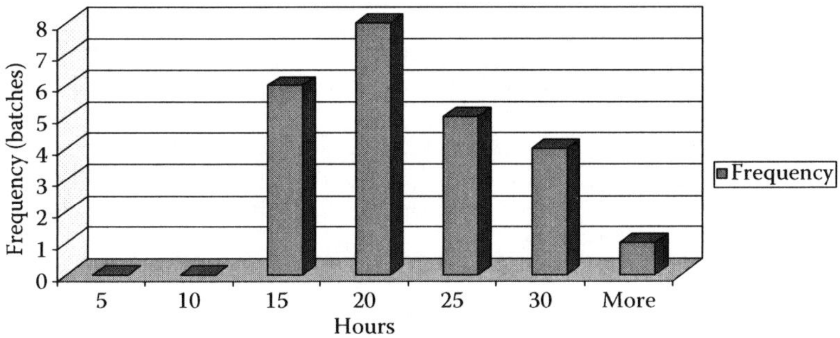

FIGURE 5.20 Histogram of payroll batch time.

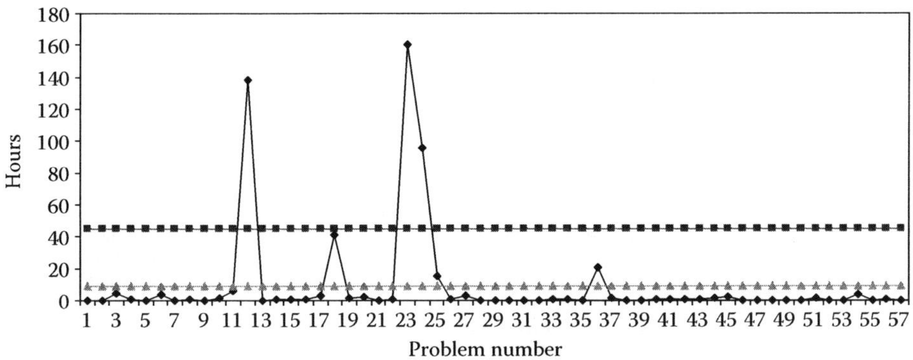

FIGURE 5.21 Mean time to resolve problems control chart.

does not appear to make sense that, as the number of invoices increases, the time decreases. There are probably other variables that are more highly correlated, or there is so much variability in the process that the error is causing an appearance of moderate relationship between the two variables.

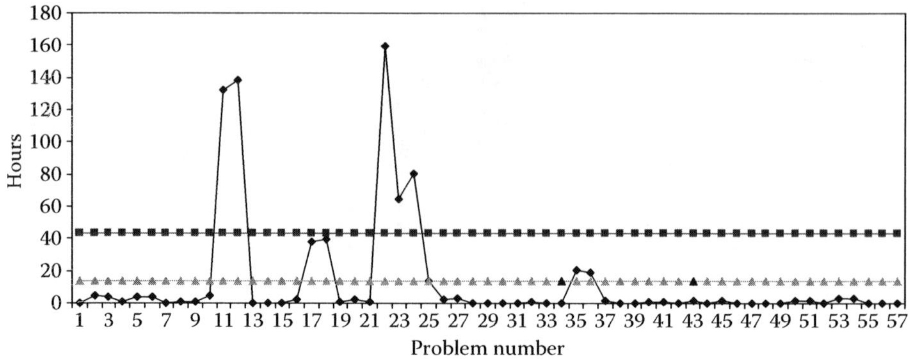

FIGURE 5.22 Moving range time to resolve problems control chart.

Waste type	Process	Waste element
Transportation	AP, AR, payroll	Moving manual checks, moving funds manually, not using direct deposit so moving paychecks
Over-production	AP, AR, payroll, monthly reconciliation	Printing reports that are not used
Motion	AP, payroll	Walking to printer in other room
Defects	AP, AR, payroll, monthly reconciliation	Matching totals, process defects (wire transfers, direct deposit errors, information system process errors, printer problems), paycheck errors, timesheet errors
Delay	AP, AR, payroll, monthly reconciliation	Waiting for AP processing, waiting to deposit AR checks, not getting to Monthly Reconciliation process, paying outside accountant to balance books, payroll late
Inventory	AP, AR, payroll, monthly reconciliation	Filing/storing reports, Purchase requisitions, purchase orders, invoices, time sheets
Processing	AP, AR, payroll, monthly reconciliation	Matching and balancing, not using direct deposit (printing checks), not moving funds automatically at bank
People	AP, AR, payroll, monthly reconciliation	No focus on process improvement, not using people's ideas

FIGURE 5.23 Waste analysis.

9. Regression Analysis

Note: The data provided are for illustrative purposes only, and were not actually collected during the Lean Six Sigma project.

The team quality facilitator performed a regression analysis to try to predict the time to perform the monthly reconciliation process based on the number of defects in each of the processes (accounts payable, accounts receivable and payroll). The coefficient of determination (R^2) was only 0.27. The team concluded that this was not a very good model to use to try to predict the time that it would take to reconcile the books.

10. Confidence Intervals

Note: The data provided are for illustrative purposes only, and were not actually collected during the Lean Six Sigma project.

The team quality facilitator calculated a confidence interval about the mean for the following variables: accounts payable information system processing time; accounts receivable elapsed time from receipt of check until deposit in bank; payroll processing time from receipt of time sheets to printing paychecks; and monthly reconciliation processing time. The accounts payable processing time only includes the time when entering information in the financial system, running reports, and printing checks. The payroll processing time only includes the time entering information in the system, printing reports and processing and printing the payroll checks and any delays related to this processing. The confidence intervals are shown in Figure 5.24.

11. Hypothesis Testing

Note: The data provided are for illustrative purposes only, and were not actually collected during the Lean Six Sigma project.

The team quality facilitator performed a t-test hypothesis test about the mean to determine if the processing time for the first 12 payroll cycles was significantly different from the last 12 payroll cycles. The null hypothesis was that the means are identical, and the alternative hypothesis was that the means are different. We used the t-test assuming equal variances. The mean for the first 12 pay cycles is 18.2 hours, and for the second 12 pay cycles is 22.3 hours. The p-value is .04, with an alpha of .05 (95% confidence level), so the null hypothesis can be rejected (p-value is less than alpha of .05). We can conclude that the means are different between the first 12 pay cycles and the second 12 pay cycles. So the average payroll process time increased in the latter payroll cycles.

Process	Lower confidence interval of the mean	Upper confidence interval of the mean	Mean	Standard deviation	Sample size
Payroll	17.99	22.51	20.3	5.7	24
AR	60.45	87.55	74	33.9	24
AP	3.33	4.59	4.0	1.6	24
Monthly reconciliation	46.91	64.84	55.9	22.4	24

FIGURE 5.24 Confidence intervals about the mean and variance.

12. Survey Analysis

Note: The data provided are for illustrative purposes only, and were not actually collected during the Lean Six Sigma project.

The results of the vendor survey are shown in Figure 5.25. The areas of opportunity for the accounts payable process are related to receiving payment for invoices in a timely manner, and receiving friendly service.

The results of the internal customer survey are shown in Figure 5.26. The areas of opportunity for the payroll process are related to receiving friendly service.

13. DPPM/DPMO

Note: The data provided are for illustrative purposes only, and were not actually collected during the Lean Six Sigma project.

Survey question	% Negative (1, 2)	% Positive (3, 4, 5)
1) I receive payment for my invoices in a timely manner.	80%	20%
2) I receive accurate payments for my invoices.	15%	85%
3) If I call or see the city for customer service related to my invoice, I receive prompt service.	10%	90%
4) If I call or see the city for customer service related to my invoice, I receive friendly service.	80%	20%
5) If I call or see the city for customer service related to my invoice, my problem gets solved completely the first time.	55%	45%

FIGURE 5.25 Vendor VOC survey results summary.

Survey question	% Negative (1, 2)	% Positive (3, 4, 5)
1) I receive my paycheck in a timely manner.	7%	93%
2) I receive an accurate paycheck.	13%	87%
3) If I call or see the finance department for service related to payroll, I receive prompt service.	5%	95%
4) If I call or see the finance department for service related to payroll, I receive friendly service.	88%	12%
5) If I call or see the finance department for service related to payroll, my problem gets solved completely the first time.	53%	47%

FIGURE 5.26 Internal customer VOC survey results summary.

We calculated the DPMO and related sigma level for the payroll process, assuming a 1.5 sigma shift, three opportunities for failure (time sheet erroneous data; pay rate error; and payroll processing error), five defects and 100 employees in the payroll batch). The DPMO is 16,667, relating to a sigma level of about 3.6 sigma, indicating a great deal of opportunity for improvement.

14. Process Capability

Note: The data provided are for illustrative purposes only, and were not actually collected during the Lean Six Sigma project.

We calculated the process capability for the accounts payable processing time with the following specifications:

- Lower specification limit: two hours
- Upper specification limit: four hours.

There was one point out of control on the moving range chart, but because we did not have an assignable cause, we left the point in for the analysis. We used Minitab to calculate the Cp and Cpk. The Cp index was .23 and the Cpk index was .01, demonstrating that the process is not capable of meeting the 2–4-hour specifications, nor is the process centered. The mean accounts payable batch hours is 3.958 with a standard deviation of 1.579 hours.

15. Analyze Phase Presentation

The Analyze phase presentation can be found in the downloadable instructor materials.

ANALYZE PHASE CASE DISCUSSION

1. **Analyze Report**
   - 1.1 Review the Analyze report and brainstorm some areas for improving it.
   - 1.2 How did your team ensure the quality of the written report? How did you assign the work to your team members? Did you face challenges of team members not completing their assigned tasks in a timely manner and, if so, how did you deal with it?
   - 1.3 Did your team face difficult challenges in the Analyze phase? How did your team deal with conflict on your team?
   - 1.4 Did your instructor and/or Black Belt or Master Black Belt mentor help your team better learn how to apply the Lean Six Sigma tools in the Analyze phase, and how?
   - 1.5 Did your Analyze phase report provide a clear understanding of the root causes of the financial processes, why or why not?

2. **Cause and Effect Diagram**
   - 2.1 How did you generate the causes related to the effect for the cause and effect diagram?

3. **Cause and Effect Matrix**
   3.1 Did many of the causes apply to many of the effects?

4. **Why-Why Diagram**
   4.1 Was it easier to create the cause and effect diagram, the cause and effect matrix, or the Why-Why diagram? Which of the tools was more valuable getting to the root causes?

5. **Process Analysis**
   5.1 Discuss how your team defined whether the activities were value-added or nonvalue-added. Was the percent of value added activities or value added time what you would expect for this type of process and why?

6. **Histogram, Graphical, and Data Analysis**
   6.1 What type of distribution do your data appear to be from a graphical analysis?
   6.2 Can you test your distribution statistically and determine a likely distribution, what is it?
   6.3 Did you have outliers in your data?

7. **Waste Analysis**
   7.1 What types of waste were prevalent in the financial processes and why?

8. **Correlation Analysis**
   8.1 Were there any significant variables that were correlated? Do they appear to have a cause and effect relationship, and why?

9. **Regression Analysis**
   9.1 Were you able to identify a model that can predict any dependent variables? Why or why not?

10. **Confidence Intervals**
    10.1 What are your conclusions from the confidence intervals that you calculated?

11. **Hypothesis Testing**
    11.1 What were your key findings for your hypothesis tests?

11.2 What conclusions can you make from a practical perspective?
11.3 How might you use these findings in the Improve phase?

12. Survey Analysis

12.1 What were the significant findings in the vendor VOC survey?
12.2 What were the significant findings in the internal customer VOC survey?
12.3 Did your survey assess customer satisfaction with the accounts payable and payroll processes?

13. DPPM/DPMO

13.1 What is your DPPM/DPMO and sigma level. Is there room for improvement, and how did you determine that there is room for improvement?

14. Process Capability

14.1 What conclusions can you draw from the process capability study? Is your process capable? Is your process stable and in control? Can you have a process that is in control, but not capable, and how?

15. Analyze Phase Presentation

15.1 How did your team decide how many slides/pages to include in your presentation?
15.2 How did your team decide upon the level of detail to include in your presentation?

IMPROVE PHASE EXERCISES

1. Improve Report

Create an Improve phase report, including your findings, results and conclusions of the Improve phase.

2. Recommendations for Improvement

Brainstorm the recommendations for improvement.

3. QFD

Develop a QFD from the VOC CTS characteristics and map them to the improvement recommendations.

4. Action Plan

Create an action plan for demonstrating how you would implement the improvement recommendations.

5. **Cost/Benefit Analysis**
   Perform a cost/benefit analysis with rough estimates for automating the payroll time sheet process with the following potential solutions:
   1) Access program; 2) financial system's remote payroll module; 3) scanning and optical character recognition (OCR) program; 4) Excel timesheets.

6. **Future State Process Map**
   Create a future state process map for the following processes:
   - Accounts payable process
   - Accounts receivable process
   - Payroll process
   - Monthly reconciliation process

7. **Dashboards/Scorecards**
   Create a dashboard/scorecard for the project.

8. **Revised VOP Matrix**
   Revise your VOP matrix from the Measure phase with updated targets.

9. **Training Plans, Procedures**
   Create a training plan, and a detailed procedure for one of the financial processes.

10. **Improve Phase Presentation**
    Prepare a presentation (PowerPoint) from the case study exercises that provides a short (10–15 minutes) oral presentation of the Improve phase deliverables and findings.

IMPROVE PHASE

1. Improve Phase Report

The goal of the improve phase is to implement the improvements, measure the impact of the improvements and document procedures and train employees on the improved procedures.

2. Recommendations for Improvement

The team identified improvement opportunities that were grouped in the following Lean categories: Standardized processes and procedures, good housekeeping, kanban and visual control, waste identification and elimination, and one-piece flow.

Standardized Processes and Procedures

The team suggested that the finance department develop standardized desktop procedures. No written procedures existed in the current state. The finance clerk would keep handwritten notes, but this did not lend to standardization and repeatability.

Another improvement area was to use an Excel spreadsheet to standardize batch calculations for matching, and dividing repeating invoice amounts across different account numbers.

The fire department had converted from an association to a city department during the improvement effort. The team encouraged the finance department to integrate the fire department into the standardized payroll and accounts payable procedures.

The team recommended that the employees who used the financial system get training from the software vendor tailored specifically to their streamlined financial processes. Initially, when the city implemented the new financial system, the software vendor would train a generic process that encouraged printing of lengthy reports that the city did not need to print. The software vendor was able to provide additional understanding on the more extensive software functionality and tailor the processes better to the city's needs.

The team recommended that the city standardize the time sheets across all of the departments to help reduce payroll data entry errors and the time to enter the timesheets. The team also recommended that the finance clerk use timesheets in Excel spreadsheets to calculate the total timesheet hours by department, to compare to the payroll reports, instead of using a calculator.

Kanban and Visual Control

The team created a kanban and used visual control for the accounts payable processing. A kanban is a Lean tool that is used as a signal to pull work. The kanban we designed was a file hanging system that was easily visible to the finance clerk and the finance director. In the current process, the invoices, purchase orders, and requisitions that needed to be assigned account numbers or approved by the finance director were frequently lost in the piles of work. The kanban was organized in the order of the process steps. The documents that needed to be assigned account numbers were placed in a red folder in the first slot of the filing system. The purchase orders that needed approvals were placed in the next slot, so that the finance director would easily see them and quickly process them. The appropriate documents for each step were placed in the bin, so that the finance clerk and the finance director would have visual cues for the work that needed to be done. This greatly reduced the purchasing and accounts payable processing times. Figure 5.27 graphically depicts the purchasing and accounts payable kanban system.

Waste Identification and Elimination

The team identified unnecessary steps in the processes, such as printing lengthy reports that were never used. The team encouraged eliminating the printing of unnecessary reports, or printing them to an electronic file, which took seconds, instead of hours.

The team encouraged the use of new accounts receivables technology that automatically transferred journal entries, instead of requiring redundant data entry.

The team identified direct deposit as an improvement opportunity to eliminate printing of payroll checks. They suggested having a payroll direct deposit contest between departments to encourage use of direct deposit. This was after identifying and eliminating problems with the direct deposit process.

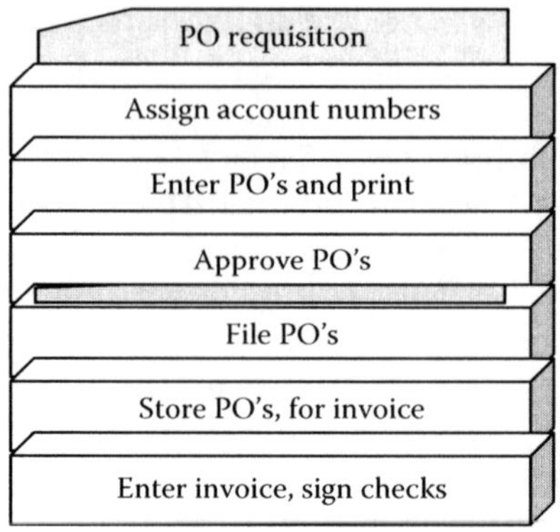

FIGURE 5.27 Purchasing and accounts payable kanban system.

The team recommended extensive information technology improvements that further streamlined the processes, and eliminated redundant data entry.

One-Piece Flow

One-piece flow is a Lean concept that tries to reduce the batch processing size to one or very small to flow work through the process more quickly. Another improvement idea that the team identified was to reduce the batch sizes of the accounts payable and accounts receivable batches. This would help to move closer to one-piece flow, and enable vendors to get their payments quicker by processing smaller batches more frequently. This was also dependent upon other improvements for both of the processes, so the batches could be processed more quickly. The team recommended the accounts receivable (revenue) batches be processed daily, instead of holding them for 1–2 weeks. This would increase the potential revenue from interest received by depositing the checks more quickly at the bank.

The team used the vendor Pareto analysis to identify duplicate vendors and recommended the number of vendors be reduced. The duplicate vendors were mainly due to each department choosing their own vendors for similar purchases across the city. This would also help the accounts payable processing to move closer to one-piece flow, or smaller batch sizes, by reducing the number of vendors and invoices.

3. QFD

A QFD house of quality was developed to map the CTS criteria to the improvement recommendations, to show alignment between the customer requirements and the improvements. The QFD is shown in Figure 5.28.

4. Action Plan

The team implemented the initial financial process improvements to the payroll and pension reporting, purchasing and accounts payable processes across a four-month period. They implemented improvements to accounts receivable, and monthly

Project name: Financial process

Customer requirements / *Technical requirements*	Importance	Standardized processes and procedures	Good housekeeping	Kanban and visual control	Waste identification and elimination	One piece flow		
Cycle time	8	9	3	3	9	9		
Accuracy of the process	10	9	3	1	1			
Customer satisfaction	6	9	9		3	3		
Absolute weight		216	108	34	100	90		
Relative weight		1	2	5	3	4		

FIGURE 5.28 QFD house of quality.

reconciliation throughout the next year, as time and resources permitted. They did not implement budgeting process improvements, because the finance director wanted to focus only on the processes performed by the finance clerk. The team first collected further information to validate the feasibility of the process improvement ideas presented in the Analyze phase. They created an implementation plan for any improvements that would take more than one week to implement or that required significant expenditures, and defined the associated costs and benefits at a finer detail than in the Analyze phase. The team gained approval from the finance director to proceed with the implementation of the improvement opportunities. The team implemented the improvements and redesigned the appropriate processes to incorporate the improvements. As part of the project management of the implementation the team quality facilitator provided weekly status reports to the team that included the tasks that were completed and the status and estimated completion date. The team quality facilitator documented any outstanding unresolved issues on items for resolution form (IFR). The IFR form included a description of the issue, the owner who was responsible for ensuring that the issue was resolved, the estimated resolution date, the priority of the issue, the status, the date the issue was opened and resolved, the impact of the issue to the project, and a description of the resolution.

5. Cost/Benefit Analysis

The team quality facilitator and process analyst identified potential costs and proposed benefits of each proposed improvement to determine if the estimated benefits are greater than the costs to implement. They also provided advantages and disadvantages to each solution, so that the finance director could make an informed and data-oriented decision. Most of the costs were related to training, and resources needed to implement and document the standardized procedures. The largest costs were related to consulting fees, and obtaining laser printers for check printing. A cost/benefit analysis for automating the processing of payroll timesheet hours is presented in Appendix B. Four alternatives were identified that could automate the timesheet payroll hours entry and verification activities. The first solution (Alternative 1) was to create a Microsoft Access program that would allow entry of timesheet data and perform automated verification and summing of hours by department. Alternative 2 was to implement an existing module from the financial information system vendor to automate the timesheet data, allow remote entry by each department, and allow automated integration of the timesheet data into the payroll system. Alternative 3 was to implement custom design and development of scanning and optical character recognition software to enable scanning or input from Excel timesheet data. Alternative 3 would require the highest cost, the longest implementation time, and the highest level of technology skills needed by the department employees.

Alternative 4 was to develop Excel timesheets that would enable automated entry of timesheet data within each department, and allow automated verification and summing of the timesheet data. The entered payroll hours data could then be compared with payroll hour reports to ensure payroll data accuracy. Alternative 4 required the lowest cost, the shortest implementation time, and a lower level of technology skills needed by department employees. An economic analysis was performed to determine which alternative was the most economically attractive alternative.

The net present worth of the costs and benefits over a five-year project life for the projects were:

- Alternative 1, Net Present Worth: –$15,349
- Alternative 2, Net Present Worth: –$12,542
- Alternative 3, Net Present Worth: –$74,961
- Alternative 4, Net Present Worth: $7,289

Only Alternative 4 had a positive net present worth, or a benefit/cost ratio greater than one. The internal rate of return for alternatives 1, 2, and 3 were all negative. The internal rate of return for Alternative 4 was 48%. The payback period for Alternative 4 was 2.02 years.

The city implemented Alternative 4, which reduced the time needed by the finance clerk to enter and verify timesheet data. It also pushed accountability of timesheet data to the originating department, who had the most knowledge about whether the data was accurate. Alternative 4 also eliminated the cumbersome, time-intensive, off-line calculator-based payroll hours verification step. This alternative also standardized the timesheet format and process across all of the city departments. The timesheet errors and payroll processing time was reduced by automating and standardizing the payroll timesheet entry and verification process.

6. Future State Process Map

The team revised the process maps to include the improvement recommendations. Many of the nonvalue-added activities were removed by focusing on removing the wasteful activities.

7. Dashboards/Scorecards

Because there were no process measures in place prior to the Lean Six Sigma project, the team developed detailed process measures and a metrics guide document. The process measures are shown in Figure 5.29. The metrics guide document includes a detailed description of each metric, how to measure it, including the data collection mechanism. The metrics can be arranged in a dashboard (Figure 5.30).

8. Revised VOP Matrix

The revised VOP matrix is shown in Figure 5.31, including incorporating a control chart to track defects in the accounts payable and payroll processes, with a more realistic target for tracking control of the process. The percentage for ratings in the positive categories (4-Agree, and 5-Strongly Agree) was also revised to be more realistically aligned to the results of the surveys.

9. Training Plans, Procedures

The Lean Six Sigma team created detailed desktop procedures for each of the financial processes, and trained the finance clerk in the procedures. The procedures were extremely detailed and even included screen shots populated with sample data, and step-by-step instructions. The procedures were very successful in helping to train the finance clerk, remove resistance to change, and eliminate problems reported to the help desk.

The procedures were developed based on our detailed knowledge of the financial information system acquired during the project. The desktop procedures were so thorough, that on several occasions when the finance clerk was not available, the finance director, and the income tax clerk were able to perform the payroll process with limited advanced training.

Proposed process measure	Data collection mechanism
Payroll and pension reporting	
Number and type of payroll problems encountered per number of employees	• Payroll check sheet • Payroll metric log • Moving range and individual control chart of problems per employee
Payroll processing time by payroll period	• Payroll check sheet • Moving range and individual control chart of payroll processing time
Purchasing and accounts payable	
Number of problems per invoice	• Accounts payable check sheet • Moving range and individual control chart of AP problems per invoice
Time per invoice	• Accounts payable check sheet • Moving range and individual control chart of time per invoice
Percent invoices without purchase orders	• Accounts payable check sheet
Percent invoices paid within discount period	• Accounts payable check sheet
Accounts receivable	
Time per receipt	• Accounts receivable check sheet • Accounts receivable metrics log • Moving range and individual control chart of time per receipt
Number of problems per receipt	• Accounts receivable check sheet • Accounts receivable metrics log • Moving range and individual control chart of problems per receipt
Monthly reconciliation	
Number of problems by type	• Monthly reconciliation check sheet • Monthly reconciliation problem Pareto chart

FIGURE 5.29 Proposed process measures for scorecard.

The finance clerk was trained on all of the improved processes using the detailed desktop procedures. She also received process-specific training on the financial information system from the software vendor.

10. Improve Phase Presentation

The Improve phase presentation can be found in the downloadable instructor materials.

IMPROVE PHASE CASE DISCUSSION

1. **Improve Report**
   1.1 Review the Improve report and brainstorm some areas for improving it.
   1.2 How did your team ensure the quality of the written report? How did you assign the work to your team members? Did you face challenges of

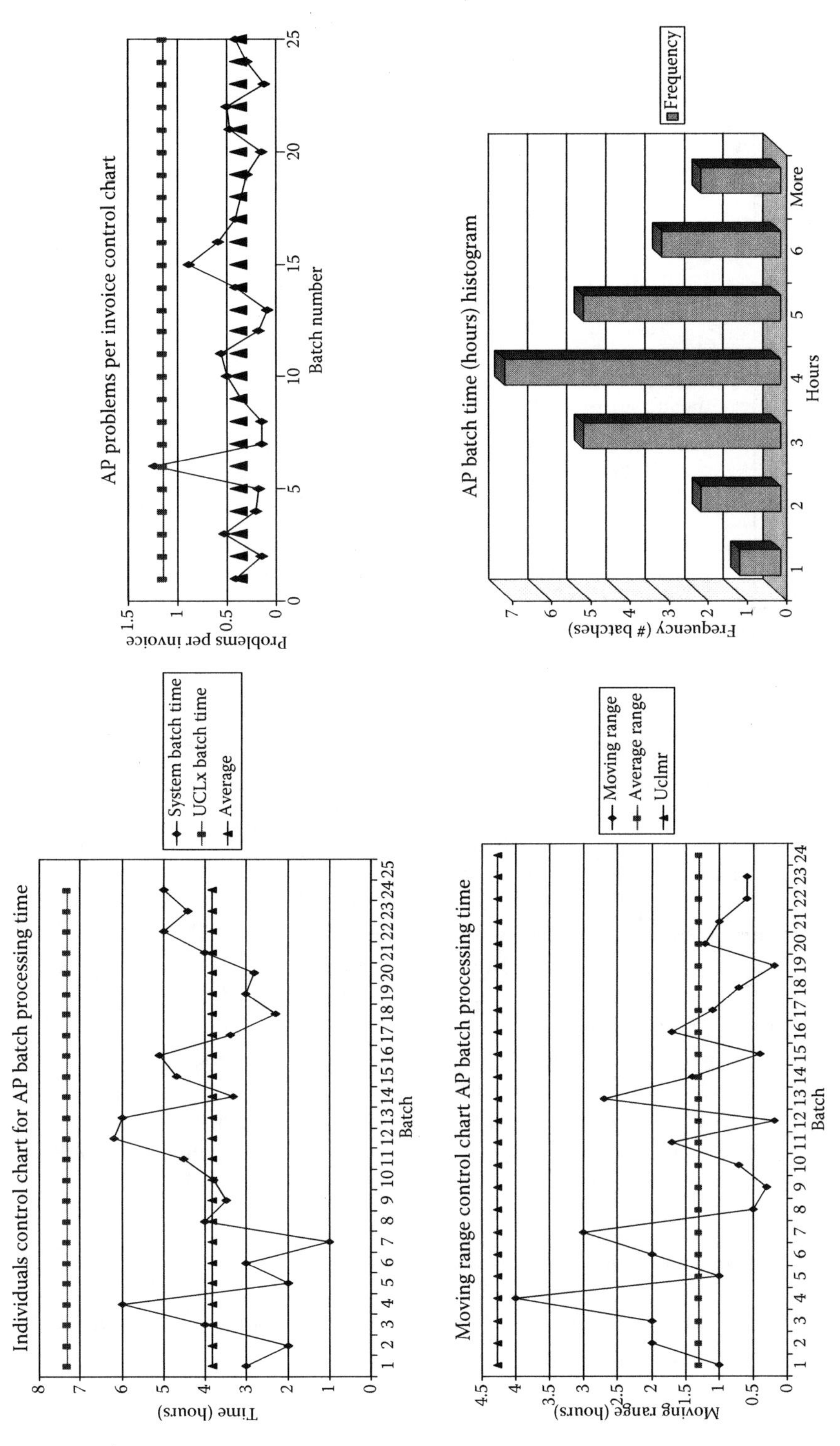

FIGURE 5.30 Dashboard example.

CTS	Process factors	Operational definition	Metric	Target
Cycle time	▪ Standard procedures ▪ Exist ▪ Streamlined processes ▪ Training ▪ Volume of invoices	Measure each process time	AP: cycle time – vendor invoice received to paid AR: time to deposit funds in bank Recon: time takes to close Payroll: Paid on time per schedule	AP: 10 business days AR: 2 days Recon: 10 days Payroll: Paid on time
Accuracy of the process	▪ Training in procedures and software	Measure each process and defect types	Defects per invoice (or paycheck) by process and type	No out of control points where assignable cause cannot be found
Customer satisfaction	▪ Repeatable process ▪ Collect and assess VOC	Measure customer satisfaction through customer and vendor surveys	% of positive responses for identified survey questions	60% of responses are rated 4 or 5 for identified questions

FIGURE 5.31 Revised VOP matrix.

team members not completing their assigned tasks in a timely manner and, if so, how did you deal with it?

1.3 Did your team face difficult challenges in the Improve phase? How did your team deal with conflict on your team?

1.4 Did your instructor and/or Black Belt or Master Black Belt mentor help your team better learn how to apply the Lean Six Sigma tools in the Improve phase, and how?

1.5 Did your Improve phase report provide a clear understanding of the root causes of the processes, why or why not?

1.6 Compare your improve report to the improve report in the book, what are the major differences between your report and the author's report?

1.7 How would you improve your report?

2. Recommendations for Improvement

2.1 How did your team generate ideas for improvement?

2.2 What tools and previous data did you use to extract information for the improvement recommendations?

2.3 How do your recommendations differ from the ones in the book?

3. **QFD**
   3.1 Does the QFD support the alignment with the CTS characteristics?
   3.2 How will you assess customer satisfaction?

4. **Action Plan**
   4.1 How did your Six Sigma team identify the timings for when to implement your recommendations?

5. **Cost/Benefit Analysis**
   5.1 How did you collect data on the potential costs, and benefits for the potential solutions?
   5.2 How did you validate the reasonableness of your cost/benefit analysis?
   5.3 Which solution would you recommend and why?

6. **Future State Process Map**
   6.1 Did your future state process map to the one eliminate all of the nonvalue-added activities? Why or why not?

7. **Dashboards/Scorecards**
   7.1 How does your dashboard compare to the one in the book?

8. **Revised VOP Matrix**
   8.1 Does the VOP matrix provide alignment between the CTSs, the recommendations, metrics and target?

9. **Training Plans, Procedures**
   9.1 How did you determine which procedures should be developed?
   9.2 How did you decide what type of training should be done?

10. **Improve Phase Presentation**
    10.1 How did your team decide how many slides/pages to include in your presentation?
    10.2 How did your team decide upon the level of detail to include in your presentation?

CONTROL PHASE EXERCISES

1. **Control Report**
   Create a Control phase report, including your findings, results and conclusions of the Control phase.

2. **Hypothesis Tests**
   Compare the before and after processing times for the following processes (hypothetical data):

   - Accounts payable
   - Accounts receivable
   - Payroll
   - Monthly reconciliation

3. **Mistake Proofing**
   Create a mistake proofing plan to prevent errors from occurring in the Monthly Reconciliation process.

4. **Control Plan**
   Develop a control plan for each improvement recommendation from the Improve phase report.

5. **Process Capability, DPMO**
   Calculate the process capability for the revised time to perform the financial processes.

6. **Control Charts**
   Create an idea for applying control charts to control the financial processes.

7. **Replication Opportunities**
   Identify some potential replication opportunities within the city to apply some of the improvement recommendations.

8. **Standard Work, Kaizen**
   Create a plan for standardizing the work.

9. **Control Phase Presentation**
   Prepare a presentation (PowerPoint) from the case study exercises that provides a short (10–15 minutes) oral presentation of the Control phase deliverables and findings.

CONTROL PHASE

1. Control Report

The goal of the control phase is to implement performance measures and other methods to control and continuously improve the processes.

2. Hypothesis Tests

The team measured the impact of the improvements after the majority of the improvement opportunities were implemented for each financial process. The payroll processing time was reduced by approximately 60%. Although the errors were not measured prior to the improvement implementation, no paycheck errors were found while migrating the fire department into the finance department procedures and financial systems, using the revised and improved payroll processes.

The purchasing and accounts payable processing time was reduced by approximately 40%, and all the vendors started getting paid on a consistent and timely basis. The accounts payable improvements also completely eliminated some of the nonvalue-added processing steps such as no longer having to verify that duplicate invoices had been paid due to paying invoices on time.

The accounts receivable processing time was reduced by approximately 90%. Revenue checks were getting deposited into the bank daily. The monthly reconciliation processing time was reduced by approximately 87%. Additionally, the monthly reconciliation process was performed on a consistent monthly basis due to providing more capacity for the finance clerk. the increased capacity was a result of the elimination of nonvalue-added tasks, and reducing the payroll, accounts payable and accounts receivable processing times.

Financial processes could be performed by one person working 40 hours per week, instead of 1.5 employees prior to the Lean Six Sigma implementation.

Another significant improvement related to the improved processes and subsequent training was the number of financial system problems reported to the software vendor greatly decreased from an average of 13 problems reported per month by the finance clerk to an average of six per month.

Figure 5.32 summarizes the estimated prior processing times, the estimated processing times after the improvements, and the percentage reduction of processing times. More specific performance measures to measure actual cycle times per batch, and quality of the processes were recommended to the city, but were not implemented prior to the end of the initial project.

The consultants encouraged the finance department to implement a continuous improvement process to continue to improve the productivity and the quality of the financial processes. This would be especially important if turnover occurred, so that the culture would change to one that continually and always improved. The good news related to changing the culture was that an upstream process in the billing department saw the value of the improvements by the reduction in the number of the billing reconciliation problems when they had to send their journal entries to finance. Sometime after the project, the finance clerk left the position, and the

Process	Average estimated processing time prior to improvements	Average estimated processing time after improvements	Percentage reduction of processing times
Payroll and pension reporting	60 hours	24 hours	60%
Purchasing/accounts payable	40 hours	24 hours	40%
Accounts receivable	60 hours	6 hours	90%
Monthly reconciliation	60 hours	8 hours	87%

FIGURE 5.32 Improved financial processing times.

utility billing clerk was able to step into the finance position and improve the financial close process to one day.

No formal hypothesis tests were used to measure the improvements because at the time of the project close there was not enough data collected to apply the statistical tests, even though there was a large amount of anecdotal evidence to suggest that the improvements were significant.

3. Mistake Proofing

As much as possible, the information system functionality was used for mistake proofing, by automating the steps that were possible to automate. This eliminated many of the manual or calculator-based activities for balancing that led to many mistakes.

4. Control Plan

One of the last (but very important) steps of the control phase is to take the time to celebrate the improvement effort, even if it was something as simple as going out to lunch to celebrate, which the team did. The finance department had not yet changed their reward and recognition system to accommodate continuous improvement and performance-based metrics.

The entire Lean Six Sigma implementation in the finance department took about 1.5 calendar years. The Define phase took three months, the Measure and Analyze phases took two months each. The Improve and Control phases took about one year together.

Through implementing a Lean Six Sigma program, the city's finance department was able to significantly reduce the time to process payroll, purchasing and accounts payable, accounts receivable and monthly reconciliation. Payroll processing time was reduced by 60%. Purchasing and accounts payable processing time were reduced by 40%. Accounts receivable processing time was reduced by 90%. Monthly reconciliation processing time was reduced by 87%.

The detailed metrics guide, summarized by the performance measures in Figure 5.29, was used as the control plan.

5. Process Capability, DPMO

A formal process capability analysis was not performed due to the sample size being low at the time of the project close.

6. Control Charts

Several types of control charts are suggested in the process metrics guide, including the following:

Payroll: Moving range and individual control charts of number of problems per employee; Moving range and individual control chart of payroll processing time.

Accounts Payable: Moving range and individual control charts of number of Accounts Payable problems per invoice; Moving range and individual control charts of time per invoice.

Accounts Receivable: Moving range and individual control charts of time per receipt; moving range and individual control charts of number of problems per receipt.

7. Replication Opportunities

The finance department migrated the fire department into the city's standardized and improved financial processes and systems when they became a city department. The migration was seamless. No paycheck errors occurred during the first pay period when the fire department's payroll was processed by the finance department using the improved procedures.

8. Standard Work, Kaizen

The monthly reconciliation process was performed on a consistent monthly basis, due to providing more capacity for the finance clerk. The increased capacity was a result of the elimination of nonvalue-added tasks, and reducing the payroll, accounts payable and accounts receivable processing times.

The financial processes were able to be performed by one person working 40 hours per week, instead of 1.5 employees prior to the Lean Six Sigma implementation.

Another significant improvement related to the improved processes and subsequent training was that the number of financial system problems reported to the software vendor greatly decreased from an average of about 13 problems reported per month by the finance clerk to about six per month.

Combining the principles and tools of Lean Enterprise and Six Sigma provides an excellent way to improve the productivity and quality of providing financial services in a local government. Although the majority of Lean Six Sigma applications have been in private industry, focusing mostly on manufacturing applications, this case study is an excellent example of how Lean Six Sigma tools can be applied in a service-oriented, transaction-based entity, such as a local government.

9. Control Phase Presentation

The Control phase presentation can be found in the downloadable instructor materials.

CONTROL PHASE CASE DISCUSSION

1. **Control Report**
   - 1.1 Review the Control report and brainstorm some areas for improving the report.
   - 1.2 How did your team ensure the quality of the written report? How did you assign the work to your team members? Did you face challenges of team members not completing their assigned tasks in a timely manner and, if so, how did you deal with it?

1.3 Did your team face difficult challenges in the Control phase? How did your team deal with conflict on your team?
1.4 Did your instructor and/or Black Belt or Master Black Belt mentor help your team better learn how to apply the Lean Six Sigma tools in the Control phase, and how?
1.5 Did your Control phase report provide a clear understanding of the root causes of the process, why or why not?
1.6 Compare your Control report with the Control report in the book, what are the major differences between your report and the author's report?
1.7 How would you improve your report?

2. Hypothesis Tests

2.1 What hypothesis tests could you perform, assuming that the data was available?

3. Mistake Proofing

3.1 How well did your team assess the mistake proofing ideas to prevent errors?

4. Control Plan

4.1 How well will your control plan ensure that the improved process will continue to be used by the process owner?
4.2 Are their additional control charts that could be used to ensure process control?

5. Process Capability, DPMO

5.1 Did you validate that your process was in control before calculating the process capability? Why is this important?

6. Control Charts

6.1 For this project did you find attribute or variable control charts to be more applicable for controlling this process?

7. Replication Opportunities

7.1 How did your team identify additional replication opportunities for the processes within the city?

8. Standard Work, Kaizen

8.1 How might you use a kaizen event to have identified process improvement areas, or ways to standardize the process?
8.2 How would you recommend ensuring that the process owners follow the standardized work procedures?

9. Control Phase Presentation

9.1 How did your team decide how many slides/pages to include in your presentation?

9.2 How did your team decide upon the level of detail to include in your presentation?

ACKNOWLEDGMENT

Portions of this case study were published in Furterer and Elshennawy, 2005.

REFERENCE

Furterer, S.L., and Elshennawy, A. K. 2005. Implementation of TQM and Lean Six Sigma tools in local government: A framework and a case study. *Total Quality Management and Business Excellence Journal*, 16 (10) 1179–1191.

6 Industrial Distribution and Logistics (IDIS) Program Recruiting Process Design—A Lean Six Sigma Case Study

Blake Hussion, Stefan McMurray, Parker Rowe, Matt Smith, and Sandra L. Furterer

CONTENTS

OVERVIEW OF THE PROBLEM

The Industrial Distribution and Logistics (IDIS) Department is part of the College of Technology and Computer Science at East Carolina University (ECU). Distribution and logistics represent professions in the workplace concerned with the movement and delivery of goods and services throughout the world. At ECU, this program provides a unique combination of coursework that prepares students for successful careers in a range of challenging areas. Courses cover areas of the distribution and logistics industry, including sales and branch operations, supply chain management, marketing, purchasing, and procurement, warehousing and materials handling, inventory management, production planning, and quality control. The goal of the Distribution and Logistics program is to provide applied distribution and logistics education as a basis for career advancement and life-long learning. A vast array of technology and simulations, as well as hands-on training, is used in the industry to prepare students for the skills required in the professional setting. The IDIS program at ECU offers a course load that provides students with an in-depth investigation into the industry and applies the material into real-world issues. By allowing students to experience and apply what they hear during lectures to a real-world setting, they are able to better understand the business and are therefore better prepared to enter the workforce.

Concerns have been voiced about the recent decline in the number of students entering the IDIS program and our group wishes to evaluate the problem and implement a solution that will help raise the number of applications to the school. There is currently no defined marketing procedure for the IDIS program and most current students are in the program due to peers or certain faculty who ignited their interest in the IDIS program. The Industrial Distribution program does not have a process to attract new students to the program.

DEFINE PHASE EXERCISES

It is recommended that the students work in project teams of 4–6 students throughout the Lean Six Sigma Case Study.

1. **Define Phase Written Report**
   Prepare a written report from the case study exercises that describes the Define phase activities and key findings.

2. **Lean Six Sigma Project Charter**
   Use the information provided in the Overview of the Problem section above, in addition to the project charter format, to develop a project charter for the Lean Six Sigma project.

3. **Stakeholder Analysis**
   Use the information provided in the Overview of the Problem section above, in addition to the stakeholder analysis format, to develop a stakeholder analysis, including stakeholder analysis roles and impact definition, and stakeholder resistance to change.

4. **Team Ground Rules and Roles**
   Develop the project team's ground rules and team members' roles.

5. **Project Plan and Responsibilities Matrix**
   Develop your team's project plan for the DMAIC project. Develop a responsibilities matrix to identify the team members who will be responsible for completing each of the project activities.

6. **SIPOC**
   Use the information provided in Overview of the Problem section above to develop a SIPOC of the high-level process.

7. **Team Member Biographies (Bios)**
   Each team member should create a short bio of themselves so that the key customers, stakeholders, project champion, sponsor, Black Belt and/or Master Black Belt, can get to know them, and understand the skills and achievements that they bring to the project.

8. **Define Phase Presentation**
   Prepare a presentation (PowerPoint) from the case study exercises that provides a short (10–15 minutes) oral presentation of the Define phase deliverables and findings.

DEFINE PHASE

1. Define Phase Written Report

The IDIS Department is a part of the College of Technology and Computer Science at ECU. Distribution and logistics represents professions in the workplace concerned with the movement and delivery of goods and services throughout the world. At ECU, this program provides a unique combination of coursework that prepares students for successful careers in a range of challenging areas. Courses cover areas of the distribution and logistics industry, including sales and branch operations, supply chain management, marketing, purchasing and procurement, warehousing and materials handling, inventory management, production planning, and quality control. The goal of the Distribution and Logistics program is to provide applied distribution and logistics education as a basis for career advancement and life-long learning. They also use a vast array of technology and simulations as well as hands-on training in the industry to prepare students for the skills required in the professional setting. The IDIS program at ECU offers a course load that provides students with an in-depth investigation into the industry and applies the material on real-world issues. By allowing students to experience and apply what they hear during lectures into a real-world setting, they are able to better understand the business and are therefore better prepared to enter the workforce.

2. Lean Six Sigma Project Charter

The objective of this project is to design student recruiting processes that will enhance the recruiting efforts for the IDIS program internal and external to the university. Concerns have been voiced about the recent fall in the number of students entering the IDIS program. Our team has been tasked with evaluating the problem and implementing a solution that will help to increase the number of applicants into the program. Working closely with the department administration, other faculty and current IDIS students, we plan to identify what is appealing about the IDIS program and what is not. By addressing these concerns, hopefully a marketing process can be executed that will help in the growth of the IDIS program, enabling the program to continue to exist. Currently, the IDIS program's faculty and administration has no organized way to attract students to the program, or understand what attracts students to the program. There are several customers and stakeholders that are part of the project. Current IDIS students are primary stakeholders because they contain the wealth of information as to what attracted them to the IDIS program. This will provide information to incorporate into the marketing plan to highlight the value of the program to potential students. Potential students are also primary stakeholders because they are the population to draw into the program. We just need to understand how to reach them. The department and program administration, as well as current IDIS faculty, are stakeholders. The department wants to keep viable programs that attract students, and the IDIS faculty wants to be secure in their jobs. The administration of the College of Technology and Computer Science also wants a viable IDIS program. The initial CTS characteristics for this project are increasing the number of students to the program while maintaining the satisfaction of IDIS students. The goal of the project is to develop a marketing process to identify voice of customer (VOC) for what attracts students to the IDIS program, and develop communication mechanisms and processes for increasing the number of students in the program. The scope of the project is to focus on attracting students internal and external to the university and the IDIS program. The financial benefits will maintain the on-going viability of the IDIS program and include potential additional tuition from new students. Some of the potential risks to the project are not attaining stakeholder buy-in (especially related to new students) and to faculty and administration not implementing the marketing plan. Additionally, it will be critical to get responses to the VOC surveys so that the team can understand the perceived value of the IDIS program to potential students. The project sponsor is Mark Angolia, an IDIS faculty member. The Master Black Belt is the course instructor, Dr. Sandy Furterer. The project deliverables are a marketing plan, VOC surveys, and analysis of data collected. The project is expected to take four months and complete the DMAIC problem solving methodology. The project charter is shown in Figure 6.1.

Project Name: IDIS Program Recruiting Process Design.

Problem Statement: The enrollment of students is decreasing in the IDIS program. The IDIS program's faculty and administration has no organized way to attract students to the program, or understand what attracts students to the program.

Customer/Stakeholders: (Internal/External) current IDIS students, future IDIS students, IDIS faculty and administration, ECU undergraduate studies, College of Technology and Computer Science.

What is important to these customers–CTS: Increasing number of students to the program; IDIS student satisfaction.

Goal of the Project: To develop a marketing process to identify VOC for what attracts students to the IDIS program, and develop communication mechanisms and processes for increasing the number of students in the program.

Scope Statement: Development of a marketing program for attracting students internal and external to the university to the IDIS program within the College of Technology and Computer Sciences at East Carolina University.

Financial and Other Benefit(s): Continue the viability of the IDIS program at ECU, increase students and related tuition and fees associated with them.

Potential Risks: Stakeholder buy-in; not getting responses to VOC surveys.

FIGURE 6.1 Project charter.

3. Stakeholder Analysis

Distribution and logistics represent professions in the workplace that are concerned with the movement and delivery of goods and services throughout the world. At ECU, this program provides a unique combination of coursework that prepares students for successful careers in a range of challenging areas. Courses cover areas of the distribution and logistics industry, including sales and branch operations, supply chain management, marketing, purchasing and procurement, warehousing and materials handling, inventory management, production planning, and quality control.

There are several primary stakeholders for the Lean Six Sigma project. Prospective IDIS undergraduate students are primary stakeholders. Prospective students are incoming freshmen, students who are undecided in their field of study, and students who are unhappy in their current major. Prospective IDIS students are concerned with having an interesting major and future career potential. Current IDIS students are also primary stakeholders. Current students are already registered in the IDIS program. They are concerned with getting additional students into the major to continue the IDIS program. IDIS faculty and program administration are another primary stakeholder group. They are concerned with the on-going viability of the IDIS program, and continuing their positions with the program. They also are concerned with preparing the students for the distribution and logistics industries, and providing the tools needed to make an immediate impact in today's competitive market.

There are several secondary stakeholders for the project. The College of Technology and Computer Science is a stakeholder. They are concerned with the on-going program viability and with satisfied students. Another secondary stakeholder is the ECU Division of Undergraduate Studies Office that receives official documents and uploads student information to the system. They are concerned with reduction of errors and resistance to change to the current procedures.

The stakeholder analysis definition is shown in Figure 6.2.

The stakeholder commitment scale is shown in Figure 6.3. Current students and the IDIS faculty and administration are extremely supportive of the project. The prospective students are neutral because they do not yet know about the IDIS program. The College and Undergraduate Studies Office are also neutral at the beginning of the project, being neither supportive nor against the project.

4. Team Ground Rules and Roles

The team ground rules were brainstormed with the Lean Six Sigma project team and included ground rules related to the team's attitudes and the processes.

Attitudes:

- Be open to all information relating to the project.
- Speak-up and be clear about all ideas.
- Member participation and questions are essential to the project.
- Respect other team members' ideas and be supportive of these ideas.
- Allow for critique of other people's ideas.
- Be open-minded.

	Stakeholders	Who are they?	Potential impact or concerns	
PRIMARY	Prospective IDIS undergraduate students	Incoming freshmen, students who are undecided in their field of study, and students who are unhappy in their current major	Interesting major with high career potential	+
PRIMARY	Current IDIS undergraduate students	Those who are already registered in the IDIS program	Recruitment of other students for IDIS program	+
PRIMARY	IDIS faculty and program administration	Current faculty who teach courses in the areas of the distribution and logistics industry including sales and branch operations, supply chain management, marketing, purchasing and procurement, warehousing and materials handling, inventory management, production planning, and quality control	On-going program viability Keeping their jobs Prepares students for the distribution and logistics industries Provides the tools needed to make an immediate impact in today's competitive market	+ – + +
SECONDARY	College of Technology and Computer Science	College administration, department chairs	On-going program viability Satisfied students	+ +
SECONDARY	ECU division of Undergraduate Studies	Office that receives official documents, upload student information to the system	Reduction of errors Resistance to change current procedures	+ –

FIGURE 6.2 Stakeholder analysis definition.

Stakeholders	Strongly against	Moderate against	Neutral	Moderate support	Strongly support
Prospective students			X		O
Current students					XO
IDIS faculty and admin					XO
College			X		O
Undergraduate studies			X		

FIGURE 6.3 Stakeholder commitment scale.

- Speak-up if anyone has a difference of opinion.
- Share and receive member experiences and/or knowledge relating to the team project.

Processes:

- Arrive at meetings on time and prepared and have a set schedule.
- Have an agenda prepared and record outcomes of scheduled meetings and progress.

- Retrieve and organize all information from facilitator and group members.
- Try to schedule all meetings and, if one team member is absent, ensure that the others pick-up his/her workload.

The team members included four IDIS students: Blake Hussion, Parker Rowe, Stefan McMurray, and Matthew Smith. Blake is the team leader and meeting facilitator. Parker developed the work plan and was the process expert on the team. Stefan is the meeting analyst and scheduled meetings. Matthew had the role of the process expert. The project champion was Dr. Leslie Pagliari, the IDIS undergraduate program director, and the project sponsor was IDIS instructor Mark Angolia.

5. Project Plan and Responsibilities Matrix

The project plan with resources responsible for each activity is shown in Figure 6.4.

6. SIPOC

The team developed a SIPOC (Figure 6.5) that described the high-level processes, suppliers, customers, inputs, and outputs that are part of the scope of the Lean Six Sigma project for developing a marketing plan. The suppliers include the faculty and staff that supply us with valuable information and insight to help with the recruitment for the IDIS program. The parents of the students are also suppliers because they influence their children's opinions on which field of study they will choose. Other universities can help to improve our own program at ECU by providing best practice information. The inputs to the processes are ECU along with its faculty and students; all put in time and effort to help promote and enhance the IDIS program. We can take successful techniques and tactics used by premiere IDIS programs at other universities and apply some of those strategies to our own program. Future students can also strive to continue the recruitment and improve IDIS. The IDIS program is an input to the processes, as well as other resources to understand potential students and the value of the IDIS program. The processes to be performed as part of the Lean Six Sigma project are: developing a marketing plan; developing VOC surveys to gain helpful information on the overall knowledge of IDIS and the value of the IDIS program for current students; and recruiting new students to the IDIS Program. Our team met with advisors as well as our sponsor to gather as many data as possible on how to market IDIS. We took successful strategies they have used and applied them to our project.

Outputs of the process are a marketing program to help recruit students to the IDIS program. New students are an output of successful recruiting, as well as recruiting information from the recruiting process.

The customers of this project are future students, existing and future faculty, who can lend a hand in continually improving our IDIS program. Undergraduate studies and the college are also customers of the recruiting process.

No.	Activity	Status	Due date	Deliverables	Resources
		Define phase			
1	Define the problem		8/20	Project charter	Blake
2	Define scope of the project		8/20	Project charter	Blake
3	Prepare project charter		8/27	Project charter	Matt
4	Customer stakeholder analysis		8/27	Stakeholder analysis chart	Matt
5	Prepare work plan		8/27	Work plan	Parker
6	Responsibilities matrix		8/27	Matrix	Parker
7	Prepare participation log		9/5	Participation log	Stefan
8	Create bios		9/5	Bios	Stefan
9	Create a SIPOC		9/5	SIPOC	Stefan
		Measure phase			
10	Draw process flow charts		9/16	Process flow chart	Blake
11	Create pareto charts		9/23	Pareto chart	Stefan
12	VOC summary		9/23	VOC summary	Matt
13	CTS measures		9/23	CTS measures	Parker
14	Key metrics		10/5	Key metrics	Blake
15	Prepare participation log		10/5	Participation log	Stefan
16	QFD		10/5	QFD	Matt
		Analyze phase			
17	Summary of problems		10/12	Summary of problems	Blake
18	Cause and effect analysis		10/15	Cause and effect analysis	Stefan
19	Summary of data collected		10/20	Summary of data collected	Matt
20	COPQ		10/20	COPQ	Parker
21	FMEA		10/20	FMEA	Parker
22	CTS- VOP matrix		10/30	CTS - VOP matrix	Blake
23	Statistical analysis		10/30	Statistical analysis	Stefan
24	Prepare participation log		11/5	Participation log	Matt
		Improve/ Control phase			
25	Action plans		12/1	Action plans	Blake

FIGURE 6.4 Work plan.

26	Recommendations for improvement		12/1	Recommendations for improvement	Blake
27	Revised process flows		12/1	Revised process flows	Stefan
28	Control plan w/ proposed control mechanisms		12/1	Control plan w/ proposed control mechanisms	Stefan
29	Significant lessons learned		12/5	Significant lessons learned	Parker
30	Prepare participation log		12/5	Participation log	Matt
		Presentation			
31	Prepare participation log		12/5	Participation log	Blake

FIGURE 6.4 (Continued)

Suppliers	Inputs	Process	Outputs	Customer
- IDIS faculty - Potential students - East Carolina - Other universities - Parents - Prospective employers	- Knowledge, time, effort - Best practice - Recruitment - IDIS programs - Resources - Students	- Marketing - Surveys - Benchmarking - Recruiting	- Marketing plan - New students - Recruiting information	- Future students - Faculty - Program admin - Undergraduate studies - College - Distribution companies

FIGURE 6.5 SIPOC.

7. Team Member Bios

Stefan McMurray grew-up in Richmond Virginia, where he participated in several activities ranging from sports around the community to school-related activities all the way up through high school. He attends ECU, where he became involved in the construction management program. After realizing his true calling, he switched majors to the industrial distribution & logistics field where he studied different aspects of transportation, logistics, quality, pricing, and business ethics.

Matt Smith is a senior at ECU, and plans to graduate with a bachelor of science in industrial distribution and logistics. After graduation, Matt is planning on working in sales for a distribution company.

Blake Hussion grew up in Cary, NC, and graduated from Cary High School in 2002. He attends ECU and will graduate with a bachelor of science in industrial distribution and logistics and a minor in business administration.

Parker Rowe was born in Bloomfield Hills, MI, where he graduated from Troy High School in 2002. Parker is a senior at ECU and is graduating with a bachelor of science in industrial distribution and logistics. He also has a business minor along with a minor in communications.

8. Define Phase Presentation

The Define phase presentation can be found in the downloadable instructor materials.

DEFINE PHASE CASE DISCUSSION

1. **Define Phase Written Report**
   1.1 How did your team ensure the quality of the written report? How did you assign the work to your team members? Did you face challenges of team members not completing their assigned tasks in a timely manner and, if so, how did you deal with it?
   1.2 Did your team face difficult challenges in the Define phase? How did your team deal with conflict on your team?
   1.3 Did your instructor and/or Black Belt or Master Black Belt mentor help your team better learn how to apply the Lean Six Sigma tools, and how?
   1.4 Did your Define phase report provide a clear vision of the project, why or why not?

2. **Lean Six Sigma Project Charter**
   Review the project charter presented in the Define phase written report.
   2.1 A problem statement should include a view of what is going on in the business, and when it is occurring. The problem statement should provide data to quantify the problem. Does the problem statement in the Define phase case study example written report provide a clear picture of the business problem? Rewrite the problem statement to improve it.
   2.2 The goal statement should describe the project team's objective, and be quantifiable, if possible. Rewrite the Define phase case study's goal statement to improve it.
   2.3 Did your project charter's scope differ from the example provided? How did you assess what was a reasonable scope for your project?

3. **Stakeholder Analysis**
   Review the stakeholder analysis in the Define phase.
   3.1 Is it necessary to identify the large number of stakeholders as in the example case study?
   3.2 Is it helpful to group the stakeholders into primary and secondary stakeholders? Describe the difference between the primary and secondary stakeholder groups.

4. **Team Ground Rules and Roles**
   4.1 Discuss how your team developed your team's ground rules. How did you reach consensus on the team's ground rules?

5. **Project Plan and Responsibilities Matrix**
   5.1 Discuss how your team developed their project plan and how they assigned resources to the tasks. How did the team determine estimated durations for the work activities?

6. **SIPOC**
   6.1 How did your team develop the SIPOC? Was it difficult to start at a high level, or did the team start at a detailed level and move up to a high-level SIPOC?

7. **Team Member Bios**
   7.1 What was the value in developing the bios, and summarizing your unique skills related to the project? Who receives value from this exercise?

8. **Define Phase Presentation**
   8.1 How did your team decide how many slides/pages to include in your presentation?
   8.2 How did your team decide upon the level of detail to include in your presentation?

MEASURE PHASE EXERCISES

1. **Measure Report**
   Create a Measure phase report, including your findings, results and conclusions of the Measure phase.

2. **Process Maps**
   Create level-1 and level-2 process maps for each of the following processes. You may need to benchmark some other similar programs and processes related to marketing, creating surveys, and recruiting.
   - Develop marketing plan
   - Develop VOC surveys
   - Recruit students

3. **Operational Definitions**
   Develop an operational definition for each of the four identified CTS criteria:
   - Awareness of program through current students
   - Awareness of program from undergraduate students at ECU
   - Program benefits, marketing techniques
   - Enrollment

4. **Data Collection Plan**
   Use the data collection plan format to develop a data collection plan that will collect VOC and voice of process (VOP) data during the Measure phase.

5. **VOC Surveys**
   Create a VOC survey to better understand the current and prospective students' requirements related to the IDIS program marketing plan and recruiting needs.

6. **VOP Matrix**
   Create a VOP matrix using the VOP matrix template to identify how the CTS, process factors, operational definitions, metrics and targets relate to each other.

7. **Benchmarking**
   Perform benchmarking of other similar programs to IDIS to understand how they perform recruiting processes for their programs.

8. **COPQ**
   Brainstorm potential COPQ for the case study for the following categories:
   - Prevention
   - Appraisal
   - Internal failure
   - External failure

9. **Measure Phase Presentation**
   Prepare a presentation (PowerPoint) from the case study exercises that provides a short (10–15 minutes) oral presentation of the Measure phase deliverables and findings.

MEASURE PHASE

1. Measure Report

The second phase of our project DMAIC is the Measure phase. Within this phase we gathered important data on how the current IDIS recruiting processes work and whether or not it is successful in informing East Carolina students about the program. The Measure phase allowed us to collect and analyze important information, which gave a better understanding of how current IDIS students view the program and which aspects they would like to see change.

Using tools such as CTS characteristics and analysis of survey results through charts and graphs, we were able to better define the aspects of the recruiting process which needed further evaluation. The following report is a description of our Measure phase findings and analysis of our survey data.

2. Process Maps

Currently, the department of Industrial Distribution and Logistics has 160 students enrolled within the program. Due to the high volume of students graduating within the next two semesters, the project champion, Dr. Pagliari, has addressed concern regarding enrollment numbers. There is currently no defined marketing procedure for the IDIS program and most current students are in the program due to peers or a certain faculty who kindled their interest in the IDIS program. Dr. Pagliari has defined a goal of maintaining around 200 students from current levels of 160 students for the IDIS program in the future, and feels that this goal is attainable.

To better understand how the current IDIS students decided on the program, we created and distributed a survey consisting of eight questions to currently enrolled IDIS students. Asking questions such as "How did you hear about the program?" and "What made you declare as an IDIS student?" gave us a better understanding of the methods that have worked in the past. Also, the survey contained open-ended questions that allowed feedback to determine the positives and negatives about the program and how the program can improve.

Although there is currently no defined process for recruiting students into the program, certain techniques are used to try to promote the program. Using brochures, seminars and the ECU website, IDIS has a basis for advertising, but they lack a concrete marketing process. By analyzing the data from the surveys, we hope to identify and implement a comprehensive marketing strategy that will also create name recognition for the IDIS program among all undergraduate ECU students, faculty, North Carolina high-school students and prospective employers of IDIS program graduates. Ultimately providing high-quality recruits to prospective employers will create a demand that will increase starting salaries and increase the profile of positions offered to IDIS program graduates upon completion of their studies.

There are no process maps for the As Is process because there is no current process.

3. Operational Definitions

The input to identifying the CTS characteristics was collected through interviews with our project sponsor (Mark Angolia) and the project champion (Dr. Leslie Pagliari), as well as through student surveys. The following characteristics were identified as being the elements that would significantly affect the output of the process as perceived by the customer. CTS criteria were identified as the following:

- Awareness of the IDIS program through current students
- Awareness of the IDIS program from undergraduate students at ECU
- Program benefits and marketing techniques
- Enrollment in the IDIS program

The operational definition for the awareness of the IDIS program through current and other undergraduate students at ECU will be measured through VOC surveys. After obtaining CTS criteria, we were able to develop two surveys that would assist in compiling data that would measure our CTS. We focused on two metrics that we tried to measure to better our understanding of students' views about the IDIS program:

A. Current IDIS students and their views on the program
B. Non-IDIS students and their understanding of the program

Program benefits were identified through the surveys and through interviews with existing students. Enrollment will be tracked by the program's administration, and will be defined as the increase in students registered in the IDIS program after piloting or implementing the new marketing plan and recruiting techniques.

4. Data Collection Plan

The data collection plan is shown in Figure 6.6. It identifies the specific question numbers from the VOC surveys that map to each CTS (awareness, and program benefits) and how enrollment will be assessed. To measure awareness of the IDIS program through current students, the team plans to perform a VOC survey. The team will analyze the survey using Pareto and chi-square analyses. Awareness of the IDIS program through undergraduate students (non-IDIS) will also be assessed using a VOC survey, and analyzed with Pareto and chi-square analyses. Program

CTS	Metric	Data collection mechanism	Analysis mechanism	Sampling plan	Sampling instructions
Awareness of program through current students	Current students views and thoughts of IDIS program Determined using questions 2, 3, 7 and 8 on current IDIS student survey	Student input of current program and its processes	Survey responses Data analysis through charts and graphs	Survey	List of questions asked
Awareness of program from undergraduate students at ECU	Undergraduate students and their familiarity with the IDIS program Determined using questions 2, 3, and 4 in survey for undergraduate students	Meetings, lectures and presentations to students who are undecided or are thinking of changing majors	Survey	Survey	Questions asked
Program benefits, marketing techniques	Current marketing procedures and how program advertises itself Determined using survey questions 3 and 4	Determine best marketing strategy by benchmarking and survey Determine best method for advertising IDIS through student responses and past successes	Survey	Survey	Which marketing technique is most successful in promoting program
Enrollment	Number of students increases program funding	Numbers in program	Summary of % of increase among student population	Students	Marketing/ Increase awareness

FIGURE 6.6 Data collection plan.

benefits and marketing techniques will also be assessed through surveys. Enrollment will be tracked through the program's administration.

5. VOC Surveys

Each of the CTS characteristics is associated with one or more key metrics that can quantify the characteristics by measuring them through the data collected in the surveys. By defining these aspects, we are able to better understand current processes used in the recruiting process and whether or not they are successful in promoting the IDIS program.

Survey Approach

The Industrial Distribution and Logistics recruitment process team's attitude toward these surveys was that of gathering relevant information toward current students within the program and the perspectives of other undergraduate students about the IDIS program.

The first survey was focused on current IDIS students, and consisted of the following questions:

1. How long have you been affiliated with the Industrial Distribution and Logistics (IDIS) program at East Carolina University?
2. How did you become familiar with the IDIS program?
3. What made you declare (your major) as an IDIS student?
4. What area within the industrial distribution and logistics field do you want to be involved with after graduation?
5. Programs that students switched from into IDIS?
6. What do you like about the Industrial Distribution program?
7. What do you dislike about the Industrial Distribution program?
8. How can the IDIS program be improved?

A response was obtained from 105 students out of 160 surveys distributed. The survey gathered information regarding how long the students have been involved in IDIS and when they declared IDIS as their major. We also tried to incorporate questions to gain further knowledge as to how they became familiar with the program and what made them want to declare IDIS as their current major. We also tried to gather likes and dislikes about the program so we knew what aspects we wanted to focus on when trying to sell the IDIS program.

The second survey that was distributed consisted of a sample size of 50 undergraduate students not affiliated with the IDIS program. This was a small sample size given the large population at ECU, but we wanted some brief feedback initially to gather some input from possible prospective students. These questions focused mainly on the current class of a student at ECU as well as a current major (if any). This enabled us to obtain a little insight as to where most students tend to declare their major, and who our target audience would be. We also focused on questions about the familiarity of the IDIS program and what concerns students had when deciding on a major. It was interesting to discover that student concerns when declaring a major were those that could be fulfilled through IDIS, but that the majority of students had no understanding of the major.

The second survey consisted of the following questions:

1. What is your current class?
2. What is your current major, if any?
3. Are you familiar with the Industrial Distribution and Logistics program at East Carolina University?
4. What do you think the Industrial Distribution and Logistics students do?

6. VOP Matrix

The VOP matrix helped the team to better understand the metrics, and the potential factors that affect the awareness of students with the IDIS program. It also helps to clearly articulate the target for the metrics. The metrics are focused on the VOC surveys, other than tracking enrollment in the IDIS program. The VOP matrix is shown in Figure 6.7.

7. Benchmarking

Benchmarking of other university's programs was not done due to lack of time.

CTS	Factors	Operational definition	Metric	Target
Awareness of program through current students	Program exists	Program exists, but marketing plan can be altered to increase awareness	Current students views and thoughts of IDIS program Determined using questions 1, 2, 3, and 5 on current IDIS student survey	Increase awareness by 100% to current students
Awareness of program to undergraduate students at ECU	Program exists	Program exists, but marketing plan can be altered to increase awareness	Undergraduate students and their familiarity with the IDIS program Determined using questions 1 and 3 in survey for undergraduate students	Increase awareness by 100% to undergraduates
Program benefits & marketing techniques	Good benefits, but unknown and poor marketing	IDIS offers good and solid benefits, but lacks in a very poor marketing strategy	Current marketing procedures and how program advertises itself determined using survey questions 6 and 7	Increase benefits and marketing techniques in the program
Enrollment	Is low	Due to poor marketing is a strong result of low enrollment	Number of students increases program funding	To increase enrollment by 25%

FIGURE 6.7 VOP matrix.

8. COPQ

A potential COPQ for prevention costs is implementing a marketing and recruiting program. Appraisal costs can include assessing the VOC through the surveys, and tracking enrollment figures in the IDIS program. Internal failure costs are students not knowing about the IDIS major, when they might find it a good match; and declaring the IDIS major, but then dropping out of it. External failure costs could be that a student graduates in a different major, but then ends up moving into the industrial distribution and logistics field after college, but does not have the educational background that he/she could have had through the IDIS program.

Measure Phase Presentation

The Measure phase presentation can be found in the downloadable instructor materials.

MEASURE PHASE CASE DISCUSSION

1. **Measure Report**
   - 1.1 Review the Measure report and brainstorm some areas for improving the report.
   - 1.2 How did your team ensure the quality of the written report? How did you assign the work to your team members? Did you face challenges of team members not completing their assigned tasks in a timely manner and, if so, how did you deal with it?
   - 1.3 Did your team face difficult challenges in the Measure phase? How did your team deal with conflict on your team?
   - 1.4 Did your instructor and/or Black Belt or Master Black Belt mentor help your team better learn how to apply the Lean Six Sigma tools in the Measure phase, and how?
   - 1.5 Did your Measure phase report provide a clear understanding of the VOC and the VOP, why or why not?

2. **Process Maps**
   - 2.1 While developing the process maps, how did your team decide what these processes might look like? If you were not a Subject Matter Expert, how did you collect information to develop the process maps?
   - 2.2 Was it difficult to develop a level-2 from the level-1 process maps? What were the challenges?

3. **Operational Definitions**
   - 3.1 Review the operational definitions from the Measure phase report; define an operational definition that provides a better metric for assessing the awareness of students with the IDIS program.
   - 3.2 Discuss why it may be important for the students to be aware of the benefits of the IDIS program.

4. Data Collection Plan

4.1 Incorporate the enhanced operational definition developed in number 3 above into the data collection plan from the Measure phase report.

5. VOC Surveys

5.1 How did your team develop the questions for the VOC surveys? Did you review them with other students to assess whether the questions met your needs?

5.2 Create an affinity diagram for the main categories on either of the VOC surveys, grouping the questions into the higher-level "affinities." Was this an easier way to approach and organize the questions of the surveys?

6. VOP Matrix

6.1 How does the VOP matrix help to tie the CTS measures, the operational definitions and the metrics together?

7. Benchmarking

7.1 Was it difficult to find benchmarking information specific to marketing and recruiting processes?

8. COPQ

8.1 Would it be easy to quantify, and collect data on the costs of quality that you identified for the case study exercise?

9. Measure Phase Presentation

9.1 How did your team decide how many slides/pages to include in your presentation?

9.2 How did your team decide upon the level of detail to include in your presentation?

ANALYZE PHASE EXERCISES

1. Analyze Report

Create an Analyze phase report, including your findings, results, and conclusions of the Analyze phase.

2. Process Analysis (Process Map)

Because there was no existing process to develop a process map in the Measure phase, we will need to develop a proposed process map, and then perform a process analysis for the following processes:

- Developing a marketing plan
- Performing recruiting for potential IDIS students

3. **Cause and Effect Diagram**
   Create a cause and effect diagram for the following effects:
   - Lack of awareness of the IDIS program for undergraduate students at ECU.

   Create a cause and effect diagram from a positive viewpoint:
   - By identifying potential factors (causes) that could help to increase enrollment in the IDIS program (the effect).

4. **Why-Why Diagram**
   Create a Why-Why diagram for why enrollment in the IDIS program has been declining.

5. **Waste Analysis**
   Perform a waste analysis for the following processes:
   - Developing a marketing plan
   - Performing recruiting for potential IDIS students

6. **Failure Mode and Effect Analysis (FMEA)**
   Develop a failure mode and effects analysis for developing the marketing plan and the recruiting process.

7. **5S**
   Identify how you might apply the 5S Lean tool in this project.

8. **Survey Analysis**
   - Perform survey analysis for current IDIS student survey data (Current Student Survey Data.xls). Include Pareto charts for each question, and chi-square analysis.
   - Perform survey analysis for the ECU undergraduate student survey data (Undergraduate Student Survey Data.xls). Include Pareto charts for each question, and Chi-square analysis.

9. **DPPM/DPMO**
   Calculate the DPMO and related sigma level for the process, assuming a 1.5 sigma shift, for the following data:
   Opportunities for failure:
      - Student does not select IDIS as a major as a freshman
      - Student drops out of IDIS as a major.

   Defects:
      - Number of students who meet with advisor but do not enroll in IDIS per month: 15.
      - Number of times a student drops out of IDIS per month: 0.25.

   Units:
      - Number of students who meet with advisor to discuss IDIS as a major: 20.

10. **Analyze Phase Presentation**
    Prepare a presentation (PowerPoint) from the case study exercises that provides a short (10–15 minutes) oral presentation of the Analyze phase deliverables and findings.

ANALYZE PHASE

1. Analyze Report

The Analyze phase allows our team to examine the current state of the IDIS program and provide feedback on which processes need further evaluation for the Improve and Control phases related to the development of the marketing plan and the recruiting processes for new students to the IDIS program. This phase is a critical part of the DMAIC process in that it offers conclusions on the current flaws of the program and allows us to create a foundation for a future recruitment process. By being able to create some strategies that will help in establishing a better marketing plan for the IDIS program, we will ultimately institute a procedure for increasing enrollment and further developing the department. Through the student surveys that we distributed within the Measure phase, we established some positives and negatives of the IDIS program that will help us to create the best possible deliverables for the Improve and Control phases of the project. Analyzing the data collected and identifying the root causes of the negative feedback are the main goals of this phase. Being able to better understand the aspects of the program which are not appealing to the current students allows for a more concrete analysis of the program's current marketing plan and the steps needed for implementing a better procedure. Analysis of the data we collected within the Measure phase is shown through the following tools: process mapping, process analysis, failure modes and effects analysis, and the 5S diagram. All these tools were used to ascertain the best possible scenarios for the deliverables we wished to implement within the next phases of the project. All of these tools help provide recommendations on how the current process could be made more efficient and which strategies need more focus.

2. Process Analysis (Process Map)

A process map is a basis of the information and activities that the Industrial Distribution program is proposing to implement for the upcoming semesters. Because the Industrial Distribution program did not have a process before the start of this project, this diagram was established through meetings with our project sponsor (Mark Angolia) and our project champion (Dr. Leslie Pagliari). Having a foundation for what the program is ultimately trying to accomplish allows for a better understanding of what deliverables are needed for successful completion of the DMAIC process. Now that a process is established (Figure 6.8), we can try to create a marketing plan that will develop better name recognition for IDIS and detail the steps needed in that operation. Establishing a process flow for Industrial Distribution and Logistics was essential for developing a more focused recruiting process and to realize the factors that an increase in enrollment will require.

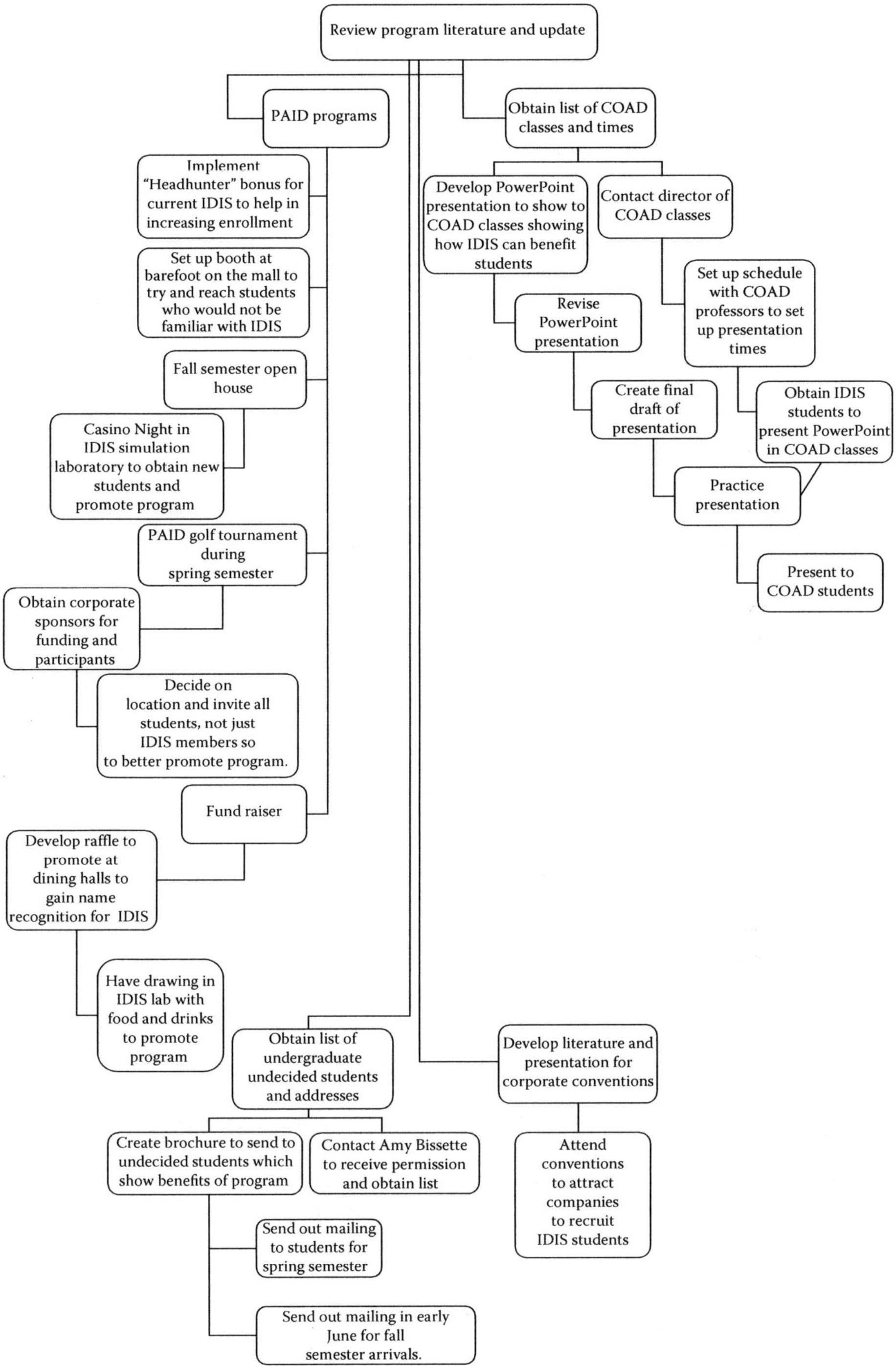

FIGURE 6.8 Proposed process map.

Process Analysis through Use of Flow Process Diagram

The flow process diagram (Figure 6.9) takes the steps from the process flow chart and breaks them down to better understand the necessary processes and whether or not they are value added or nonvalue-added to the ultimate goal of increasing enrollment within Industrial Distribution. Having already started implementing some of the processes, we were able to decipher all the required tasks that were associated with the given process. After we realized what exactly was involved with the procedures, calculating the nonvalue-added steps helped in trying to reduce the work that must be delegated within recruitment development for Industrial Distribution and Logistics. The percentage of value-added steps is 60% versus 40% of the activities being identified as nonvalue-added. There is still room for reducing the percentage of nonvalued-added activities in the future.

3. Cause and Effect Diagram

The team created a cause and effect diagram (Figure 6.10) from a positive perspective to identify the potential causes or factors that could contribute to increasing enrollment in the IDIS program. The categories that they used to group the root causes are people, information, methods and materials. In the people category, we can have more faculty market and promote IDIS to more potential students. With a better program there would be better pay to help attract more faculty and staff. Marketing will lead to higher enrollments by incoming freshmen and undecided students. In the information and materials category, we can continue to use surveys and talking with current IDIS students, as well as talking to students who are undecided on their major. We came up with a strategic set of questions that we felt would give us the knowledge and understanding we needed to understand customer requirements.

4. Why-Why Diagram

The data we gathered during the Measure phase and from our surveys led us to conclude that increasing enrollment is the main objective for the IDIS program related to this project. Further analysis of our surveys and data brought to our attention that most students are not aware of the IDIS program, which is obviously a major contributing factor to the lack of students in the program. We used a Why-Why analysis to find the probable root causes of the decline of students in IDIS (Figure 6.11). The main reason that the students are not aware of IDIS is that the program did not have a focused marketing and recruiting program to attract students. When the IDIS program faculty or students did give presentations at the class for undecided majors (COAD), the presentations were not geared to the interests of students because formal surveys had not been performed prior to this project.

5. Waste Analysis

The eight wastes were used as a guide to identify wasteful activities in the marketing plan development and recruiting processes to develop a more streamlined process. There is an over production waste when recruiting presentations are given to students who have no interest in the IDIS major. Defects in the process are when students never hear of the IDIS major, but would have been interested had they become aware

Location: Industry and technology building		Summary			
Activity: Recruitment		Event	Present	Proposed	Savings
Date: 11/8		Operation	4	4	
Operator: Mr. Angolia	Analyst: Dr. Pagliari	Transportation	3	1	
Method: Proposed		Delay	3	2	
		Inspection	2	1	
Remarks:		Storage	0	0	
		Time (mins.)	595	180	
		Costs			

Event description	Symbol					Time	Value added	Nonvalue added
Review program literature	O	→	▼	D	□	180		√
Go to Registrars Office	o	→	▼	D	□	20		√
Obtain list of undecided students	o	→	▼	D	□	10	√	
Contact project sponsor, Mr. Angolia	O	→	▼	D	□	15		√
Obtain list of COAD classes	o	→	▼	D	□	10	√	
Develop PP to show COAD classes	o	→	▼	D	□	90	√	
Meet w/ dept. head, Dr. Pagliari	o	→	▼	D	□	30		√
Create brochure to show undergraduate students	o	→	▼	D	□	120	√	
Revise PP	O	→	▼	D	□	45		√
Contact project sponsor to obtain professors COAD classes	o	→	▼	D	□	10	√	
Set up time with Professors to present	o	→	▼	D	□	35	√	
Present PP to COAD	O	→	▼	D	□	20	√	
Obtain list of incoming students	o	→	▼	D	□	??		√
Send out mailings in June To incoming students	o	→	▼	D	□	??	√	
Contact coordinators of barefoot on the mall	o	→	▼	D	□	??		√
Have PAID setup booth at barefoot on the mall	o	→	▼	D	□	??	√	
Promote PAID golf tour. PAID member + guest	o	→	▼	D	□	??	√	
Try to have every PAID member bring 1 nonIDIS student to golf tour.	o	→	▼	D	□	??	√	
Promote raffle to IDIS & nonIDIS students through dinning halls & other social areas	°	→	▼	D	□	??	√	
Have project sponsor & dept. head review ppt, program literature, & marketing ideas	°	→	▼	D	□	10		√
							60%	40%
							Percentage of value added vs. nonvalue added	

FIGURE 6.9 Process analysis through use of flow process diagram.

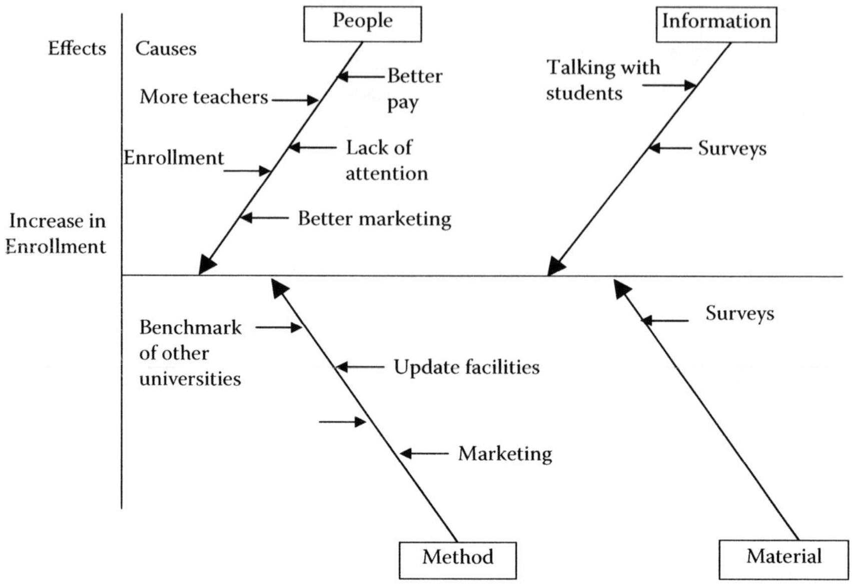

FIGURE 6.10 Cause and effect diagram.

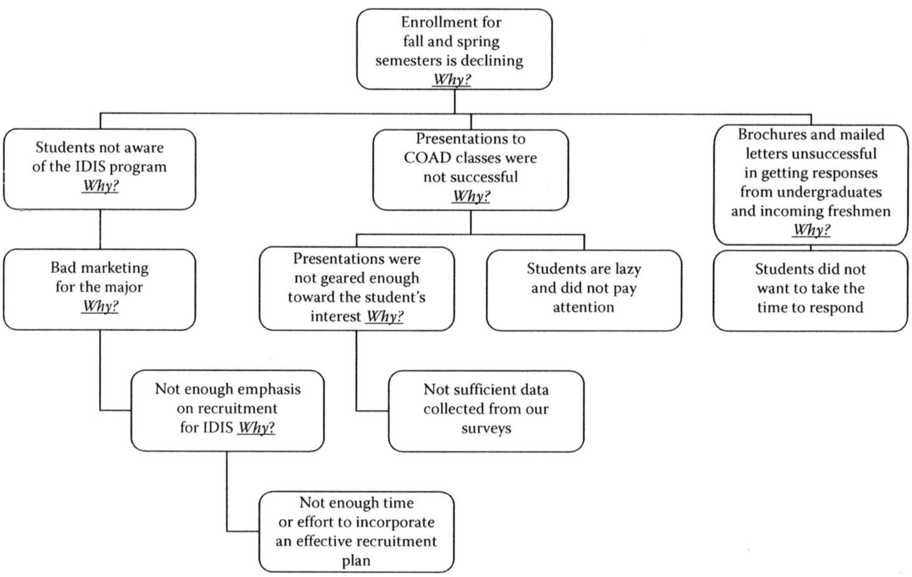

FIGURE 6.11 Why-Why diagram.

of the major, and students who become IDIS majors thinking that it would be a good fit, but find out that it is not. A delay waste is a student who does not hear about the IDIS major until his/her second or third year, switches to the major, but then is behind in classes, thereby delaying his/her graduation date. An inventory waste includes students who register for IDIS classes, but then drop the classes, potentially preventing

Waste type	*Process*	*Waste element*
Transportation	Not applicable	Not applicable
Over Production	Marketing, recruiting	Giving presentations to students who are not interested in IDIS
Motion	Not applicable	Not applicable
Defects	Marketing, recruiting	Students who are never aware of IDIS as a major; Students who are IDIS students who then drop out
Delay	Marketing, recruiting	Students who do not hear about IDIS until their second or third year and then are behind in their classes
Inventory	Recruiting	Students who register for classes, but then drop out
Processing	Marketing, recruiting	Students who seek advising who don't choose the IDIS program
People	Marketing, recruiting	No focus on process improvement, not using people's ideas

FIGURE 6.12 Waste analysis.

other students who need the class from getting registered. A processing waste is those students who seek advice, taking up the advisor's time, but never change their major. The people waste is not having a focus on process improvement and not using people's ideas to improve the processes. The wastes are summarized in Figure 6.12.

6. FMEA

Summary of Problems

All of the analysis performed in the previous sections is shown below in the problems we encountered with the IDIS program. We felt that these issues affected us the most in our research in the further development of the IDIS program at ECU.

Lack of enrolled students: Due to the low number of students in the Industrial Distribution program, we have taken new steps in revamping the program's marketing scheme. This year there will be around 70 IDIS majors graduating between December and May. So now more than ever the IDIS program is in need of new students to join our program. We have around 165 people in the program; with so many graduating it will really affect the program.

Lack of knowledge about the IDIS program: We recently took a survey of students not in the IDIS program, to find out what they know and do not know about IDIS. Seventy-two percent of the people surveyed have no idea what IDIS is. So we knew we had to get our name out there, so we developed a new benchmarking scheme and a new marketing strategy.

Awareness of the program to new students: We also surveyed new students to ECU. In this survey we tried to find out what the students thought IDIS was with the following question: "What do you think the Industrial Distribution and Logistics students do?" We had a huge variety of answers, but none were correct. This presented a significant issue. How can enrollment be increased if nobody out there has even heard of or understands the program?

Poor marketing techniques: The IDIS program has a very weak marketing plan. No where around campus do you see anything about IDIS or where you can even find us. This is a big problem we are facing, which makes it challenging to increase enrollment. We need to act fast due to the loss in numbers that will hit us by the end of the year.

Orientation: During orientation we are set-up in the Bate building on the second floor. We are out of the way of the new students. We need a new booth spot so the freshmen can come and talk with us and find out who we are and what we are all about. By us not being in a high traffic area, it potentially prevents us from properly recruiting new students.

The FMEA and the 5S diagrams give recommended actions that can be adopted to overcome the problems mentioned above.

A FMEA was conducted for our project to recognize and evaluate the process steps for the marketing plan and recruiting procedures that are proposed for implementation for the upcoming semesters. We reviewed these proposed events and decided on possible failure modes as well as potential effects that these failures could have on the Industrial Distribution program. We used the following analysis in our FMEA diagram to ascertain some conclusions on how the processes could fail and what is needed to eliminate the possible failure modes.

We reviewed key process steps by reviewing the flow process diagram and process maps to determine the most important factors to the recruitment process. We identified potential failure modes and analyzed how these steps could ultimately affect the outcome of our proposed process. We determined potential effects that the failure modes could have on the Industrial Distribution process. We concluded that the ultimate goal was increasing enrollment and geared the effects of failure mainly on the outcomes that would present themselves if the program failed to market itself properly to the student body. We identified potential causes of the failures and how those failures occur by identifying the root causes that can be corrected or controlled. The potential causes of failure were ultimately a decrease in enrollment within the Industrial Distribution program due to the fact that the process steps were recommending actions for implementation in the future recruitment process. For each of these effects, we assigned a likelihood of severity, occurrence and detection. These probabilities were assigned based on the relative importance of all those effects and given a degree of severity based on a 1–10 scale (10 being a high severity and occurrence, and 10 being a low ability to detect the failure). A risk priority number (RPN) is calculated by multiplying the severity, occurrence and detection values to yield a combined value. High RPNs indicate immediate actions to resolve the failure modes. The higher the RPN number the more severely it affects the outcome of the process and therefore needs to be resolved immediately. Because all of our potential failure modes were taken from our proposed process map or flow process diagram, we concluded that the RPN for the majority of the events ultimately was high due to the fact that every process was important in the recruitment process. The FMEA chart is shown in Figure 6.13. The highest RPN failure

Process step	Potential failure mode	Potential effect(s) of failure	Sev	Potential cause(s) of failure	Occ	Det	RPN
Review marketing plan and update literature for upcoming semesters	Old literature is used and new information is not included. This will ultimately not show how the program is advancing and creating more opportunities for its students	Decreasing number of students within the program due to the literature not successfully showing benefits of program that IDIS offers students	9	Marketing plan and literature does not contain necessary information to successfully promote program	5	3	135
Develop PowerPoint presentation to show to COAD classes showing how IDIS can benefit students	Teachers not responding to our request to present the IDIS material to their respective classes	Undecided students who attend COAD classes cannot become familiar with the program which lessens their chances for enrollment within IDIS	6	PowerPoint is unsuccessful in allowing potential students to obtain knowledge of program which in turn does not help in increasing enrollment	4	3	72
Obtain list of undergraduate undecided students and addresses	Director of undergraduate studies is not available or does not wish to release list of undergraduate undecided majors	IDIS program is not able to send out mailings and therefore cannot reach it's most valuable customer, undecided students at ECU. Enrollment will ultimately decline during the upcoming semesters	8	The Industrial Distribution and Logistics program declines in enrollment due to the fact that any undeclared student cannot fully see benefits of program through the literature that would be provided in the mailings	7	5	280
Send out mailing to undergraduate, undecided students and incoming freshman for Spring and Fall semester arrivals	Literature sent to students is not successful in showing benefits and positive qualities of IDIS	If literature that is compiled by IDIS program to show to prospective students is not appealing, the undecided students will not wish to engage in the program	8	Incoming freshman are unaware of the program when they arrive on campus and therefore do not wish to enroll within the program	8	5	320

FIGURE 6.13 Failure mode and effects analysis (FMEA).

Set up booth at Barefoot on the Mall to promote program and present literature showing benefits of program	No student visitors to the booth that is set up by the IDIS department	The IDIS program is unable to reach potential clients that attend the event	6	The Industrial Distribution and Logistics program will not be able to gain necessary recognition for the recruitment process	7	3	126
Fundraiser for program. Create a raffle and promote at dining halls	Students do not wish to participate in the fund raiser or become involved with the IDIS program	The IDIS and PAID programs will ultimately lose money and resources due to the fact that the prizes they buy for the raffle will not generate funding and potential future students	6	Decrease in funds will not allow the IDIS and PAID programs to hold future events and therefore cannot market the program in a successful manor	6	6	216
Have open house at beginning of fall semester to gain recognition for IDIS program. Create "Casino Night" to be held in the IDIS simulation lab to show benefits of program to incoming freshman and undecided students	No interested students attend the open house and therefore IDIS cannot promote itself to incoming students as well as current ECU students who are either undecided or are unhappy with their current major	The IDIS program looses name recognition on campus due to the fact that there are no interested students that wish to join Industrial Distribution	7	If the IDIS program looses name recognition then the enrollment will eventually decrease due to the fact that students are unfamiliar with the program	7	6	294
Promote yearly PAID golf tournament to any ECU student who wishes to learn more about the IDIS program and wants to participate	Corporate sponsors do not volunteer to donate funds and prizes and therefore the PAID golf tournament cannot be held	Yearly fund raiser for the program is unsuccessful and does not generate any funds for the upcoming semester	7	If the IDIS and PAID programs cannot increase their funding, then materials necessary to the promotion of the program cannot be obtained	5	7	245
	Golf courses in the area do not wish to participate in promotion of the golf tournament and therefore a venue will not be available	Fund raiser for program is unable to be held	7	Companies that usually participate in the tournament to recruit students will loose interest in the program and ultimately not become involved with future graduates	5	7	245

FIGURE 6.13 (Continued)

is that the literature sent to students may not be successful in showing benefits and positive qualities of IDIS. Another failure is that no interested students attend the open house and therefore IDIS cannot promote itself to incoming students as well as current ECU students who are undecided or are unhappy with their current major. The third and fourth highest-rated failures are corporate sponsors not volunteering to donate funds and prizes, and therefore the Professional Association of Industrial Distribution (PAID) golf tournament cannot be held and golf courses in the area do not wish to participate in the promotion of the golf tournament. Therefore, a venue will not be available.

7. 5S

This diagram addresses the factors of the current marketing plan and recruitment process that need to be improved and tries to make suggestions on how to better develop those factors for the upcoming semesters. The main focus of our project is the ultimate increase in enrollment for the IDIS program and the factors that directly affect that outcome. Implementing a Lean approach to the recruitment process will better allow the program to focus on exactly what will be needed in order to achieve an increase in the student body within Industrial Distribution. Having collected a variety of data through our student survey and interviews, we were able to make conclusions on what exactly needs to be done in order to gain more recognition throughout the ECU student body. Implementing the 5S diagram and the improvement recommendations contributes to a more focused approach to the recruitment process as well as addresses the factors that were used in the past that were not successful. The 5S diagram is shown in Figure 6.14.

5s	Issues	Recommendation
Sort	Unaware of our target audience, and what would appeal to them about the IDIS program	Obtain a list of undecided and incoming freshmen students; establish a well-organized marketing plan to attract more students to IDIS
Systematize	Most students are unaware of what the IDIS program is	Develop better marketing strategies within the department that better inform students of our program and what is has to offer
Sweep N' Clean	Students not responding to presentations or surveys	Analyze what the students did like and what they did responded to, and target those areas
Standardize	Students not enrolled in IDIS do not know about the jobs and benefits associated with industrial distribution and logistics	Inform students on the high job placement directly out of college, as well as all of the different fields they could pursue in distribution and logistics
Self-discipline	Lack of time interacting with undecided and freshmen students	Find a more efficient plan to allow time to meet with undecided and freshmen students

FIGURE 6.14 5S.

8. Survey Analysis

Current IDIS Student Survey

There are currently 160 IDIS declared students and of those we were able to survey 105. We first analyzed the current IDIS student VOC survey to understand what made them aware and interested in IDIS as a major. As shown in Figure 6.15, Question 1, "How long have you been affiliated with the IDIS program?" the majority of the students in the IDIS program have been in the program from two to three years (28 and 32%, respectively). This is a sign that the department is at its peak. With the one year (18%) being lower than the four years (22%), it is not showing a promising future for the program. However, performing a chi-square analysis, this was not significantly different ($p=.173$). By creating a better marketing campaign we will be able to reach undeclared and future freshmen on the ECU campus.

Question 2 (Figure 6.16) was "How did you become familiar with the IDIS program? It was a very relevant question to our survey because it shows us how people became associated with the IDIS program. The top-rated response were friends/peers (28%), faculty (21%), and other (17%). The "other" category consisted of many responses, including the ECU website, family and freshman orientation as being the

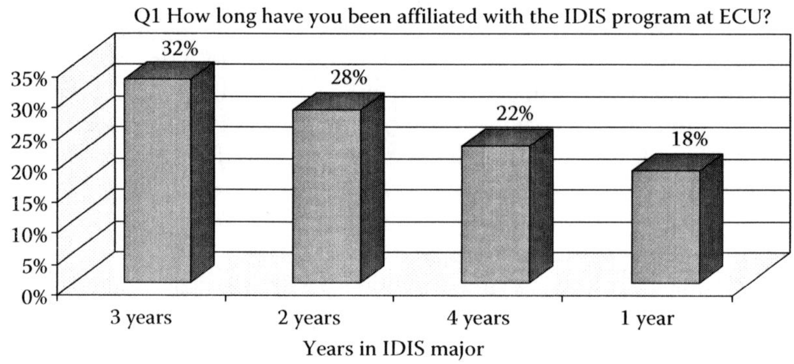

FIGURE 6.15 Pareto chart of IDIS student survey, question 1.

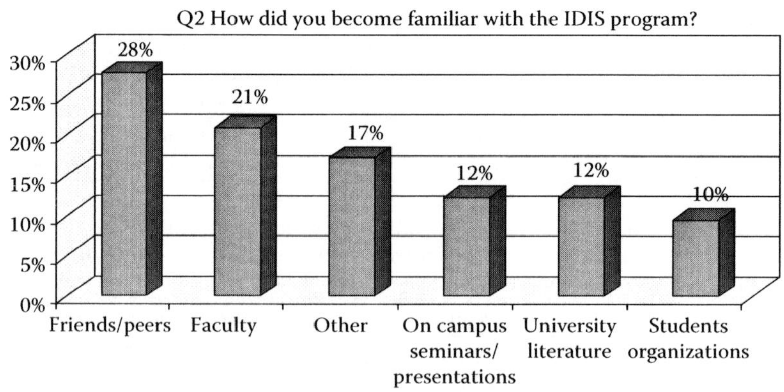

FIGURE 6.16 Pareto chart of IDIS student survey, question 2.

reason for finding out about the program. This information is helpful in realizing other aspects that familiarize students with IDIS and therefore try to take advantage of those resources that they would have otherwise overlooked.

The responses also show us where we need to improve. Our three lowest responses, student organizations (10%), on-campus seminars/presentations (12%) and university literature (12%) should be our highest, but they are not. We need to be getting our name out during freshmen orientation, setting-up booths in better locations, putting on seminars similar to the school of business and have the PAID members pass out pamphlets all over campus. Chi-square analysis showed the ratings were significantly different ($p = .014$).

The next question (Figure 6.17) was "What made you declare (your major) as an IDIS student?" It was by far our favorite survey question because we received the answers that we had hoped to receive. Most people who declared IDIS as a major did so because of job placement (26%), faculty (20%), and interesting subject matter (19%). These are the reasons that every program wants their students to declare their major. It is also nice to see students listening to their friends/peers (12%) and their parents (12%). The majority of the "other" category in this response consisted of Jim Toppen and Dr. Leslie Pagliari as being the reasons they declared their major as an IDIS student. Jim Toppen has since left the program, but played a vital part in recruiting students into the IDIS program. This statistic shows that the faculty of a program is vital in increasing the enrollment within the particular program. The p-value for the chi-square analysis was .053, so the responses were not significantly different.

Of the 105 survey responses, 29 transferred from other programs at ECU. Forty-five percent of students transferred from business, 21% from construction management, 17% from communications, 10% from political science, and 7% from design (Figure 6.18). Chi-square analysis p-value was .537, so the results were not significantly different.

For the first open-ended question, students were asked "What do you like about the Industrial Distribution Program?" Many responses were given for this particular question, such as the high job placement, professors, and the hands-on experience they receive in the classes. When students were asked what they dislike about the program, an astounding number replied that the constant changing of faculty is a downfall of

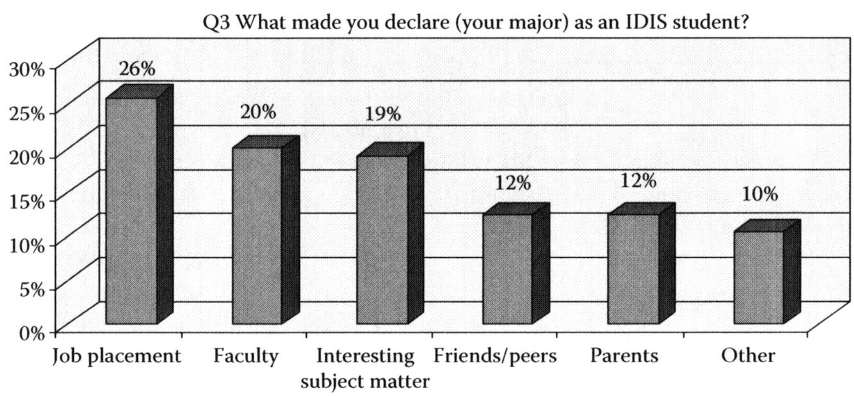

FIGURE 6.17 Pareto chart of IDIS student survey, question 3.

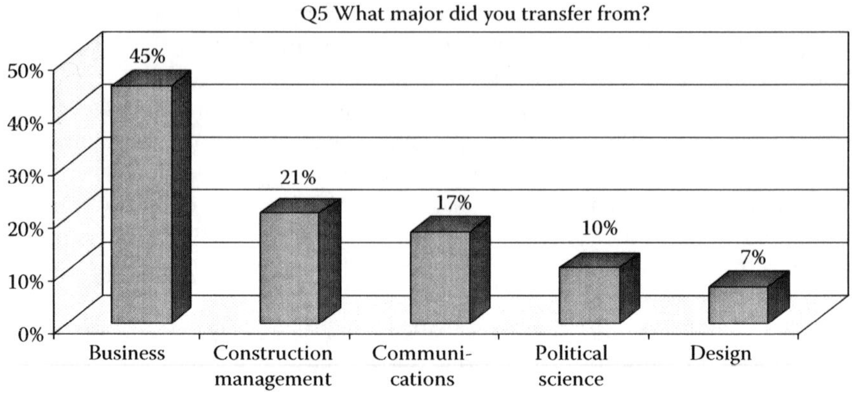

FIGURE 6.18 Pareto chart of IDIS student survey, question 5.

IDIS. Also, many students expressed a dislike in the similarity of the companies that IDIS promotes to recruit at ECU. They would like to see more than plumbing and construction suppliers, and more of a variety of distribution companies.

Many students also shared ideas on how the IDIS program can improve. Many of the responses stated a dislike for the constant change of professors in the program, as well as a need for more diversification in the recruiting of companies.

Non-IDIS Undergraduate Student Survey

For the first question, "What is your current class?", out of 50 students surveyed who were not associated with the IDIS program at ECU, most students were freshman and sophomores (48% and 36%, respectively). Though this is a small sample size, it gives insight as to who our target audience is in the recruitment process. The Chi-square *p*-value is 0, which supports the larger percentage of freshmen and sophomore students.

For the next question, "What is your current major, if any?", the predominant major is business (42%), undecided (22%), other (20%,), and 16% are in construction management. This was an interesting statistic because the IDIS program is very much business related, and if the majority of students fall under a business degree, why are they unfamiliar with the IDIS program at ECU? The Pareto chart is shown in Figure 6.19. Chi-square analysis *p*-value is .044, which supports the theory that the highest percentage of students is business students.

Question 3 was "Are you familiar with the Industrial Distribution and Logistics program at East Carolina University?" We discovered that a staggering 72% of the students surveyed had little or no knowledge about the program offered at ECU. Chi-square analysis *p*-value is .002, supporting the high response rate of students who had no knowledge of the IDIS program.

9. DPPM/DPMO

Note: This is a hypothetical example for the IDIS DPMO calculation. The DPMO and related sigma level for the marketing and recruiting processes assuming a 1.5 sigma shift for the following data is 381,250 for a sigma level of about 1.8, showing the large opportunity for recruiting more students. The opportunities for failure are

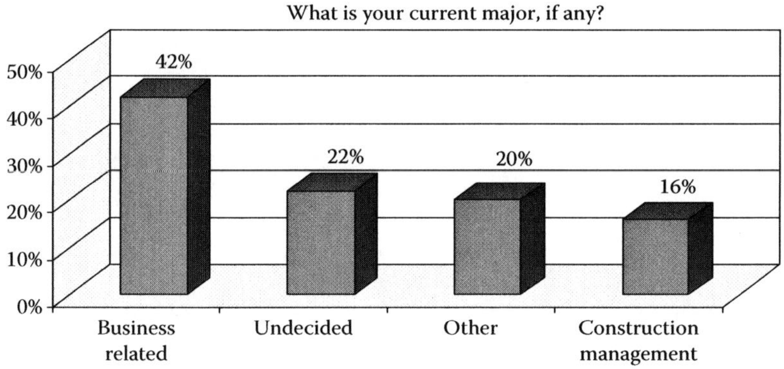

FIGURE 6.19 Pareto chart of non-IDIS student survey, question 2.

twofold: a student does not select IDIS as a major as a freshman, or a student drops out of IDIS as a major. Defects are identified as the number of students who meet an advisor but do not enroll in IDIS per month as 15, and the number of times a student drops out of IDIS per month as 0.25 (or one student every four months). The number of students (units) who met with an advisor to discuss IDIS as a major was 20.

10. Analyze Phase Presentation

The Analyze presentation can be found in the downloadable instructor materials.

ANALYZE PHASE CASE DISCUSSION

1. **Analyze Report**
   1.1 Review the Analyze report and brainstorm some areas for improving the report.
   1.2 How did your team ensure the quality of the written report? How did you assign the work to your team members? Did you face challenges of team members not completing their assigned tasks in a timely manner and, if so, how did you deal with it?
   1.3 Did your team face difficult challenges in the Analyze phase? How did your team deal with conflict on your team?
   1.4 Did your instructor and/or Black Belt or Master Black Belt mentor help your team better learn how to apply the Lean Six Sigma tools in the Analyze phase, and how?
   1.5 Did your Analyze phase report provide a clear understanding of the root causes of the process, why or why not?

2. **Process Analysis**
   2.1 Did your to be or future state process map help you to analyze the process, and how?
   2.2 Discuss how your team defined whether the activities were value-added or nonvalue-added. Was the percentage of value-added activities or value-added time what you would expect for this type of process and why?

3. **Cause and Effect Diagram**
   3.1 How did your team determine the root causes, and how did they validate the root causes?

4. **Why-Why Diagram**
   4.1 Was it easier to create the cause and effect diagram or the Why-Why diagram? Which of the tools was more valuable getting to the root causes?

5. **Waste Analysis**
   5.1 What types of waste were prevalent in the process and why?

6. **FMEA**
   6.1 What were your main failure modes and how do you plan to reduce the failure?

7. **5S**
   7.1 Did you find the 5S tool helpful for this project?

8. **Survey Analysis**
   8.1 What were the significant findings in the IDIS student survey?
   8.2 What were the significant findings in the non-IDIS student survey?
   8.3 Did your survey assess customer satisfaction with the marketing and recruiting processes?
   8.4 Was there consistency in the responses between the two surveys?

9. **DPPM/DPMO**
   9.1 What is your DPPM/DPMO and sigma level. Is there room for improvement, and how did you determine that there is room for improvement?

10. **Analyze Phase Presentation**
    10.1 How did your team decide how many slides/pages to include in your presentation?
    10.2 How did your team decide upon the level of detail to include in your presentation?

IMPROVE PHASE EXERCISES

1. **Improve Report**
   Create an Improve phase report, including your findings, results, and conclusions of the Improve phase.

2. **Recommendations for Improvement**
   Brainstorm the recommendations for improvement.

3. **Revised QFD**
   Create a QFD to map the improvement recommendations to the CTS characteristics.

4. **Action Plan**
   Create an action plan for demonstrating how you would implement the improvement recommendations.

5. **Future State Process Map**
   Create a future state process map for the following processes:
   - Developing a marketing plan process
   - Recruiting process plan

6. **Revised VOP Matrix**
   Revise your VOP matrix from the Measure phase with updated targets.

7. **Training plans, procedures**
   Create a training plan, and a detailed procedure for the process.

8. **Improve Phase Presentation**
   Prepare a presentation (PowerPoint) from the case study exercises that provides a short (10–15 minutes) oral presentation of the Improve phase deliverables and findings.

IMPROVE PHASE

1. Improve Report

Issues have now been identified and associated with potential improvement strategies. We can now develop an overall plan for the improvement of enrollment and recruiting for the IDIS program.

Comparison of Improvement Strategies

Upon identification of improvement strategies, through the tools and methods in the Analyze phase, we group the strategies to create an affinity diagram to compare and ascertain their relationship to the CTS elements that were originally developed in the Define phase. The affinity diagram allows us to do a side-by-side comparison of the improvement strategies so they may be consolidated and later grouped according to whether they are short-term, long-term, global, or local in nature. Figure 6.21 lists the potential improvements identified by each of the tools enlisted during the Measure and Analyze phases. These items relate to the CTS items that were originally identified in the Define phase and further refined in the Analyze phase. This affinity diagram is shown in Figure 6.20.

2. Recommendations for Improvement

The information shown in Figure 6.21 summarizes the improvement recommendations that support the CTS measures. Students have already begun to make use of our new marketing techniques by going to COAD classes and recruiting undeclared students into the IDIS program. So far our presentations have brought about nine new students to the program. It is not a lot, but it is a start and shows us we are on the right page. We are hoping that the ideas and improvements that we have developed over this past semester will be in full effect by the beginning of the next academic year.

5S	House of quality	FMEA
Develop effective recruiting strategies and better marketing for IDIS Improve awareness of the IDIS program to undecided and incoming students Improve emails and presentations given to students Inform students on the high job placement directly out of college, as well as all of the different fields they could pursue in distribution and logistics	Discover the wants and needs of our potential new students Focus on the strong relationships and strive to improve the weaker ones Increase awareness to undecided students and incoming freshmen Improve recruitment and marketing to increase enrollment in IDIS	Implement the necessary literature and marketing plan to better promote IDIS Increase the number of presentations given to COAD classes Improve communication with incoming freshmen and undecided students Develop more attractive brochures to appeal more to the students interest

FIGURE 6.20 Affinity diagram.

CTS	Improvement recommendations
Awareness of program through current students	• Communication to current students to speak of our program to others • Have students get more involved in PAID. (Professional Association of Industrial Distribution) • Encourage students to participate in recruiting processes
Awareness of program from undergraduate students at ECU	• Improve ease of use of website/access to IDIS program • Set up booths during orientation • Develop a better slide show/ presentation for orientation • Go to the COAD (undecided major) classes and give presentations
Program benefits, marketing techniques	• Employ new marketing strategy for next semester • Inform new students of the benefits that our program has to offer • Put up flyers in freshmen dorms • Continue COAD presentations
Enrollment	• Encourage PAID members to be active in the recruiting process • Stay in contact with COAD professors • Make PAID meetings mandatory for active IDIS students • Tell students to bring their friends to the PAID meetings

FIGURE 6.21 CTS and improvement recommendation mapping.

3. Revised QFD

QFD is a tool to ensure alignment between the customers' needs (CTS measures) or requirements and the improvement recommendations. We were able to determine the CTS needs through interviews with the IDIS faculty and surveys that were distributed to current IDIS students as well as non-IDIS undergraduate students. These surveys established the programs positive and negative attributes and allowed us to ascertain how the customers' requirements needed to be met. After gathering all of our research, we then converted the data into a house of quality, and we used this to assess where the strong and weak relationships existed between the CTS and our proposed improvements. We then were able to conclude where we need to focus our attention in the future recruitment process. By doing this, it is easier for faculty and current IDIS students to read and evaluate the areas that needed improvement for the improvement of the IDIS program. Overall, we are on the right path for fixing our weak areas within the program. We feel strongly that we will have great success in the future with Industrial Distribution at ECU.

The highest priority recommendation is to improve the mailing list. Next is to implement an email notification regarding IDIS information. Freshmen orientation and updating the IDIS website are also high-priority recommendations. Interacting with non-IDIS undergraduate students is another important recommendation. The QFD house of quality matrix is shown in Figure 6.22.

Project name: IDIS recruiting process

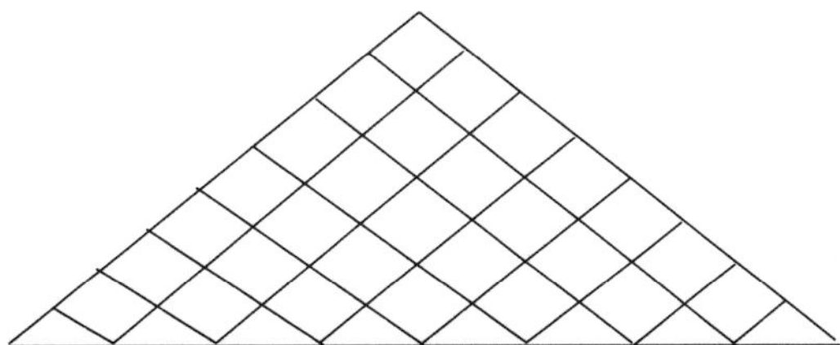

Technical requirements / *Customer requirements*	Importance	Email notification	Update website	Improve mailing list	COAD presentations	Lecture meetings	Freshmen orientation	Advertisement	Surveys	Benchmarking	Undergraduate students	Families
Awareness of program through current students	6	9	9	9		3		3			3	3
Awareness of program to undergraduate students at ECU	5		9	3	3	3	9	3	1		3	3
Program benefits & marketing techniques	9	9		9	3		3				3	3
Enrollment	10	3		3	3		3	3	3	3	3	3
Absolute weight		165	99	180	72	33	102	63	35	30	90	90
Relative weight		2	4	1	7	10	3	8	9	11	5	5

FIGURE 6.22 QFD house of quality.

4. Action Plan

Once the improvement strategies are consolidated, we determined what the level of difficulty (risk) and importance of the strategies are, along with whether they will be implemented in the short- or long-term time frame, and the area of responsibility for the recommendations. To do this, we rated each improvement category on a scale according to level of difficulty (1–5, 5 being the highest) and importance to the overall success of the project (1–5, 5 being the highest). The improvement strategies are grouped according to whether they can be classified as short-term, relatively low-cost improvements, and longer-term improvements that require a more significant investment of time and resources. Once the improvements are prioritized, we can establish a sequence of implementation. Finally, the anticipated responsible partner for implementation is identified. The short- and long-term recommendations are shown in the action plan in Figure 6.23.

Improvement	**Level of difficulty (Risk)**	**Importance**	**Schedule**	**Responsibility**
Short-term	**(1–5)**	**(1–5)**		
Marketing and recruitment strategies to better promote IDIS	3	5	Immediate (spring 07')	Increase the enrollment in the IDIS Program
Increase awareness of IDIS program to undecided and new students	4	4	Immediate (beginning next semester)	Current IDIS students and faculty need to meet the goals set in place
Long-term	**Risk**	**Importance**	**Schedule**	**Responsibility**
Increase funding and assets for the IDIS program and facilities	3	4	Phase in over the next 5 to 10 years	Develop ways to raise money for IDIS (fundraisers, sponsored events, etc.)
Increase outside contacts that want to be affiliated with the ECU IDIS program	2	4	Steadily increase over the next few years	Develop good relationships with employers interested in IDIS
Keep up to date with the changing literature and technology associated with IDIS	3	5	Discussions with faculty and other professionals to determine when a change is necessary	Stay on top of the new developments and strategies that are being implemented into the distribution industry

FIGURE 6.23 Action plan.

Short-Term Improvements:

1. Meet with faculty, sponsors, and other professionals in the distribution industry to come-up with an effective marketing plan. Take strategies that have been successful in the past and implement them into our program.
2. Develop well-organized presentations to show to undergraduate and new students that could possibly influence them to join the IDIS program. We also developed surveys to gain insight that will hopefully help in the recruitment of IDIS students.

Long-Term Improvements:

1. We can have more sponsored events that will raise money for the program, which will allow us to expand on what we can offer to students in the future.
2. If we continue to have good attendance at the career fairs and PAID meetings, employers will want to stay involved in the IDIS program. Keeping good relationships with these employers is a key element to the success of IDIS.
3. Keep in contact with professionals in the distribution industry that can provide valuable insight on improvements and new technology for the near future. We can also benchmark other universities to see if our literature and technology are up to par with some of the more prestigious schools around the country.

5. Future State Process Map

The future state process map presented in the Analyze phase was analyzed to reduce the nonvalue-added activities to streamline the recruiting and marketing processes further.

6. Revised VOP Matrix

It is necessary to institute performance targets to establish the level of performance needed for the process to operate well. Our group has established the necessary performance targets corresponding with the respective CTS characteristics. It has been established that the Industrial Distribution program recruiting process has certain characteristics that will ensure an increase in students for future semesters. These characteristics have been addressed and are shown in Figure 6.24.

The metrics corresponding to the CTS measures in the Measure phase have been slightly modified upon further investigation. The updated metrics corresponding to the CTS measures along with parallel performance targets are summarized in the abbreviated VOP matrix.

7. Training Plans and Procedures

The future process map will serve as the training plan, along with the IDIS presentations. Most of the students and faculty who would be giving these presentations are

CTS	Metrics	Performance targets
Awareness of program through current IDIS students	Current students views, thoughts and suggested improvements of IDIS program	Use suggestions made by current IDIS students and try to accommodate student needs so that IDIS is more appealing to future students
Program awareness through undergraduate students at ECU	Undergraduate students not affiliated with the program and their familiarity with IDIS	Implement new marketing strategy to better inform students of the benefits of the IDIS program. Allow them to see curriculum, fields of study and opportunities upon graduation
Program benefits and marketing techniques	Current marketing procedures and how program advertises itself to undergraduate students at ECU	Mailings showing what the IDIS program offers students has already been established and sent to students' permanent addresses
Enrollment	Number of students enrolled in IDIS increases program funding	Can be determined upon start of next semester. Some of our suggestions have already been implemented and current undergraduate students have already expressed interest in program. Increase enrollment from 160 to 200 by end of next academic year

FIGURE 6.24 Abbreviated revised VOP matrix.

familiar with the program, and already have the knowledge required to give effective presentations on IDIS.

8. Improve Phase Presentation

The Improve presentation can be found in the downloadable instructor materials.

IMPROVE PHASE CASE DISCUSSION

1. **Improve Report**
    - 1.1 Review the Improve report and brainstorm some areas for improving the report.
    - 1.2 How did your team ensure the quality of the written report? How did you assign the work to your team members? Did you face challenges of team members not completing their assigned tasks in a timely manner and, if so, how did you deal with it?
    - 1.3 Did your team face difficult challenges in the Improve phase? How did your team deal with conflict on your team?
    - 1.4 Did your instructor and/or Black Belt or Master Black Belt mentor help your team better learn how to apply the Lean Six Sigma tools in the Improve phase, and how?
    - 1.5 Did your Improve phase report provide a clear understanding of the improvement recommendations of the process, why or why not?

1.6 Compare your improve report to the improve report in the book, what are the major differences between your report and the author's report?
1.7 How would you improve your report?

2. Recommendations for Improvement

2.1 How did your team generate ideas for improvement?
2.2 What tools and previous data did you use to extract information for the improvement recommendations?
2.3 How do your recommendations differ from the one's in the book?

3. Revised QFD

3.1 Does the QFD support the alignment with the CTS characteristics?
3.2 How will you assess customer satisfaction?

4. Action Plan

4.1 How did your Six Sigma team identify the timings for when to implement your recommendations?

5. Future State Process Map

5.1 Compare your future state process map to the one in the book. How does it differ? Is yours better, worse, or the same?

6. Revised VOP Matrix

6.1 Does the VOP matrix provide alignment between the CTS measures, the recommendations, metrics and target?

7. Training Plans, Procedures

7.1 How did you determine which procedures should be developed?
7.2 How did you decide what type of training should be done?

8. Improve Phase Presentation

8.1 How did your team decide how many slides/pages to include in your presentation?
8.2 How did your team decide upon the level of detail to include in your presentation?

CONTROL PHASE EXERCISES

1. Control Report

Create a Control phase report, including your findings, results, and conclusions of the Control phase.

2. **Hypothesis Tests**
   Compare the improvement in the number of students enrolled in the IDIS program before and after improvements were implemented. The enrollment before improvements was 160, and after was 169.

3. **Control Plan**
   Develop a control plan for each improvement recommendation from the Improve phase report.

4. **Control Charts**
   Create an idea for applying control charts to control the recruiting or marketing Plan processes.

5. **Replication Opportunities**
   Identify some potential replication opportunities within the college or university.

6. **Standard Work, Kaizen**
   Create a plan for standardizing the work.

7. **Dashboards/Scorecards**
   Create a dashboard or scorecard for tracking and controlling the recruiting process.

8. **Control Phase Presentation**
   Prepare a presentation (PowerPoint) from the case study exercises that provides a short (10–15 minutes) oral presentation of the Control phase deliverables and findings.

CONTROL PHASE

1. Control Report

To successfully track the progress of the Improvement steps, a strong control plan needs to be established. The IDIS program recruiting process Lean Six Sigma team has implemented certain procedures that will create a baseline for maintaining and increasing the number of students within the program. To successfully monitor the progress of the team, guidelines have been established and recommendations set to allow for a concrete system in which to measure student enrollment.

2. Hypothesis Tests

There is already a significant difference in the number of students that declared IDIS as a major as a result of IDIS presentations to the undecided majors class (COAD).

We performed a two-proportion test to compare the additional nine students that joined the program since some of the changes were implemented. Prior to the changes, there were 160 students, and after the COAD presentations nine additional students joined the program based on those presentations. The p-value was 0, so we conclude that this is a significant increase in the number of students entering the program.

3. Control Plan

Recommendation #1

Marketing Strategy—Develop better marketing strategy to help promote the IDIS program.

Proposed Control

A more efficient marketing strategy would gain more students' attention and recognition about the IDIS program through better promotion techniques.

Goal: Gain more recognition through the university and student body.

Counter Reactions

If positive: More students will show increasing interest in the IDIS program.

If negative: No reaction from prospective students, leads to trying other marketing techniques.

Data Available

Available data come from past surveys handed-out to current and prospective IDIS students. Such valid survey responses include "How current students joined the program," and "What did prospective students know about the program." Marketing techniques can be generated from those responses given by students.

Recommendation #2

COAD Presentations—Continue to present to COAD classes to help inform prospective students about the IDIS program.

Proposed Control

Presentations to COAD classes would be beneficial in trying to increase enrollment within the program because that is where a number of undecided students reside. The faster we can inform undecided students (mainly freshman), the better chance the program has of gaining recognition and increasing enrollment.

Counter Reactions

If positive: We should see an increased interest in the program from COAD students and continuation of COAD presentations.

If negative: We need to look at what better ways we can present the program and/or revise COAD presentations.

Data Available

Available data comes from current PowerPoint presentations given to COAD classes as well as responses that the students gave that might help better the presentations in the future.

Recommendation #3

Orientation—Prepare presentation for freshman/new student orientation.

Proposed Control

Presentations given to incoming freshman would better help students coming to ECU to start thinking about what they want to major in before they actually arrive. Setting-up booths, and speaking at orientation would help to inform those undecided students what the IDIS program has to offer.

Counter Reactions

If positive: We should see an increase in declared IDIS students before new comers arrive at ECU and/or feedback from those interested in the program.

If negative: We would see neither change in prospective students nor feedback from those who come through orientation.

Data Available

Available data is poor location of current IDIS set-up during orientation. Current data show us that we need more visible locations for incoming students so that they can recognize the program and hopefully be able to gain better insight into what IDIS has to offer.

Recommendation #4

Acquisition of undergraduate student list. Distribute mailings to prospective students and better inform students of the IDIS program and the benefits that are offered.

Proposed Control

Distribution of mailings to undergraduate students would allow the IDIS program to reach a much broader spectrum of students and in turn use a strategic marketing process to inform students of the major.

Counter Reactions

If positive: Student enrollment will increase within the IDIS program.

If negative: Time and money will have been wasted and the IDIS program will not gain the name recognition that it was hoping for.

Data Available

Until this strategy has been implemented, there is not a way to determine if this recommendation will be successful in the recruiting process.

4. Control Charts

A proposed control chart could be to use an individuals and moving range chart to track and control the number of students that hear the IDIS presentation in the undecided majors class (COAD). Another idea would be to use the individuals and moving range chart to track the number of students that attend orientation and discuss IDIS with the faculty and students offering information at the orientation booth.

5. Replication Opportunities

Almost any other program in the university could use similar recommendations to enhance their marketing plan, recruiting efforts, and VOC surveys and analysis. Additionally, other IDIS type programs at other universities could use similar recommendations.

6. Standard Work, Kaizen

The standardized work can be attained by having a standard IDIS presentation, brochures and marketing materials. Also, the college can adopt a common look and feel to the website. The current website requires a minimum of seven clicks to get to the IDIS program website page, which is quite excessive and confusing. This would be a great area to standardize.

7. Dashboards/Scorecards

A sample dashboard (Figure 6.25) summarizes the improvements in enrollment, the increase in the number of presentations to the COAD undecided major's classes, improved awareness based on the VOC survey, and increased PAID attendance. The IDIS program has already added nine additional students through the new COAD presentations. The other improvements will be tracked in the future, but are only hypothetical at this point and are used for illustrative purposes.

8. Control Phase Presentation

The Control phase presentation can be found in the downloadable instructor materials.

CONTROL PHASE CASE DISCUSSION

1. **Control Report**
   1.1 Review the Control report and brainstorm some areas for improving the report.
   1.2 How did your team ensure the quality of the written report? How did you assign the work to your team members? Did you face challenges of team members not completing their assigned tasks in a timely manner and, if so, how did you deal with it?
   1.3 Did your team face difficult challenges in the Control phase? How did your team deal with conflict on your team?

Metric	Baseline	Improvement level	Improvement (%)
Enrollment	160	169	5.6
Number presentations	1	4	300
Awareness with IDIS and non-IDIS students	28%	32%	4%
PAID attendance	40	100	150

FIGURE 6.25 Sample dashboard.

1.4 Did your instructor and/or Black Belt or Master Black Belt mentor help your team better learn how to apply the Lean Six Sigma tools in the Control phase, and how?
1.5 Compare your Control report with the Control report in the book, what are the major differences between your report and the author's report?
1.6 How would you improve your report?

2. Hypothesis Tests

2.1 How did you assess the improvement for the CTS?

3. Control Plan

3.1 How well will your control plan ensure that the improved process will continue to be used by the process owner?

4. Control Charts

4.1 For this project did you find attribute or variable control charts to be more applicable for controlling this process.
4.2 Are their additional control charts that could be used to ensure process control?

5. Replication Opportunities

5.1 How did your team identify additional replication opportunities for the marketing and recruiting processes?

6. Standard Work, Kaizen

6.1 How might you use a kaizen event to have identified process improvement areas, or ways to standardize the process?
6.2 How would you recommend ensuring that the process owners follow the standardized procedures or presentations?

7. Dashboards/Scorecards

7.1 How would your dashboard differ if it was going to be used to present the results of the marketing and recruiting to the entire college or university?

8. Control Phase Presentation

8.1 How did your team decide how many slides/pages to include in your presentation?
8.2 How did your team decide upon the level of detail to include in your presentation?

7 CECS Inventory and Asset Management Process Improvement—A Lean Six Sigma Case Study

Felix Martinez, Varshini Gopal, Amol Shah, Robert Beaver, Russ D'Angelo, Miguel Torrejon, and Sandra L. Furterer

CONTENTS

PROJECT OVERVIEW

The College of Engineering and Computer Science (CECS), along with all other colleges at the University of Central Florida (UCF), as a public institution, is entrusted with state-owned assets. Jose Murphy is the property manager for Engineering Buildings

I and II, as well as the physical plant. He is responsible for the safeguard, tracking, and managing of said assets, as specified in Chapter 80–380 of the Florida Statutes.

The departments housed in these buildings are:

- Industrial Engineering and Management Systems (IEMS)
- Electrical and Computer Engineering (ECE)
- Engineering Technology (ENT)
- Mechanical, Materials, and Aerospace Engineering (MMAE)
- Civil and Environmental Engineering (CEE)

All personnel at CECS are responsible for notifying Murphy of any and all relocations of state-entrusted property assigned to them or otherwise in their possession.

CURRENT PROCESS

Before we can proceed with defining the business objectives, we need to get to the meat of the matter, i.e., study the current process to answer questions, such as, what is the process that you are improving and why is it important enough to spend time improving it? The following paragraph provides a description of our current process.

It is a statewide policy for universities to maintain control of all nonconsumable items worth more than $1000. CECS has a series of custodians specifically in charge of more than 4000 items spread across the engineering buildings and Research Park. At the beginning of the year, they are given an inventory list of items which they must account for by the end of the fiscal year. During this period, they follow a series of "passes" in which people from the UCF property office scan a specific tag placed on all the items that need to be accounted for.

From our kick-off meeting with Jose Murphy, we discovered they perform three passes. The fiscal year begins on July 1. The first pass is conducted during the first three months, the second pass is conducted during the next three months, and the final pass is done during the last six months. Any items not located during the first two passes are searched for in the third pass, which begins around January 1, at the beginning of the following semester. Items not located by the end of the year are reported to the police at the end of the fiscal year. These items may later be recovered or never be found. Items may also reach the end of their useful life and therefore must be surplused or "cannibalized" following strict guidelines set forth by the UCF property office.

Even though custodians are responsible for the safekeeping of these items, they are not held accountable for items that are declared lost at the end of the year. Thus, there is no sense of ownership or responsibility for strict tracking of items. Based on this issue, the UCF property office has declared that it will now charge each department for the value of those items lost by that department at the end of the fiscal year.

Tracking of items is conducted using specific software installed on scanners that are taken across UCF and its satellite campuses. After a scanning session, the data collected are uploaded to the computers and to their PeopleSoft financial software system. The UCF property office does not upload the data from the scanners on a daily basis, thus custodians experience a short delay in the retrieval of information.

DEFINE PHASE EXERCISES

It is recommended that the students work in project teams of four to six students throughout the Lean Six Sigma Case Study.

1. **Define Phase Written Report**
   Prepare a written report from the case study exercises that describes the Define phase activities and key findings.

2. **Lean Six Sigma Project Charter**
   Use the information provided in the Project Overview and Current Process sections above, in addition to the project charter format, to develop a project charter for the Lean Six Sigma project.

3. **Stakeholder Analysis**
   Use the information provided in the Project Overview and Current Process sections above, in addition to the stakeholder analysis format, to develop a stakeholder analysis, including stakeholder analysis roles and impact definition, and stakeholder resistance to change.

4. **Team Ground Rules and Roles**
   Develop the project team's ground rules and team members' roles.

5. **Project Plan and Responsibilities Matrix**
   Develop your team's project plan for the DMAIC project. Develop a responsibilities matrix to identify the team members who will be responsible for completing each of the project activities.

6. **SIPOC**
   Use the information provided in the Project Overview and Current Process sections above to develop a SIPOC of the high-level process.

7. **Team Member Bios**
   Each team member should create a short bio of themselves so the key customers, stakeholders, project champion, sponsor, Black Belt and/or Master Black Belt can get to know them, and understand the skills and achievements they bring to the project.

8. **Define Phase Presentation**
   Prepare a presentation (PowerPoint) from the case study exercises that provides a short (10–15 minutes) oral presentation of the Define phase deliverables and findings.

DEFINE PHASE

1. Define Phase Report

We begin by stating the *business objectives* related to the asset management process.

- Increase efficiency needed to track registered assets
- Increase effectiveness of inventory tracking to prevent losses
- Improve stewardship of federal and state-funded acquisitions

These objectives were further defined in terms of proposed deliverables:

- A refined process for asset tracking throughout the asset life cycle
- Recommended technology (if applicable) for tracking assets to determine location
- Consolidated communications between stakeholders as to result in a completely integrated system
- Periodical reports outlining progress and recommendations

Our project goals provide a clearer statement of our visions, specifying the accomplishments to be achieved if the vision is to become real. The target objectives are clearer statements of the specific activities required to achieve the goals, starting from the current state. Our *primary goal* was to improve the overall performance of the inventory management system. To achieve that goal we identified our *target objectives* as the following:

- Determine a new set of procedures to be abided by the stakeholders
- Identify areas of improvement for the current organization; develop recommendations for the system and those who interact with it
- Study alternative solutions/technologies
- Reduce inefficiencies and redundancies in the process of asset tracking

2. Lean Six Sigma Project Charter

The effort needed to ascertain location and condition of assets, for inventory control and use by our students and faculty, requires an efficient process to track the items. This project seeks to discover issues affecting the efficiency of the tracking process and recommend ways and technology to improve or streamline the process, which will result in better asset utilization and reduction in property loss, as a secondary effect.

The DMAIC Six Sigma approach will study the asset management system of the CECS. It will focus only on nonconsumable items of a physical nature. The project will not focus on the financial aspects of the item management. The monetary value of items will be used only if it is determined to bring a benefit to the process. The project will develop solutions that involve only the stakeholders mentioned in the stakeholder analysis section. The detailed project charter is shown in Figure 7.1.

Project name: College Asset Inventory/Management Process Improvement

Problem statement: The College Property Management department has provided an opportunity to analyze their asset management system, identify problems, and design solutions to improve their current situation.

Customer/Stakeholders: (internal / external) Executive associate dean; property manager; department chairs; Office of Property and Inventory Control; CECS faculty, staff and students; Government

What is important to these customers – CTS: Faculty/Staff awareness of process; documented location of assets; identification of assets; efficiency of yearly scanning; values of assets lost; number of assets lost; undocumented assets; efficiency of list update; sorting efficiency of lists; loss avoidance.

Goal of the Project: Streamline the process of asset tracking to enhance control and reduce the effort needed to manage them. Use the DMAIC Six Sigma approach to understand the system and develop an overall improvement on the process.

Scope statement: The scope of the project is focused on asset management for the College of Engineering and Computer Science (CECS).

Financial and other benefit(s): Reduced effort (labor) in tracking assets; More efficient utilization of assets due to better location management; Reduced losses; Better communication among stakeholders

Potential risks: Project contact unavailable; Difficulty applying DMAIC strategy; Conflict team schedule; Contradictions between theory and practice; Change of customer requirements

FIGURE 7.1 Project charter.

All projects entail risks. A risk analysis will help identify those risks and mitigate them before they can occur. The following provides a brief description of the Risk Analysis we performed. The potential risks that could occur were brainstormed and are listed in column (1). For each of these, the probability of occurrence, severity, and detection were determined and rated on a scale from 1 to 10. These values are listed in columns (2), (3), and (4), respectively. Occurrence determines the likelihood of the risk to occur. So "1" indicates that the risk is very unlikely to occur and "10" indicates that the risk is very likely to occur. Severity measures the seriousness of the effects of a risk. Severity scores are assigned only to the effects of the risk but not the risk itself. A "1" on the rating scale indicates that the effect will be almost unnoticeable, but a "10" indicates that it could result in a total lack of function. Detection determines the likelihood of the risk being detected before it reaches the customer. The rating on this scale decreases as the chance of detecting the problem increases. Therefore "1" indicates that risk is almost certain to be detected, and "10" indicates that it is impossible to detect. After the ratings were assigned, the risk priority number (RPN) for each risk was calculated by multiplying occurrence, severity, and detection. The focus should be on the risk with the largest RPN. The risk matrix is given in Figure 7.2.

Risk (1)	Occurrence (2)	Severity (3)	Detection (4)	Risk priority number (5)	Risk mitigation strategy (6)
Change in customer requirements	1	10	9	90	Always keep up-to-date with the customer requirements
Unclear objectives	2	6	4	48	Reiterate objectives and goals very clearly so that everyone understands them
Changes in schedule	2	9	5	90	Provide a flexible Gantt chart that is dynamic and can absorb the changes in schedule
Project contact unavailable	3	6	1	18	Contact other personnel already acquainted
Difficulty applying DMAIC strategy	3	8	5	120	Consult with Black Belts
Conflict team schedule	5	5	1	25	Develop collaboration plan and commitments
Contradictions between theory and practice	9	8	3	216	Quick adaptation strategy

FIGURE 7.2 Risk analysis.

3. Stakeholder Analysis

A stakeholder analysis was performed to enlist the stakeholders, conduct an assessment of their interests, and identify the ways in which these interests affect project riskiness and viability of the basic process. The stakeholders analysis recognizes our stakeholders, their role in our project and what they expect from our project. We classified our stakeholders as primary and secondary based on the level of effect the project would have on them. The stakeholder definition is shown in Figure 7.3.

4. Team Ground Rules and Roles

Felix Martinez will be the project leader. He will be responsible for delegation of job tasks and acts as liaison between the Six Sigma Team and the project contact. He will also be responsible for ensuring that all deliverables are reviewed and approved by the project Black Belt, project contact, and the project champion. Robert Beaver will be the project expert. He will be responsible for the general overall maintenance of the team, as well as ensuring we are on schedule and heading in the right direction. Varshini Gopal will serve as the technology specialist. She will be responsible for in-depth analysis and maintenance of data bases and risk, analysis, as well as

	Stakeholders	Role	Impact / Concern	+ / –
Primary	Executive associate dean	Administrative responsibility for asset management	▪ Ensure effective asset management ▪ Satisfy the interests of department heads	+ +
	Property manager	Manages assets	▪ Reduce effort needed to manage assets ▪ Improve efficiency of tracking property and equipment	+ +
	Department heads	Responsible for property movement/ disposition	▪ Reduce number of lost/ stolen items ▪ Track all items more effectively	+ +
Secondary	Office of property and inventory control	Property custodians	▪ Reduce effort needed to manage assets ▪ Improve efficiency of tracking property and equipment	+ +
	CECS faculty, staff, students	Faculty, staff, students of CECS	▪ Implement new procedures to report items that have been transferred to a different location	+
	Government	Local and federal governments	▪ Reduce cash outflow on item recovery and replacement	+

FIGURE 7.3 Stakeholder definition.

the fabrication of graphs, charts, and relationship matrices. Miguel Torrejon will be the research analyst for this project. His responsibilities will include forecasting and scheduling all components of each phase of the project. Amol Shah will be the process analyst. His responsibilities will include maintaining participation logs and analyzing statistical data to include measures of variances, probability distributions, and hypothesis testing. Russell D'Angelo will be the quality assurance specialist. His responsibilities will include the seamless compilation of deliverables, as well as verifying that all deliverables are accounted for (including graphs, charts, and matrices).

5. Project Plan and Responsibilities Matrix

The detailed project plan and responsibilities matrix is shown in Figure 7.4.

6. SIPOC

With the SIPOC, we identify all the critical elements of the current state and therefore the elements that can be addressed during the process improvement.

Responsibility	Project leader	Project expert	Process analysts	Technology specialist	Research analyst	Quality assurance specialist	Project contact	Project black belt
Define phase								
Form team	X	X	X	X	X	X	X	
Define ground rules	X	X	X	X	X	X	X	
Define team roles	X	X	X	X	X	X	X	
Define responsibilities	X	X	X	X	X	X	X	
Define project objectives	X	X					X	
Inspect procedural manual						X		
Prepare work plan					X			
Tool assessment			X					
Identify milestones	X	X						
Create participation log								
Stakeholder analysis				X			X	
Compile project charter	X	X						
Prepare define report						X		
Provide expert guidance								X
Inspection and approval of project charter							X	
Measure phase								
AS IS process chart	X							
Variation		X			X	X	X	

FIGURE 7.4 Project plan and responsibilities matrix.

Six Sigma DPMO		X	X	X	X			
Yield			X		X	X	X	
Cost of quality (Poor)	X	X					X	
Metrics				X			X	
CTS		X			X		X	
Baseline	X		X					
Benchmarking				X				
Statistical measures			X	X		X		
Probability distributions	X					X		
Hypothesis testing	X	X	X					
Analyze phase								
Assess key processes	X	X			X		X	
Identify value added processes				X			X	
Improve phase								
Stakeholder buy-in of new procedures	X	X			X			
Implement suggested improvements	X		X					
Evaluation/final assessment					X			X
Alternative solutions								
Cost analysis								

FIGURE 7.4 (Continued)

Suppliers

Suppliers for this process are the faculty and staff that need to use an item. When they require a new item, they purchase it personally or through a "purchase card." This request for an item serves as input for the process of asset management.

Inputs

If the item is transferred between departments, the location record of the item is updated on the PeopleSoft system. If the requested item is purchased, it is tagged by UCF property control only if it is a nonconsumable item costing more than $1000. Each item gets a unique bar code with specific data that are uploaded on the PeopleSoft system and used for tracking through every fiscal year.

Process

The process of asset management starts when the item is received by the faculty or staff. The detailed tracking process is explained in the flow chart.

Outputs

When the third scanning pass is completed, item status is declared. Three passes are required to find the items or to verify that the item is lost and cannot be found. An item is declared as found, lost/stolen, surplus, or cannibalized. The record of found items enters the list of assets for the next fiscal year. All other items are managed according to its new status. If an item is declared as surplused or cannibalized, it is disbursed appropriately.

Customers

The customers of this process are UCF Property Control, department property manager, department heads, faculty and staff. The SIPOC is shown in Figure 7.5.

7. Team Member Bios

Felix Martinez is a graduate student in quality engineering at UCF; he obtained his bachelor of science in industrial engineering in Spring 2005. Felix works as a graduate research assistant in the Housing Constructability Laboratory, where he is

Start boundary: Receipt of an item.

End boundary: Declaration of an item status.

Supplier→	**Input→**	**Process→**	**Output→**	**Customer**
• Faculty • Staff	• Tagging • Recording • Transfer of location on PeopleSoft	• Receipt of an item • Yearly scan	• Items found • Lost/stolen items • Surplus items • Cannibalized items	• UCF Property Control • Dept. property manager • Dept. heads • Faculty and staff

Input Indicator: Approval of the request, value of $1000 or more.
Output Indicator: Declaration of an item status.

FIGURE 7.5 SIPOC

leading a project regarding water intrusion in masonry walls. Previous work experience includes a year-long internship with United Parcel Service, where he helped implement new package-tracking systems and conducted time studies on personnel.

Bob Beaver is a graduate of UCF (1977) with a master's degree in engineering. He has 28 years of experience in private sector planning and engineering infrastructure design work around Central and South Florida, including facilities management and project management. He manages Walt Disney World's Civil/Structural Engineering SME group and is working toward his candidacy in the doctoral program at Industrial Engineering and Management Systems program at UCF.

Varshini Gopal is originally from Bangalore, India. She is a full-time student at UCF pursuing a master's degree in engineering management. Gopal attained her bachelor's degree in industrial engineering and management in India. She is working as a graduate research assistant under Dr. Pet-Armacost in the Department of Information, Analysis and Assessment. Gopal did an internship at MICO (member of the Bosch Group) in Bangalore and also worked there as a graduate trainee for one year.

Amol Shah is a graduate student pursuing a master's degree in industrial engineering. He has a bachelor's degree in production engineering from the University of Mumbai, India. Shah has worked as an in-plant trainee in Mahindra & Mahindra Limited (manufacturing company of General Purpose Vehicles) and completed the project titled "Application of 'value analysis' to reduce major rework on vehicles."

Miguel Torrejon was born in Peru. Miguel is a senior, pursuing an industrial engineering major at UCF. He came to the U.S. five years' ago to attend UCF, and is planning to follow his studies with a doctorate in an ergonomics discipline at UCF. Torrejon works in the Housing Constructability Laboratory doing research related to water intrusion in houses in the Central Florida area among other issues that could be solved by applying industrial engineering tools in the construction field. Torrejon has worked in several group projects and always likes to give alternative solution ideas to be analyzed within the group for a better outcome of the project.

Russell D'Angelo is pursuing a bachelor's degree in industrial engineering from UCF. He led an ergonomics process improvement team for Boeing at Cape Canaveral. D'Angelo has also provided team support for a project involving the redesign of the layout for the shipping and receiving department at Lockheed Martin. He has seven years of experience managing and overseeing the complete design and fabrication of the Removable Prosthesis department for the Nelson Dental Laboratory. D'Angelo's future plans are to acquire a master's degree in quality engineering.

8. Define Phase Presentation

The Define phase presentation can be found in the downloadable instructor materials.

DEFINE PHASE CASE DISCUSSION

1. **Define Phase Written Report**
   1.1 How did your team ensure the quality of the written report? How did you assign the work to your team members? Did you face challenges of

team members not completing their assigned tasks in a timely manner and, if so, how did you deal with it?

1.2 Did your team face difficult challenges in the Define phase? How did your team deal with conflict on your team?

1.3 Did your instructor and/or Black Belt or Master Black Belt mentor help your team better learn how to apply the Lean Six Sigma tools, and how?

1.4 Did your Define phase report provide a clear vision of the project, why or why not?

2. Lean Six Sigma Project Charter

Review the project charter presented in the Define phase case study example written report.

2.1 A problem statement should include a view of what is going on in the business, and when it is occurring. The problem statement should provide data to quantify the problem. Does the problem statement in the Define phase case study example written report provide a clear picture of the business problem? Rewrite the problem statement to improve it.

2.2 The goal statement should describe the project team's objective, and be quantifiable, if possible. Rewrite the Define phase case study example's goal statement to improve it.

2.3 Did your project charter's scope differ from the example provided? How did you assess what was a reasonable scope for your project?

3. Stakeholder Analysis

Review the stakeholder analysis in the Define phase case study example.

3.1 Is it necessary to identify the large number of stakeholders as in the example case study?

3.2 Is it helpful to group the stakeholders into primary and secondary stakeholders? Describe the difference between the primary and secondary stakeholder groups.

4. Team Ground Rules and Roles

4.1 Discuss how your team developed your team's ground rules. How did you reach consensus on the team's ground rules?

5. Project Plan and Responsibilities Matrix

5.1 Discuss how your team developed their project plan and how they assigned resources to the tasks. How did the team determine estimated durations for the work activities?

6. SIPOC

6.1 How did your team develop the SIPOC? Was it difficult to start at a high level, or did the team start at a detailed level and move up to a high-level SIPOC?

7. Team Member Bios

7.1 What was the value in developing the bios, and summarizing your unique skills related to the project? Who receives value from this exercise?

8. Define Phase Presentation

8.1 How did your team decide how many slides/pages to include in your presentation?
8.2 How did your team decide upon the level of detail to include in your presentation?

MEASURE PHASE EXERCISES

1. Measure Report
Create a Measure phase report, including your findings, results and conclusions of the Measure phase.

2. Process Maps
Create level-1 and level-2 process maps for the asset management process.

3. Operational Definitions
Develop an operational definition for each of the identified CTS criteria:

- Faculty/Staff awareness of process
- Identification of assets
- Efficiency of yearly scanning
- Characteristics of assets managed (value and number of assets lost)

4. Data Collection Plan
Use the data collection plan format to develop a data collection plan that will collect voice of customer (VOC) and voice of process (VOP) data during the Measure phase.

5. VOC Surveys
Create a VOC survey to address one of the main concerns, which was whether or not the faculty are presently aware of the policies when relocating or discarding state-entrusted assets.

6. Pareto Chart
Create a Pareto chart using the following scanning data:
Number of items scanned in first pass is 1935; number of items scanned in the second pass is 1577; and number of items scanned in the third pass is 647.

7. VOP Matrix
Create a VOP matrix using the VOP matrix template to identify how the CTS, process factors, operational definitions, metrics and targets relate to each other.

8. **Benchmarking**
   Perform a benchmarking of how other organizations manage their assets.

9. **Statistical Analysis**
   Use the asset management data to calculate the average value of an asset.

10. **COPQ**
    Brainstorm potential COPQ for the case study for the following categories: prevention; appraisal; internal failure; external failure.

11. **Measure Phase Presentation**
    Prepare a presentation (PowerPoint) from the case study exercises that provides a short (10–15 minutes) oral presentation of the Measure phase deliverables and findings.

MEASURE PHASE

1. Measure Report

The second phase of our project DMAIC is the Measure phase. In this sector, we evaluate and quantify certain performance levels to help us establish a clear picture of current events. The Measure phase allows us to baseline the current process and system's capabilities, obtain a VOC and identify key metrics.

We applied the Measure phase for the CECS inventory and asset management project, using tools such as process maps and data mining. We were able to scientifically model the current flow of events and get an "holistic" view of the process.

2. Process Maps

It is a statewide policy for universities to maintain control of all nonconsumable items worth more than $1000. CECS has a series of custodians specifically in charge of more than 4000 items spread across the engineering buildings and Research Park. At the beginning of the year, they are given an inventory list of items which must be accounted for by the end of the fiscal year. During this period, a series of "passes" in which people from the UCF property office scan a specific tag placed on all the items that need to be accounted for is completed.

From our kick-off meeting with Jose Murphy, we discovered that they perform three passes. The fiscal year begins on July 1. The first pass is conducted during the first three months, the second pass is conducted during the next three months, and the third pass is done during the last six months. Items not located during the first two passes are searched for in the third pass, which begins around January 1, at the beginning of the following semester. Items not located by the end of the year are reported to the police at the end of the fiscal year. These items may later be recovered or never found again. Items may also reach the end of their useful life, and therefore must be surplused or cannibalized following strict guidelines set forth by the UCF property office.

Even though custodians are responsible for the safekeeping of these items, they are not held accountable for items declared lost at the end of the year. Thus, there is no sense of ownership or responsibility for strict tracking of items. Based on this issue, the UCF property office has declared that it will now charge each department for the value of those items lost by that department at the end of the fiscal year.

Tracking of items is conducted using specific software installed on scanners that are taken across UCF and its satellite campuses. After a scanning session, the data collected is uploaded to the computers and to the PeopleSoft financial software system. The UCF property office does not upload the data from the scanners every day, thus departmental custodians experience a small delay in the retrieval of information. The process map is shown in Figure 7.6.

3. Operational Definitions

It is generally accepted that customers are the most important part of a business. No customers means no business. It is they who define what the quality of the product or service is going to be. CTSs represent the important measurable characteristics of a process whose performance standards must be met to satisfy the customer. It essentially involves getting the VOC.

The input to identifying the CTS characteristics was collected through interviews with our project sponsor (Jose Murphy), project champion (Dr. Debra Reinhart) and the senior property manager (Tereasa Clarkson), as well as through faculty surveys. The following characteristics were identified as being the elements that would significantly affect the output of the process as perceived by the customer. Each of these characteristics is associated with one or more key metrics that can quantify the characteristics by measuring them. There were several CTSs we deemed to be important, but several we could not measure with the current measurement system in place. The following CTSs were important, but not measurable: documented location of assets; undocumented assets; efficiency of list update; sorting efficiency of lists; and loss avoidance.

The following metrics were deemed as important and measurable with the current measurement system: faculty/staff awareness of process; identification of assets; efficiency of yearly scanning; and characteristics of assets managed (value and number of assets lost)

CTS: Faculty/Staff Awareness of the Process

During our interviews with the CECS senior property manager and the UCF Property Office scanners, they complained that their biggest concern was the lack of understanding by faculty and staff on procedures regarding relocation of an item. Appropriate procedures indicate that relocation of items must be reported to the custodian, an activity seldom conducted. We set out to measure faculty levels of awareness by conducting a quick nine-question survey.

The Six Sigma team's approach to conducting a survey was to first take measures to achieve 95% confidence. We searched the databases of the CECS website and determined the population of faculty who would potentially make use of the state entrusted assets to be 163. With an anticipated variance in response not more than 10% and interpolation of the values; the proper sample size for a population of 163

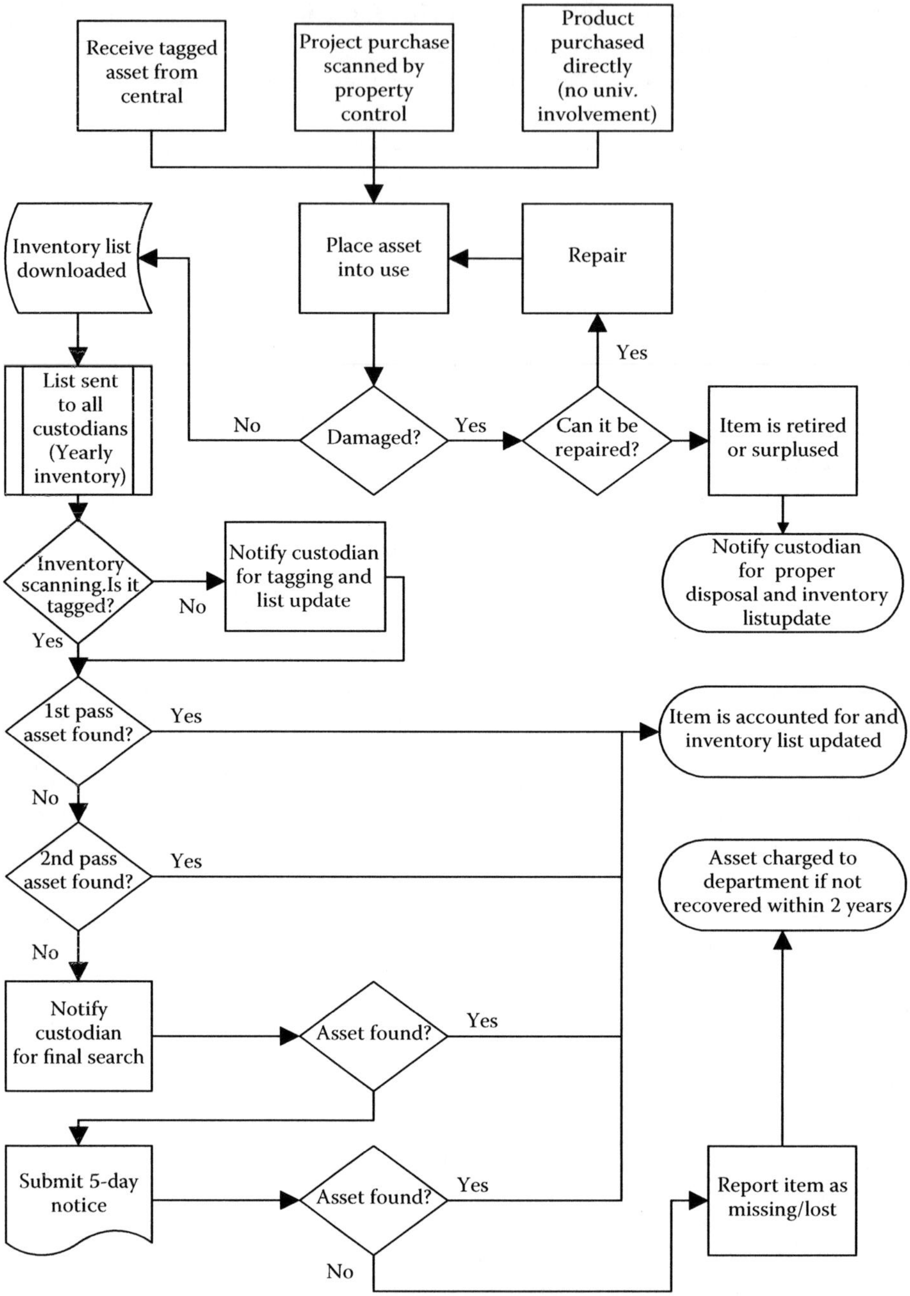

FIGURE 7.6 Process map.

was calculated to be 43. The Six Sigma team was able to survey 44 faculty members which allowed us to maintain a 5% margin of error.

Each of the questions designed had to be answered in one of the following scales:

1. Never (1); Rarely (2); Sometimes (3); Most of the time (4); Always (5)
2. Yes – No – Don't Know

The survey questions were shuffled and arranged in a random order so as not to lead the respondent into a particular response due to pattern recognition. This approach was intended to establish basic awareness of the policies and procedures.

CTS: Identification of Assets

Another key metric is the number of items lacking a proper description. The inventory list of items contains a section labeled "Description." In this section, the purchasing department has the ability to write a small phrase that helps identify the asset. It is very important that the item contains a good description because scanners use this information to look for the items that were missed at the first pass. Looking at the inventory list, we find that many items cannot be identified based on their description because they have an ambiguous description or because they contain information of no use to the scanner.

CTS: Characteristics of Assets Managed (Value and Number of Assets Lost)

Our project contact gave us their most recent inventory list, as well as the list of assets lost since record-keeping began. We used the data to make some calculations that allowed us to establish patterns and make inferences about the assets managed by CECS.

CTS: Efficiency of Yearly Scanning (Rate of Items Scanned Throughout the Year)

Data were collected on the dates assets were scanned by CECS in the prior fiscal year. The UCF property office divides their scanning periods into three phases:

1. First Pass (July 1–September 30)
2. Second Pass (October 1–December 31)
3. Third Pass (January 1–June 30)

Each item was placed into its respective category, and we used these data to create a Pareto chart showing comparative numbers of scanned items for the first, second and third (final) scanning attempts. According to interview information, the final scan attempt (which includes a physical search of items by the property custodians and staff) takes as much time as the first and second scan efforts combined. Therefore, the 16% found on the final scan attempt takes more than five-times the effort (in terms of elapsed time) to locate and scan on a per item basis than the first 84%.

4. Data Collection Plan

The data collection plan is summarized in Figure 7.7. Data were to be collected by interviews with key stakeholders, surveys with faculty and staff, and reviewing the asset management database. A great deal of the VOP was collected through the interviews to understand the current process for asset management and scanning. VOC surveys were developed to understand the faculty and staff awareness of asset management procedures and the process. The asset management database provided a wealth of knowledge to understand how many items exist, how many were lost over the last ten years, and the dollar value of items.

5. VOC Surveys

The voice of customer survey developed to understand the awareness of the faculty and staff with the current asset management process and procedures. Following are the questions on the survey:

Critical to Satisfaction (CTS)	Metric	Data collection mechanism (survey, interview, focus group, etc.)	Analysis mechanism (statistics, statistical tests, etc.)	Sampling plan (sample size, sample frequency)	Sampling instructions (who, where, when, how)
Faculty/staff awareness of procedures	Faculty awareness of asset management procedures	Survey, Interviews	Survey analysis	Population size = 163; 95% confidence; 10% desired precision; 5% margin of error, with sample size of 44	Questions presented in a random order
Nondescriptive items	To be able to identify the item from the description in the system field	Review data in system; interviews	Pareto analysis; data analysis	Current items	Not applicable
Characteristics of assets managed	Total numbers of items; Total number of lost items	Review data in system; interviews	Pareto analysis; data analysis	Current items	Not applicable

FIGURE 7.7 Data collection plan.

Questionnaire for Faculty and Staff: CECS Asset Management Practices

The following questions are intended in the spirit of providing a more efficient and effective property management system so faculty and staff can gain better use of the assets available to them.

1. I use state/federally purchased equipment in my work (including items obtained from grants).
   Y/N (if N, then no need to proceed further).
2. I have had a situation in which an item I needed for class or research was lost and not recovered.
   Never – Rarely – Sometimes – Most of the time – Always.
3. I can easily locate the equipment I need for classes/research.
   Never – Rarely – Sometimes – Most of the time – Always.
4. Existing equipment that I need is where I need it.
   Never – Rarely – Sometimes – Most of the time – Always.
5. I require the services of the property custodian to help find items I cannot locate.
   Never – Rarely – Sometimes – Most of the time – Always.
6. I know what department assets and equipment are available to me.
   Never – Rarely – Sometimes – Most of the time – Always.
7. I am aware of the SUS policy on care and reporting of state- and federally funded assets.
   Yes – No – Don't know.
8. I am aware of the SUS policy on discarding state- and federally funded assets
   Yes – No – Don't know.
9. Availability of assets and equipment affects my ability to conduct classes and research.
   Never – Rarely – Sometimes – Most of the time – Always.

6. Pareto Chart

For the yearly scan, each item was placed into its respective category of when it was found (first, second or third pass). We used these data to create a Pareto chart showing comparative numbers of scanned items for the first, second and third (final) scanning attempts. According to interview information, the final scanning attempt (which includes physical search of items by the property custodians and staff) takes as much time as the first and second-scanning efforts combined. We found that about 46% of the items are scanned during the first pass, about 38% of the items are scanned during the second phase, and about 16% of the items are scanned during the third pass.

Therefore, the 16% found on the final scanning attempt takes more than five times the effort (in terms of elapsed time) to locate and scan on a per item basis than the first 84%. The Pareto chart is shown in Figure 7.8.

We examined the list of assets (current and lost items) for costs and type of assets, as well as assigned departments. Pareto analyses were conducted to determine types of assets involved in loss, loss by department, and item unit costs. The Pareto charts showed that most of the items managed/lost fall in the range of \$1000 to \$3000

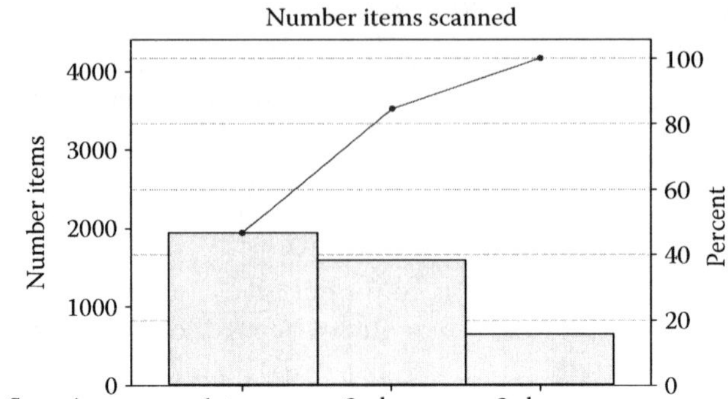

Scanning pass	1st pass	2nd pass	3rd pass
Number items	1935	1577	647
Percent	46.5	37.9	15.6
Cum %	46.5	84.4	100.0

FIGURE 7.8 Pareto chart of items found by scanning period.

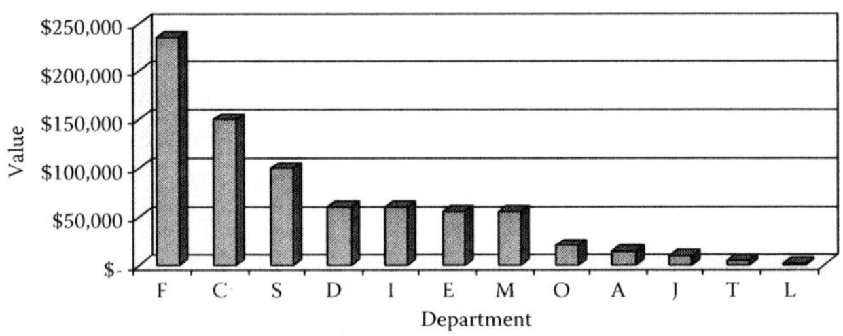

FIGURE 7.9 Pareto chart of lost item value over ten year period by department.

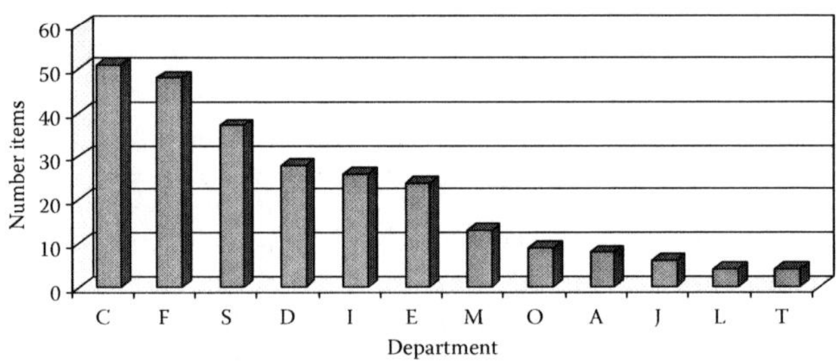

FIGURE 7.10 Pareto chart of number lost items over ten year period.

(which will be discussed further in the summary of the Analyze phase), and the number of items in both lists decreases as the value range increases. The types of items that were lost were mostly computers and/or printers.

Analyses of ten years' worth of data of the lost items showed that, based on assets assigned, Departments F, C, and S have experienced the greatest losses. This may be attributable to the relatively high dollar value of assets assigned to these departments as well as the out-of-doors/field-remote nature of specialized equipment used in Department C. Figure 7.9 shows the relative dollar volume of lost items over a ten-year period by department and by total item value. Figure 7.10 shows the number of items lost over a ten-year period by department. The relative dollar value and the number of items by department are quite similar. Another significant result obtained was from classifying the acquisition value of the items. It was identified that only five items constituted 14% of the value of items lost over this period. This specifies the importance of closely keeping track of expensive items.

7. VOP Matrix

A compilation of issues has been made from the interview results. These interviews were conducted with Dr. Debra Reinhart, Mr. Jose Murphy, Ms. Theresa Clarkson, and Chris Vu. The issues developed are tabulated as follows:

Property Management

- Function of property control is to track and monitor status of assets.
- Perceived most difficult aspect seems to be making faculty aware of policy and procedures.
- Staff seems to have more knowledge of system and policy than faculty.
- Descriptions on purchase orders (POs) can be fixed, but takes a while (post-procurement).
- Most time in the process is taken-up with checking lists and uploading the results.
- Would like to see an education program for faculty to get them familiar with system.
- Only two years of detailed data due to change over to the PeopleSoft system.

Champion

- No one in the dean's office actively tracks assets, departments have custodians and there is a property manager for the entire college.
- Biggest concern from the dean's office is for lost property and implication for back charges to departments on capital losses.
- Also concern over lost time to classes and research if items cannot be located.

Contacts

- Custodians tier up to departments (no direct report to property manager unless problems develop).
- Purchasing is sometimes done directly and some items (cannot quantify) are not tagged.
- Scanners are unfamiliar with and do not know what many items look like. Descriptions from POs are lacking in some cases.
- Lists have to be checked manually to get missing items identified.
- When scanners enter a room, they scan everything, location does not matter. Recognizing items is more important.
- The local department custodian is more likely to be contacted by faculty to find items.
- Recent first-scan results increased in efficiency from 46 to 76%. Scanners are more familiar with objects.

CTS	Process factors	Operational definition	Metric	Target
Faculty/Staff awareness of process	Procedures exist	Procedures exist and are auditable	Number of departments with procedures	100% of departments have procedures by Jan. 1
	Training in procedures	All faculty will take 1 hour training session within 3 months of hire	Number of faculty trained	100% of faculty are trained within 3 months of hire or Jan. 1
Documented location of assets	Procedures exist	Procedures exist and are auditable	Number of departments with procedures	100% of departments have procedures by Jan. 1
	Training in procedures	All faculty will take 1 hour training session within 3 months of hire	Number of faculty trained	100% of faculty are trained within 3 months of hire or Jan. 1
Identify of assets	Description on PO	All purchasers will input detailed description of asset on PO	Number of POs with detailed description	95% of POs sampled have detailed descriptions
	Description in system	PO description will transfer to asset management system	Number of asset descriptions in asset mgt.	95% of POs sampled have detailed descriptions
Efficiency of yearly scanning	Training	All property managers will be trained in process	Number property managers trained	100% of property managers trained within 3 months
	Process	Quality of process	Proportion of items found on first try	95% of items found on first scan

FIGURE 7.11 Voice of process (VOP) matrix.

- The property manager never sees the purchase order. Needs clearance for that.
- If items are disposed of, the property manager rarely sees the request.

The VOP matrix is shown in Figure 7.11.

8. Benchmarking

Benchmarking was used to determine best management practices at other schools. The asset management experiences at two other Florida universities (University of Florida and Florida Atlantic University) reflect similar attitudes by users, but slightly different approaches and levels of maturity in their programs. The benchmarking summary is shown in Figure 7.12.

Based on interviews with representatives of the two universities considered above, the following issues become apparent in the "gap analysis:"

Aspect	UCF	UF	FAU
P.O. process	Grants Projects Yearly capital	Grants Projects Yearly capital	Grants Projects Yearly capital
Scan process	Scan twice first 6 months Final scan after 5 day letter. Latest scan (2005) found 76% in first 3 months. Process explained on website.	Scan once then send letter. First scan typically picks up 70 – 80% of items. Process explained on website.	Scan twice in first 6 months Final scan after letter notification. Recent “crash” project resulted in 70% found in 5 weeks.
Tagging of items	Paper UPC tag	Paper UPC tag Optional tag for “attractive” items	Paper UPC tag Optional tag for “attractive” items
Identification of assets	UPC barcode applied at receiving. Identity of item determined by purchaser.	UPC barcode applied when received. Identity of item determined by purchasing	UPC barcode applied when received. Contact PM for barcode application with help by procurement
Communications issues	50–52% of faculty have little / no knowledge of system	Some faculty not aware or consider it a priority. Lack of due care is considered a “significant” problem.	Estimate that 50% of faculty/staff either do not know or “care” about system
Untagged items	Sometimes found then call is placed to have them tagged	Sometimes found then call is placed to have them tagged	Sometimes controller “finds” them in PO’s and lets PM know of item that will need tagging.
Disposition of “Old” items	No record of what happens to old items, except for missing report.	Claims good recovery rate so that items can be recycled into other programs.	Estimate 50% of obsolete or old items are disposed not reported. High field items loss rate especially Ocean Engineering.
Use of supplementary tag	No system currently in use	Optional – some use it. More of a deterrent as there is no monitoring. Departments are directly responsible for attractive items and must do internal inventory.	Optional – Many use it. Opinion of PM is that it is an “effective” deterrent.

FIGURE 7.12 Benchmarking summary.

Replacement of lost items	Proposed value system in development for charging departments for items missing more than 2 years.	Replacement value would be used to back charge departments for lost items but not considered "enforceable."	No system under proposal currently.
Staff and reporting	Local department custodians are appointed but no direct reporting relationship to Property managers.	Local as well as central property managers. Departments held responsible to account for items.	Few or no local custodians to assist.
System database	PeopleSoft	PeopleSoft	Banner
Availability of forms	Part of a larger financial website. Property Control has exposure at University–wide level. Website recently redesigned.	Extensive website with transfer, property transfer, and off–site transport permission forms, along with description of procedures.	Basic information website and forms available on line for property transfer, removal and surplus.

FIGURE 7.12 (Continued)

- The scanning process at UCF could be "shortened" by reducing the number of "forgiveness" visits to one. One of the universities benchmarked requires the departments/colleges to contact them for scanning once they are "ready" and give notice to staff and faculty of upcoming inventory. The effectiveness of the first scan at UCF CECS last year was 46%. However, this year, it increased to 76%.
- No system takes full advantage of enabling control of item descriptions during the purchase process. This is apparently a common problem.
- No system employs remote sensing tracking technology of tagged items. One school interviewed had examined the use of RFID technology, but opted not to pursue due to the expected costs of implementation.
- The problem most often mentioned from a "people" perspective was lack of awareness of the property management requirements and lack of communication to property managers of items missing or disposed of. A common feeling expressed among property managers is that others in the university community "do not care."
- Optional tagging of "attractive" items (smaller items usually worth less than $1000 in value) is claimed to be effective by universities using the system, which employs highly visible tags on items that are not allowed to leave the premises.
- Time to download and convey the scanned list takes only a day according to the UF manager. This is shorter than the UCF system, in which download is done on Tuesdays and Fridays. Therefore, timing of receipt of scanned list could take two to three days to get an updated list.
- One university reports value in having the controller's office examine purchase orders and notifying property manager of purchases that qualify as capital expenditures.

Based on the above, the UCF system may wish to consider the following actions:

- A program to educate staff and faculty of the needs and reasons for good property stewardship. Lost property results in having to repurchase or opportunity loss in not having the asset available when needed.
- The inventory process should be examined to determine if it can be made more efficient. As a part of this, better identification of items for scanning should be considered. Revising and standardizing the purchase descriptions and training to watch for capital purchases so that the appropriate managers can be notified, should be considered.
- The software should be examined to determine if updated lists can be developed as soon as data is uploaded from the scanners.

9. Statistical Analysis

Current Inventory List

The team obtained the most recent inventory list of items for the CECS. The purpose of this list is to give an idea of the most current number of items that should be in the

CECS and use them as guidance for the scanning process. The most recent inventory list shows 4865 items. It was also noted that the increment of items from the prior year was 691 items.

This inventory list enabled the Lean Six Sigma team to conduct a statistical analysis that portrayed patterns about the assets managed by CECS. Careful analysis of the inventory list revealed that a majority of the assets fell in the range of $1000 to $3000 due to the large number of computers around the college. The quantities of items within other groups decreased as their value increased.

The inventory list contains a description of the item in which the purchase department documents a small phrase that identifies each item. It is extremely crucial that each item has a good depiction because scanners use this information as guidance to look for items. It would be safe to assume that a contributing factor in the scanners' inability to locate assets is partially due to the poor level of descriptions. When analyzing the most updated inventory list, the Lean Six Sigma team noticed that of the 4865 items, 134 were unidentifiable from the current descriptions. In other words, 2.75% of the total list could not be identified.

10. COPQ

An estimate is made below of the costs associated with errors and losses that can be attributed to "poor" quality. It is these costs that we will address through our improvement plan. In our case we will assume a "zero" cost for prevention, given the present state and methods of maintaining the existing system, such as the website maintenance for finance and accounting, and administrative costs of normal operations.

Prevention

The prevention costs can be summarized as costs spent to prevent losses by employing education, training, or processes set-up to avoid losses or inefficiencies. In our case, we shall assume a zero cost for these costs.

Appraisal

Appraisal costs are those costs associated with the first scan phase. This activity is intended to verify presence of items. Under ideal conditions, all items would be located and scanned in the first-phase scan. Given costs of scanning and a three-month window, we can roughly estimate these costs as: $19,584 (cost of scanning over entire university based on four part-time and two full-time personnel over a 12-week period) and CECS represents 14.1% of the total items. If we assume that we can prorate the scanning costs to number of items, then prorated cost of scan = $19584 × 0.141 = $2761. This is a cost that cannot be avoided, but may be further reduced if scanning efficiency is increased.

Failure

These costs are made up of *internal* and *external* components. Internal costs can be summarized by the second-scan phases because they represent failure to find objects and scan them in the first phase. External costs can be summarized only as the potential cost of lost or missing items. In our case, we can only use acquisition costs of those items because we do not yet have a replacement cost formulae. Internal costs = \$2761 (second scan); external costs = \$ 66,000/year (this is calculated using the lost/missing items list over a ten-year period, so dividing by 10).

Therefore, the total cost of quality = \$ 68,761/year.

11. Measure Phase Presentation

The Measure phase presentation can be found in the downloadable instructor materials.

MEASURE PHASE CASE DISCUSSION

1. **Measure Report**
    1.1 Review Measure report and brainstorm some areas for improving the report.
    1.2 How did your team ensure the quality of the written report? How did you assign the work to your team members? Did you face any challenges of team members not completing their assigned tasks in a timely manner, and how did you deal with it?
    1.3 Did your team face difficult challenges in the Measure phase? How did your team deal with conflict on your team?
    1.4 Did your instructor and/or Black Belt or Master Black Belt mentor help your team better learn how to apply the Lean Six Sigma tools in the Measure phase, and how?
    1.5 Did your Measure phase report provide a clear understanding of the VOC and the VOP, why or why not?

2. **Process Maps**
    2.1 While developing the process maps, how did your team decide how much detail to provide on the level-2 process maps?
    2.2 Was it difficult to develop a level-2 from the level-1 process maps? What were the challenges?

3. **Operational Definitions**
    3.1 Review the operational definitions from the Measure phase report, define an operational definition that provides a better metric for assessing the faculty awareness of the asset management process and procedures.
    3.2 Discuss why it may be important for the faculty students and staff to be familiar with the asset management process and procedures.

4. **Data Collection Plan**
    4.1 Incorporate the enhanced operational definition developed in number 3 above into the data collection plan from the Measure phase report.

5. VOC Surveys

5.1 How did your team develop the questions for the VOC survey? Did you review them with other students to assess whether the questions met your needs?

5.2 Create an affinity diagram for the main categories of the VOC survey, grouping the questions into the higher level "affinities." Was this an easier way to approach and organize the questions of the surveys?

6. Pareto Chart

6.1 Discuss how the Pareto chart provides an assessment of the asset management data.

7. VOP Matrix

7.1 How does the VOP matrix help to tie the CTSs, the operational definitions and the metrics together?

8. Benchmarking

8.1 Was it difficult to find benchmarking information specific to asset management processes.

9. Statistical Analysis

9.1 What additional statistical analysis could be performed on the asset data.

10. COPQ

10.1 Would it be easy to quantify, and collect data on the costs of quality that you identified for the case study exercise?

11. Measure Phase Presentation

11.1 How did your team decide how many slides/pages to include in your presentation?

11.2 How did your team decide upon the level of detail to include in your presentation?

ANALYZE PHASE EXERCISES

1. Analyze Report

Create an Analyze phase report, including your findings, results and conclusions of the Analyze phase.

2. **Cause and Effect Diagram**
   Create cause and effect diagrams for the following effects:

   - Why are assets lost?
   - Why are assets not found on the first pass?

3. **Why-Why Diagram**
   Create a Why-Why diagram for why assets are lost.

4. **Process Analysis**
   Prepare a process analysis for the asset management process.

5. **Histogram, Graphical, and Data Analysis**
   Perform a histogram and graphical analysis to categorize the asset items into three categories by dollar values: $3,000 and above, $1,000 to $2,000, and $2,000 to $3,000 in the Asset Management Data.xls file.

6. **5S Analysis**
   Perform a 5S Analysis for the asset management process.

7. **Survey analysis**
   Perform survey analysis using Pareto analysis and chi-square analysis for each of the questions for the VOC survey. The data can be found in the "Asset Management VOC Survey.xls" file.

8. **FMEA**
   Perform an FMEA for the asset management process, using the process map from the Measure phase.

9. **DPPM/DPMO**
   Calculate the DPMO and related sigma level for the asset management process, assuming a 1.5 sigma shift, for the following data:
   Opportunities for failure
      First pass items not found; second pass items not found; and third pass items not found (one opportunity for each pass).
   Defects
      Defects in first pass are 12; defects in second pass are 10; defects in third pass are 4.
   Units scanned
      The total number of units scanned in the first pass is 1,935, the second pass is 1,577 and the third pass is 647.

10. **Analyze Phase Presentation**
    Prepare a presentation (PowerPoint) from the case study exercises that provides a short (10–15 minutes) oral presentation of the Analyze phase deliverables and findings.

ANALYZE PHASE

1. Analyze Report

The Analyze phase is the mid-point in the DMAIC cycle. It is a critical part of the process because, based on the conclusions from the Analyze phase, we furnish the best possible deliverables to our customer in the Improve & Control phases. We start the Analyze phase with an assessment of the root causes that contribute to problems in the asset management process. We perform system-wide analysis of the data gathered by applying various tools and techniques. The process analysis allows us to look at every process step to identify potential defects and the affinity diagram provides a linkage between the common issues that came up through the interviews, surveys and benchmarking processes.

Next, we commence the problem specific analysis through hypothesis testing to infer results from a certain proposed hypothesis, study of items moved by comparing the records for the prior fiscal year with that of the current year, and ABC inventory analysis which helps prioritize the scanning according to the value of the items. We give a brief synopsis of the interviews and surveys as well as results that were obtained by studying the current inventory list and making conclusions about the summary of lost/missing items. We conclude this section by providing a summary of the problems we identified.

The Analyze phase helps establish a core set of principles and undeniable facts about the system that assist in making a smooth transition into the Improve phase. For this we use tools such as the 5S, failures modes and effect analysis and the Lean waste approach. All of these tools help provide recommendations on how the current process could be made more efficient.

2. Cause and Effect Diagram

We use the cause and effect diagram to cite the obvious causes that are leading to an inefficient asset management process. The potential drivers we identified are people, method, material and information. The cause and effect diagram is shown in Figure 7.13. A discussion of the potential costs follows.

People

Relocation of an item: The policy states that the user is supposed to complete a form to inform the custodian while moving an item from one place to another. However, the item is moved without informing the custodian and the record of an item shows the old location while scanning.

Item stolen: There is a time delay from when an item is considered stolen to the time that it is reported to the police.

Improper disposal of an item: Some of the items are cannibalized without completing the respective form. So the record on the PeopleSoft system shows an item located at its old location. However, when the scanners scan the item, it is not found at its place and it leads them to search for an item at different places.

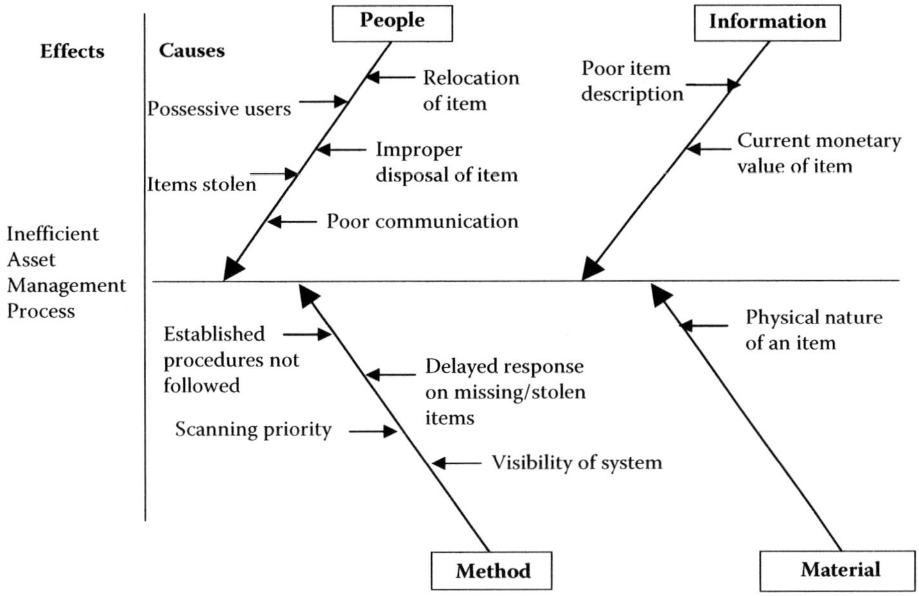

FIGURE 7.13 Cause and effect diagram.

Poor communication: The faculty and staff using the item are not informed about the policies. For example, some of the faculty or staff did not know they are supposed to fill out the respective forms and inform the custodian before moving or cannibalizing an item.

Possessive users: Once the users obtain the item, they tend to be vigilant by safeguarding their items. This makes it difficult for scanners to obtain that item while scanning.

Method

Visibility of system: There are forms to be filled-out before moving or cannibalizing an item, but most of them are not aware of the procedure that needs to be followed when doing this. Hence, users sometimes do not spend their time looking for the form on the website before moving or cannibalizing the item.

Scanning priority: When the item is not found in the first pass, scanners continue searching for an item in the second pass, but no special consideration is given to the items considering the dollar value of an item. Scanners select items randomly and start searching for them. Thus, they sometimes spend a significant amount of time searching for items that have low monetary values and less time searching for items that are more expensive. This increases the cost of lost items.

Delayed response on missing/stolen items: Some of the users do not inform the custodians if the items are missing or stolen. When these items are not found until the second pass of scanning, the custodian contacts the user and at that time they inform the custodian the item is lost. This causes a considerable loss of time spent in looking for them in the second pass of scanning.

Established procedures not followed: There are standardized procedures which are to be followed while moving or cannibalizing an item or when the item is lost or stolen. However, they are not followed by some of the users.

Material

Physical nature of an item: There are tags on items which are used to track them throughout the fiscal year. The physical nature of some items is such that tags cannot be attached to them. Therefore the card is located in a different place than that of the item. This can sometimes be misleading because it does not give definitive information about the actual location of the item.

Information

Poor item description: The description provided to scanners while scanning includes acquisition date and cost, location and barcode. However, they are not given detailed information about items which would make it easy to find them. Sometimes scanners have the title and barcode of an item, but they do not know what it looks like while looking for them. It makes the searching process more difficult. For example, when scanners are looking for laptops, giving the information about the manufacturing company or a model of the laptop would make the searching process much easier.

Current monetary value of an item: Starting this fiscal year, the UCF property office will charge each department for the value of the items lost during the respective fiscal year. However, there is no consensus on whether they will charge their acquisition value or the current depreciated value.

3. Why-Why Diagram

We introduced the cause and effect diagram in the previous section. Further brainstorming of this diagram allowed us to conclude that one of the most significant causes leading to the inefficient asset management process is items are not found at the correct location while scanning. We used the Why-Why analysis to find the probable root causes of not finding the items at the right location while scanning. The Why-Why diagram is shown in Figure 7.14.

We found five root causes for not finding the items at the correct location:

- Poor communication
- Poor visibility of the system
- Items stolen
- Description procedures are not standardized
- Custodian does not have authority to update the description of an item

These are the most basic reasons a problem has or could occur. In the improvement phase, we will prepare the action plan so that the probable root causes will be eliminated or corrected and it would prevent the problem from existing or reduce significantly the occurrences of the problem.

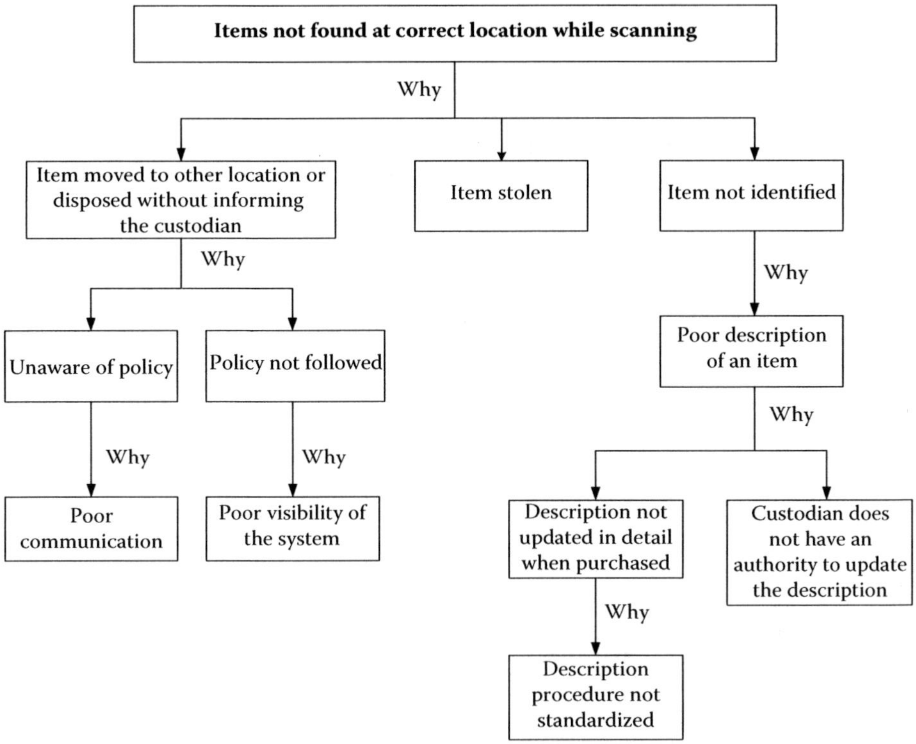

FIGURE 7.14 Why-Why diagram.

4. Process Analysis

Issues are developed from the process map from the Measure phase by breaking down and analyzing each step. We examine two "subprocesses" that involve the item or asset itself, and addresses the information about that item or asset. Issues that result from this examination are:

- Visibility of PO to concerned property manager
- How is an item identified (or made identifiable) to those not familiar with the object?
- Decision to locate an item in a certain location or room is a group, leadership, or individual decision?
- Is location "optimal" for its use? How easy is it to move or relocate?
- Who checks or how is item checked for repair needed?
- Notification to property manager of item to be disposed or surplused.
- Identity of item on list for scanning
- Does route taken by scanners minimize double-back? No defined route taken to minimize walk time
- Scanners sometimes held up by classes in progress
- List comparison and missing list generation methodology
- No theft prevention methodology (visible stickers or continuous monitoring)

We created an affinity diagram relating issues found in the process map, interviews, surveys and benchmarking. If we examine the "common threads" amongst the issues in the previous matrix, we find the following:

- Scanners and property custodians have problems identifying items, but faculty, for the most part, know where to find the items needed.
- Description of the item is an issue for property managers in being able to find it.
- Location of the item is not as much an issue as first thought. If faculty knows where it is and scanners can identify it and scan it, it has a location.
- Decisions to retire or surplus an asset are made without knowledge of the property manager.
- Missing items rarely get reported to the property manager.
- There may be a lack of involvement by faculty in the scanning process (making items kept under lock and key available for scanning).
- Is the scanning process itself susceptible to repetition, added walk time or inefficient routing?
- Time to download updated lists appears to add to feedback time.
- Lack of involvement by faculty and staff may contribute to lack of communication and reporting difficulties.

The process analysis is shown in Figure 7.15. As a result of the comparison of issues arising from the information gathered from the measurement phase, we can ensure alignment to the issues and the CTS characteristics (Figure 7.16).

Upon review of issues developed in the analysis of the information collected, it appears all the CTS elements previously developed are well supported, except perhaps for "Documented location of assets." Though it is important to know where the asset location of the asset may be, faculty do not seem to think this is a very important issue and regardless of location of the asset, wherever it is found, it will be scanned and the new location documented. Regardless, we will elect to keep these CTS in the matrix because it may be important to the efficiency of scanning efforts. This was demonstrated by this year's first scan, which displayed an increase in efficiency as a possible result of increasing familiarity with item location on the part of the scanners. It is important to distinguish this characteristic from "Identification of Assets," which concerns itself with the ability to recognize the item and describe it to someone. Also, this is differentiated from "Undocumented assets," which concerns itself with the procurement stage. Though it shows-up only once in the CTS', it represents a key part of *prevention* costs of quality.

The CTS known as "Sorting efficiency of lists" is repetitious with respect to "Efficiency of list update" and can be eliminated from further consideration.

The foregoing analysis is meant to associate issues identified in each of the measurements conducted to the items considered critical to satisfaction. However, some CTSs (documented location of assets, undocumented assets, efficiency of list update,

Activity in the "value stream"	Tasks involved	Issues/Potential defect "producer"
Obtain asset	Purchase through project Obtain through grant Obtain by yearly capital	P.O. not "visible" to property manager
Assign location	Pick up from receiving and tag/take to room Direct delivery and take to room	Not clear who decision maker is for where item goes
Use of item	Item transported Taken off-campus or at remote campus	Item location may change. How much mobility is required?
Damage/obsolescence	Must be checked for wear. Examine for repair or recovery.	Who checks or calls for repair? If disposed, is Property Control called?
Retire due to obsolescence.	Decision to retire	Who makes decision to surplus item? Is it reported to property manager?
Inventory	Find item. Scan item.	Is it in room? What if in wrong room? Is it identifiable (in the open)?
Download inventory list to property manager.	List compiled and sent to college PM	How is list sent? (physical or email)
Scanning	Walk to room – scan doorway Identify and scan objects	Are items identifiable as assets? What if in wrong room? What if no UPC label? Does route taken throughout engineering complex minimize "double-back"?
Stop and generate new list.	Stop scanning/download to property management.	Time to download, review and send new list?
Second-pass scanning	List checked against newly compiled list and differences noted. New list compiled and sent to property manager. Scanners sent when available to scan and collect data.	What areas are visited? Time to develop list? Scanners repeat same route as before?
Generate final list five day letter.	Second scan data compiled and "missing" list is modified. List is sent to college with letter requiring action in five days.	How is final list developed (by spreadsheet sorting)? Time spent in download and sending list out to departments.
Final search by custodian/property manager.	Search conducted in response to letter. Items found must be scanned.	One scan effort or several visits in response to found items?

FIGURE 7.15 Process analysis.

CTS	Associated issue(s)
Faculty/Staff awareness of process	4, 5, 7, 8
Documented location of assets	2, 3
Identification of assets	2
Efficiency of yearly scanning	5
Value of assets lost	4, 7
Number of assets lost	4, 7, 8
Undocumented assets	1
Efficiency of list update	5, 6
Sorting efficiency of lists	6
Loss avoidance	7, 8, 9

FIGURE 7.16 CTS and issue alignment.

sorting efficiency, and loss avoidance) could not be specifically quantified. In these cases, we performed interviews and conducted questionnaires to compare answers and draw conclusions based on perceptions of the stakeholders.

5. Histogram, Graphical, and Data Analysis

Using ABC inventory data analysis, we concluded that security priorities can be targeted to certain items that have a higher unit value (based on acquisition value). Figure 7.17 summarizes the data of that analysis. What this table shows is that any technology or tagging security system to be considered may be targeted toward the "A" items, which comprise 27.1% of the items, but over 68% of the value.

6. 5S Analysis

Through the 5S' analysis we begin identification of specific improvements and prioritization.

The 5S analysis stresses the need for an ABC inventory method to place emphasis on items worth substantially more than the typical 2–3-year-old laptop. It also flags the need for faculty and staff education in the needs and importance of the asset management system (from a regulatory as well as more practical standpoint), creation of a centralized and easily accessible information system on how to use the various forms and procedures, and better item description. The 5S matrix is shown in Figure 7.18.

7. Survey Analysis

The VOC captured through the surveys told us essentially that from the perspective of the faculty there are no problems locating state-entrusted assets. Questions 2, 3, and 4 were created with the purpose of obtaining consistency in response as to whether or not the availability of assets creates a problem. From the survey we saw that 80% of the faculty responded they never or rarely had a situation in which an

Classification	Value range of items ($)	Number of items	Percentage of number of items	Cumulative percentage of number of items	Value of items ($)	Percentage of total value of items	Cumulative percentage of value of items
A	3,000 and above	1,191	27	27	11,853,570	68	68
B	1,000–2,000	2,227	51	78	3,193,012	18	86
C	2,000–3,000	970	22.10	100	2,328,056	13	100
	TOTAL	3,197			17,374,639		

FIGURE 7.17 Data analysis: ABC inventory analysis.

5S	ISSUE	RECOMMENDATION
SORT	High-value items are being lost, incurring a considerable cost to the university	Establish an ABC inventory method. Place emphasis on items worth a substantial amount.
SYSTEMATIZE	In the process of scanning, items may be unavailable because they are in a locked cabinet or off–campus (faculty may have taken it home, etc.)	Establish a system in which faculty is warned beforehand on the scanning visit. Order faculty to bring all items and unlock all cabinets on that day.
SWEEP AND CLEAN	Information regarding property office and all necessary forms are dispersed across different websites	Create a centralized center of information for all faculty/staff.
STANDARDIZE	Information provided on the description section for each item can vary drastically from one item to another, even when they are actually very similar. Items with poor descriptions are difficult to find.	Set a standard for the information provided in the description. Brand name, color, use, and size are very helpful characteristics when searching for a difficult item.
SELF DISCIPLINE	Professors are not following the procedures set forth by the property office and the state.	Reeducate the professors, discuss the issues and create a culture of concern toward state property.

FIGURE 7.18 5S analysis.

item they needed for class was lost and not recovered. Another 82% said that needed equipment is in fact where they need it to be. Furthermore, 89% said they always or most of the time can easily locate needed assets.

From this information it is easy to infer they know where the assets are because presumably they know where they moved them. Therefore, the faculty does not see it to be problematic when they move an asset to another room. Furthermore, the survey revealed that 47.73% of the faculty are not aware or have no knowledge of the policies concerning the care and reporting of the relocation of state-entrusted assets.

However, the faculty is unaware of the implications that relocating an asset has on inventory management. The faculty does not notify property management or the custodian of their department upon relocating an asset due to this lack of awareness. They have no idea they are contributing to a problem. If they were aware of the problem they are creating and the remedy according to proper procedures is to simply notify the custodian of their department upon relocation of an asset; they would probably more than likely do so. An easy remedy would be a memorandum from the dean educating the faculty and making them aware of the "Request for transfer and receipt of state-owned property" form that is available on the finance and accounting section of the UCF website. However, searching the finance and accounting website seems a bit counterintuitive for a person looking for a form to notify property management of an asset being relocated. A better approach would be to construct a property and asset management website directly linked from the CECS website.

We used Pareto and chi-square analysis to analyze each question of the VOC survey as follows.

1. I use state/federally purchased equipment in my work (including items obtained from grants).
   Y/N (if N, then no need to proceed further).
2. I have had a situation in which an item I needed for class or research was lost and not recovered.
   Most of the people surveyed (43.18%) have never had an asset lost. Although the percentage of faculty who responded that they often cannot find what they need is relatively low at 6.81%, it is worth noting that the cumulative amount of the faculty who has had some problem locating an item needed for class or research is significantly high at 56.82%. More than half of the faculty being hindered due to the inability to locate an asset is something that needs to be improved upon. The chi-square p-value was 0.001, which supports a significant difference in ratings.
3. I can easily locate the equipment I need for classes/research.
   The surveyed sample has relative ease in locating the equipment needed for classes or research. As seen on the graph, the frequency of locating the items ranges from "Sometimes" to "Always." The majority of the samples responded "Most times" with 71% occurrence. The chi-square p-value is 0, so the ratings are significantly different.
4. Existing equipment that I need is where I need it.
   This histogram shows that only 18.18% of the faculty sometimes experience equipment not being where it should be. The other 82.82% responded that equipment is usually where it should be. The chi-square p-value is 0, supporting the difference in ratings.
5. I require the services of the property custodian to help find items I cannot locate.
   The responses to question 5 show that the faculty rarely invokes the help of the custodians in locating assets. In fact, 38.63% responded that they never contact the property custodian to help them find the missing items. This does not necessarily mean they are capable of finding the asset on their own, but simply they do not request the assistance of the department custodian.
   The chi-square p-value is 0.001, supporting the difference in ratings.
6. I know what department assets and equipment are available to me.
   Faculty has different levels of awareness of assets available to them. Only 9.09% of the surveyed faculty seems to be unaware of all available assets. This means that 90.91% of the faculty surveyed is knowledgeable of the items they are accessible to. The chi-square p-value is 0.001, supporting the difference in ratings.
7. I am aware of the SUS policy on care and reporting of state- and federally funded assets.
   More than half (52%) of the population surveyed is aware of the SUS policy on care and reporting of state and federal funded assets. More important is the fact that 47.73% are not aware or have little knowledge of the policies concerning the care and reporting of state-entrusted assets. The chi-square p-value is 0.012, supporting the difference in ratings.

8. I am aware of the SUS policy on discarding state- and federally funded assets.

   Half (50%) of the population surveyed is aware of the SUS policy on discarding state- and federally funded assets. The remainder of the survey sample is either not aware, or has partial knowledge of the policy. The chi-square p-value is 0.06, which is not a significant difference in ratings.
9. Availability of assets and equipment affects my ability to conduct classes and research.

   Most of the population surveyed has a diverse response regarding the availability of assets affecting their ability to conduct classes and research. An overall average of 22.72% responded affirmatively to this question, meaning that availability of assets is a key component for teaching and researching purposes. In lieu of these findings, it would seem imperative that assets be managed properly to ensure fluent teaching and research within CECS.

   The chi-square p-value is 0.141, which is not a significant difference in ratings.

8. FMEA Analysis

An FMEA analysis is conducted for our project to recognize and evaluate the failures of the process and identify actions that could eliminate or reduce the chance of the potential failure. We performed a process FMEA because one of our primary business objectives was to make the process more efficient. The detailed FMEA is shown in Figure 7.19.

9. DPPM/DPMO

The DPMO and related sigma level for the asset management process, assuming a 1.5 sigma shift, for the following data is 2083 DPMO, equating to a 4.3 sigma level, assuming: opportunities for failure are three (first-, second- and third-pass items not found); defects in first pass are 12; defects in second pass are 10; defects in third pass are 4. The total units scanned in the first pass are 1935, in the second pass are 1577, and in the third pass are 647.

10. Analyze Phase Presentation

The Analyze phase presentation can be found in the downloadable instructor materials.

ANALYZE PHASE CASE DISCUSSION

1. **Analyze Report**
   1.1 Review the Analyze report and brainstorm some areas for improving the report.
   1.2 How did your team ensure the quality of the written report? How did you assign the work to your team members? Did you face any challenges of team members not completing their assigned tasks in a timely manner, and how did you deal with it?

Process step	Potential failure mode	Potential effect(s) of failure	Sev	Potential cause(s) of failure	Occ	Det	RPN
Receive tagged asset from office	Tag not readable	Scan will not work	4	Barcode not good quality	2	2	16
	Multiple tags on item	Inventory wont match up	1	Operator error	0	1	0
	Damaged tag	Scan wont read tag	3	Items rubbed against another item	2	4	8
				Poor quality tag			
Place asset into use	Item could be placed in wrong location	Inventory match up difficult	4	Improper communication	5	4	80
	Item stolen	Item will not be found leading to detailed investigation	10	Not good security for rooms with items	5	9	450
	Item damaged	Cannot be put into use	6	Careless handling by users	3	2	36
	Item not put into use	Item not available for use	4	Item misplaced or custodian did not know where to place it	2	1	8
				Item was ordered because it was thought to have been lost			
	Item returned to vendor	Loss of inventory	3	Item arrived damaged or wrong item arrived	2	1	6
				Did not order this item			
	Poor description of items	Time spent by scanners in looking for items they are not sure of	10	Lack of standardized procedures when providing description of an item	10	9	900

FIGURE 7.19 Failure mode and effect analysis.

Download inventory list	Additional items on list not accepted by Dept	Mismatch inventory leading to detailed investigation	4	List not maintained and updated correctly	2	5	40
	Wrong list downloaded	Items wont match list hence items will not be found	3	Operator error, software problems	1	8	24
	List not up-to-date	Delays in scanning	6	Update interval not followed	7	4	168
	Item value less than 1000	More time spent in inventory on items less than set limit	1	Procedure not followed on what items get on the list	1	1	1
Send list to custodians	Wrong list downloaded	Items on list will not be locatable	5	Operator error	2	8	80
	No list obtained	Delay in inventory sweep	3	Operator error	1	1	3
	Partial list printed	New additional items will be noted	2	Database could have given only partial list or operator error	2	7	28
	Updated list not obtained	Old list will be used and items mismatch will occur	6	Operator did not have updated list or database not updated with latest list	5	8	240
Scan items	Scanner not working properly	Inventory delay and no data captured	4	Malfunctioning equipment	2	1	8
	Tags not readable by scanner	List will not get updated	2	Illegible tag or damaged tag	2	1	4
	Scanned wrong barcodes	Item will not show in list and will be noted as lost	2	Many similar looking tags on part, operator not aware of serial number tag or format	1	6	12
	Missed items from scanning	Time spent in second and third passes scanning them	9	Operator not systematic and carefull, no sequence followed	8	10	720

FIGURE 7.19 (Continued)

Process step	Potential failure mode	Potential effect(s) of failure	Sev	Potential cause(s) of failure	Occ	Det	RPN
	Inability to scan items either because they are locked away or inaccessible	Delays in scanning those items	10	Faculty members keep items locked away in cabinets or take it home	10	9	900
	Items moved between departments or locations without approval	Time and effort wasted in looking for those items	9	Faculty members not aware of process and procedure that needs to be followed	9	10	810
	Items surplused or cannabalized without informing	Time spent in looking for such items	9	Lack of awareness among faculty members of the process	9	10	810
		Financial loss	7		6	3	126
Update database	Incorrect data uploaded to database	List will not be accurate	6	Operator error, problems with software (PeopleSoft)	3	7	126
	New data not uploaded to database	List will not be accurate	5	Operator error, problems with software (PeopleSoft)	5	4	100
Report lost items	Items not reported at all	Process not followed and police will not have report on lost items	8	Lack of responsibility among reporting authorities	1	2	16
	Item list not reported consistently every year	Process breakdown, police not aware of lost items	5	Lack of responsibility among reporting authorities	1	1	5
Charge Dept. after 2 years	Department not charged for lost items	Property department eats loss	8	Improper procedures	2	1	16
	Department incorrectly charged too much	Loss of department funds for items not lost	7	Poor finance and accounting methods	4	4	112
	Department incorrectly charged too little	Property department eats loss	7	Poor finance and accounting methods	2	4	56

FIGURE 7.19 (Continued)

1.6 Did your team face difficult challenges in the Analyze phase? How did your team deal with conflict on your team?
1.7 Did your instructor and/or Black Belt or Master Black Belt mentor help your team better learn how to apply the Lean Six Sigma tools in the Analyze phase, and how?
1.8 Did your Analyze phase report provide a clear understanding of the root causes of the asset management process, why or why not?

2. Cause and Effect Diagram

2.1 How did your team determine the root causes, and how did they validate the root causes?

3. Why-Why Diagram

3.1 Was it easier to create the cause and effect diagram, or the Why-Why diagram? Which of the tools was more valuable getting to the root causes?

4. Process Analysis

4.1 Discuss how your team defined whether the activities were value-added or nonvalue-added? Was the percentage of value-added activities or value-added-time what you would expect for this type of process and why?

5. Histogram, Graphical, and Data Analysis

5.1 What other type of data or graphical analysis could you perform with the data that you have?
5.2 What other data could you suggest collecting to perform additional histogram or data analysis?

6. 5S Analysis

6.1 How did the 5S analysis help you to streamline and standardize the future asset management process?

7. Survey Analysis

7.1 What were the significant findings in the VOC survey?
7.2 Did your survey assess your CTS criteria for the asset management process?

8. FMEA

8.1 What types of potential failures did you identify in your FMEA?
8.2 How did you identify mitigation techniques to detect and avoid these failures?

9. DPPM/DPMO

9.1 What is your DPPM/DPMO and sigma level. Is there room for improvement, and how did you determine that there is room for improvement?

10. Analyze Phase Presentation

10.1 How did your team decide how many slides/pages to include in your presentation?

10.2 How did your team decide upon the level of detail to include in your presentation?

IMPROVE PHASE EXERCISES

1. Improve Report

Create an Improve phase report, including your findings, results and conclusions of the Improve phase.

2. Recommendations for Improvement

Brainstorm the recommendations for improvement.

3. QFD

Create a QFD to map the improvement recommendations to the CTS characteristics.

4. Action Plan

Create an action plan for demonstrating how you would implement the improvement recommendations.

5. Future State Process Map

Create a future state process map for the asset management process.

6. Revised VOP Matrix

Revise your VOP matrix from the Measure phase with updated targets.

7. Cost/Benefit Analysis

Perform a cost/benefit analysis for implementing RFID to track assets.

8. Training Plans, Procedures

Create a training plan, and a detailed procedure for the asset management process.

9. **Improve Phase Presentation**
   Prepare a presentation (PowerPoint) from the case study exercises that provides a short (10–15 minutes) oral presentation of the Improve phase deliverables and findings.

IMPROVE PHASE

1. Improve Report

Issues have now been identified and associated with potential improvement strategies. We can develop an overall plan for the improvement of the asset management process along with a strategy for implementation and control, once it is established. As a part of this phase, we propose to explain the following elements:

- Action plans—comparison of improvements
- Verification with CTSs
- Design of future state
- "To Be" process flow diagram under the new state
- Responsibilities for implementation/change management
- Benefit/Costs
- Anticipated training
- Metrics and performance targets

2. Recommendations for Improvement

Upon identification of improvement strategies, through the tools and methods in the Analyze phase, we group the strategies to create another "affinity diagram" to compare and ascertain their relationship to the CTS elements that were originally developed in the Define phase and later refined. The affinity diagram allows to us do a side-by-side comparison of the improvement strategies so that they may be consolidated and later grouped according to whether they are short-term, long-term, global or local. Once grouped, we also compare them to the CTS items as a means of verification of the improvement strategies to the performance model represented by the CTS. Figure 7.20 makes a comparison of the improvement strategies developed in the Analyze phase.

Figure 7.21 shows the mapping of the improvement recommendations to the CTS criteria.

3. QFD

The QFD maps the CTS criteria to the improvement recommendations to ensure alignment between the customer requirements and the process technical requirements. It is shown in Figure 7.22. The prioritization of the improvement recommendations from the QFD house of quality is shown in Figure 7.23.

4. Action Plan

Once the improvement strategies are consolidated, the level of difficulty (risk) and importance of the strategies are defined, along with whether short- or long-term and area of responsibility (process owner) are assessed. To do this, we rated each

5 S's	House of quality	Deployment matrix	FMEA	Lean approach
▪ Use of ABC inventory method ▪ Communication to faculty of upcoming inventory ▪ Improve/ease of use of system ▪ Education of faculty on significance of system ▪ Improved description of items through PO form	▪ Educate and Inform faculty of importance of system ▪ Inform property manager of asset relocation ▪ Improve system of item description ▪ Ease of use of FA website for forms ▪ Elimination of 2nd-pass for scans ▪ Visible tags ▪ More efficient scanning route ▪ Documented surplus or disposal	▪ Improve item description on lists ▪ Educate /Involve faculty members ▪ Make items available for inventory ▪ Immediate reporting of lost/missing items	▪ Improved item description ▪ Make items available for inventory (inform of inventory date) ▪ Document relocation of items ▪ Document/ inform of items to be disposed ▪ Scanning methods – need systematic approach to avoid wasted travel ▪ Time to update list and send out with missing item list	▪ Increase efficiency by reducing scan opportunities from 3 to 2. ▪ Increase access to items on first-pass of inventory scan ▪ Improve identification of items.

FIGURE 7.20 Improvement strategies.

improvement category on a Likert scale according to level of difficulty (1–5, 5 being the highest) and importance to the overall success of the project (1–5, 5 being the highest). The improvement strategies are grouped according to whether they can be classified as short-term, relatively low-cost improvements, and longer-term improvements requiring a more significant investment of time and resources. Once the improvements are prioritized, we can establish a sequence of implementation. Finally, the anticipated responsible partner for implementation is identified.

The short-term improvements are shown in Figure 7.24.

Through our informal "survey" of improvements, our recommendations would be to first perform the features described below.

1. *Notify faculty and staff in advance of upcoming inventory efforts.* Distribute cards or flyers coupled with email notification. Distribute cards thanking faculty for their cooperation in making items in their areas available for scanning. Provide ample warning of upcoming inventory so that faculty can make items available for scanning or otherwise provide feedback to the scanners. Provide "thank you" cards to place in offices to express appreciation for the cooperative effort.
2. *Reinforce proper use of P-card and notification of capital purchases* As of the first week in December, The finance and accounting (F&A) department

Faculty / staff awareness of process	Communication to faculty, Education in use of website and access to forms. Education in new P-card policy. Involvement in inventory process through advance notification.
Documented location of assets	Improve ease of use of website/access to forms. Inform Property manager or assets needing repair / surplus. Proper use of new P-card policy.
Identification of assets	Improved P-card system. FA notification of property manager through PO system. Add fields to PO form for exact item identifiers, brand/unit name/size/other identifying characteristics.
Efficiency of yearly scanning	Reduce to 2 from 3 scans. Apply more visible tags. Give faculty advance notification of inventory to allow access to items.
Value of assets lost	Emphasize care and security for more valuable assets. Employ RFID/camera or scanner technology on more expensive items.
Number of assets lost	Educate faculty and staff in care and safekeeping of items to prevent loss.
Undocumented assets	Improved PO System using new P-card policy.
Efficiency of list update	More efficient scanning route. Employ more visible tags. Improve item description on inventory list (through modified PO)
Loss avoidance	Education and awareness of faculty and staff, of asset policies. Hold accountable officers responsible for lost items. More visible tags on attractive items.

FIGURE 7.21 Mapping of improvement recommendations to critical to satisfaction criteria.

came out with a new policy on use of P-cards that enables capital purchases as long as certain standards are maintained with respect to notification. It is suggested F&A reinforce this with reminders and offer assistance with proper P-card use.

3. *Improve item description through changes to fields on PO form and photographs.* CECS should, through the property manager and custodians, help "tighten" descriptive information and in cases in which the item is too small or difficult to describe, require a photo of the item to be purchased so that it can be added to a database. The property manager will obtain the photos and create the database to maintain. F&A's item list should be linked to the database.
4. *Improve the F&A website to provide more direct access to the forms needed by the faculty to record relocations, and dispose or surplus items.* By adding a "hotlink" button to the website, enable faculty members to go directly to a forms section that can be selected from and have forms filled-out and sent directly to F&A, the property manager and the officer of the department accountable. It should be noted that recently this site was updated. Therefore, education of faculty of availability of the site may be all that is needed.

Customer requirements \ Technical requirements	Importance	Email notification	Improved website	Change request for transfer	Memo from the dean	Sticker on assets	Improved PO form	P-card usage	Consistency in identification	More efficient routing	RFID tags	Highly visible tags	Elimination of 2nd pass	Methodology of list update	Timing of download	Purchasing procedure	Properly documented surplus	Properly documented transfers	Properly documented cannabilism	RFID triggered cameras	Enter description
Faculty awareness of policy	5	5	5	3	5	5															
Faculty's ease of notification	5		5	5																	
Documented location of assets	5	5	3	5	5	5	5		3												3
Identification of assets	4								5				5								5
Efficiency in yearly scan	4	5				5				5			5								
Safeguarding of assets	5	3	3	3	3	5					5	5								5	
Number of items lost	2	5	5	5	5	5	3	3	5	3	5	5				3				5	5
Efficiency of list updates	4								3				5	5	5						
Undocumented asset avoidance	3			3		5	5	5			3	5		5	3	3	5	5	5		5
Loss of asset avoidance	5	5			5				5	3	5	3					3	5	3	5	
Absolute weight		120	90	99	100	120	46	21	82	41	69	65	60	35	29	15	30	40	30	60	60
Relative weight																					

FIGURE 7.22 Quality function deployment house of quality.

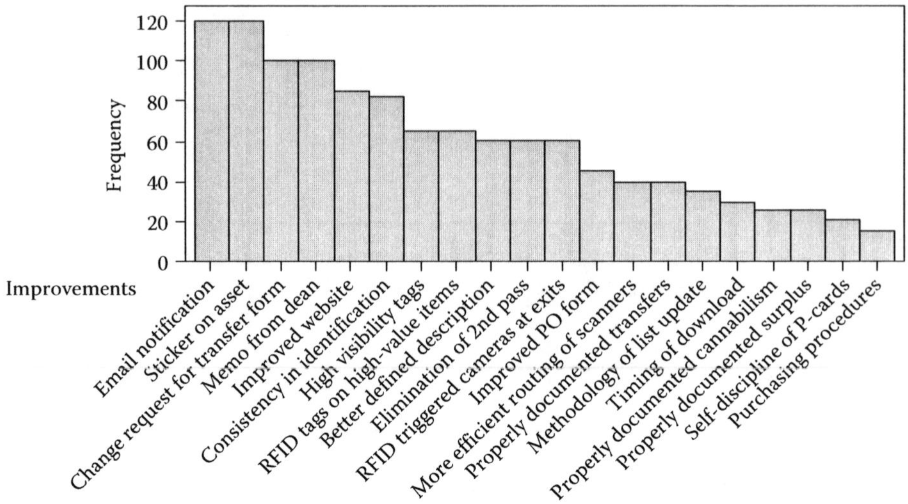

Frequency	120	120	100	100	85	82	65	65	60	60	60	45	40	40	35	29	25	25	21	15
Percent	10	10	8	8	7	7	5	5	5	5	5	4	3	3	3	2	2	2	2	1
Cum %	10	20	29	37	44	51	56	62	67	72	77	81	84	87	90	93	95	97	99	100

FIGURE 7.23 QFD prioritization of improvement recommendations.

5. *Educate the faculty of the asset management process by stressing the importance from legal (regulatory) and stewardship standpoints* (Note: it is important that the improved P-card process and website hot-links be in place before the education and information sessions). Email from the dean's office and a memorandum are needed to give credibility to an announcement reinforcing the asset management policy. Stress the importance not just from a regulatory standpoint, but from a stewardship perspective that as the college grows, competition for resources (capital) will become greater. Loss of assets will hurt everyone. In the future, the college will be back-charged for them or penalized by the grant source reducing the college's ability to obtain needed assets.
6. *Use visible tags for the more valuable items.* This will require investment in the tag system, recording, and time for application and item selection. Look into purchase of a visible tag system that can be applied to the more expensive items or more "attractive" items that can disappear. As suggested in the NPMA Manual, visible tags act as a deterrent. The property manager at one of the schools benchmarked also held this view.

Long-term improvements are shown in Figure 7.25. Long-term improvements should be prioritized as detailed below.

1. Eliminate the second scan by trial effort, in concert with training of scanners and improving efficiency of the scan process. This should include development of more systematic room-to-room coverage to avoid double

Improvement	Level of difficulty (risk)	Importance	Schedule	Responsibility
Short-term	(1 – 5)	(1 – 5)		
Education of Faculty: General Policy and Procedures (memo and guidelines for FA site use)	3	5	Immediate (Sp 06)	College administration in cooperation with property management
Use of P-card	4	4	Immediate (announcement made on 11/30/05)	Finance and accounting with support by college administration office
Notification to faculty of upcoming inventory efforts (by email or posted flyers)	1	4	Immediate – before next scan is scheduled (Sp 06)	Property manager with assistance of dept. custodians
Identification of items through precise description or photographs	2	5	Immediate implementation with photos and phase in descriptors on PO forms (Sp 06)	Purchasing (F&A) and faculty w/ support by property Manager to develop photo record
Make more visible tags available to PM and departments to place on the more "attractive" items	3	2	Phase in use of tags on more visible or attractive items over course of year (Fa 06)	College administration with approval of F&A of tag system
Improve website for faculty use. Use short cuts to forms.	2	4	Immediate: performed along with faculty education and information campaign (Fa 06)	FA and property management offices

FIGURE 7.24 Short-term improvement recommendations.

coverage and providing photographic information on items that are too small to have tags. Spend time with scanner trainees to develop a systematic pattern of room-to-room investigation, what to look for, and how to find it. Have a custodian representative of the department acquaint them with the types of objects they are likely to find. Once trained, scale back or eliminate the second scan. Use the metrics from the first pass to determine if the second scan is worth the effort. Make sure the scanners have photographic information from the database that was developed in Item 2 of the short-term improvements.

2. Invest in an RFID tagging system to provide additional security for more expensive items (ABC inventory). Develop an RFID tagging system for the fewer "expensive items" that comprise the upper end of the system. The

Long-term	Risk	Importance	Schedule	Responsibility
ABC inventory method (employ RFID tagging and sensors or cameras to identify more costly items that leave premises)	4	5	Phase in over two years, beginning in fall semester	College and departments, with support from property manager
Elimination of second scan*	3	4	Trial in next fiscal year	FA and property managers.
More efficient scanning process	3	4	Trial in next fiscal year	FA and property managers with scanner staff. Training of scanners required
Process improvement in updating and reissuing scanned list	?	3	Discussion needed with FA to determine best implementation practice	FA and property managers: Item for resolution

FIGURE 7.25 Long term improvement recommendations.

RFIDs will have to be keyed to a sensor or monitor at various doorways to record whether an item leaves the building and when. The information should then be relayed to the property manager for verification.

3. Work with F&A to develop a quicker turnaround for list updates so that missing items can be identified immediately and brought to the attention of the property manager and accountable officers (department chairpersons, etc.) Investigate with F&A as to whether a change in process or procedure can result in quicker download of lists after scanning information is uploaded. The algorithm should enable easy search and identification of items not recorded since the last scanning cycle. This will involve the cooperation of other F&A departments and may impact other colleges at the university as well.

5. Future State Process Map

As a result of the proposed changes to the process, we revisit the process diagram to understand what impact these changes will have on the flow, complexity or timing of the process. The deployment matrix revealed that during the Measure phase, there was little interaction in the process and that the property manager was not involved "until there was a problem." Our revised process puts into affect, policies that affect the issues before the inventory process takes place, so maximum benefit can be obtained for the least cost. That is, we take advantage of prevention costs as much as possible lest we allow ourselves to rely on inspection and internal failure costs. Additionally, one of the "internal failure" costs (second scan) is proposed for elimination or reduction. The revised process flow is shown in Figure 7.26.

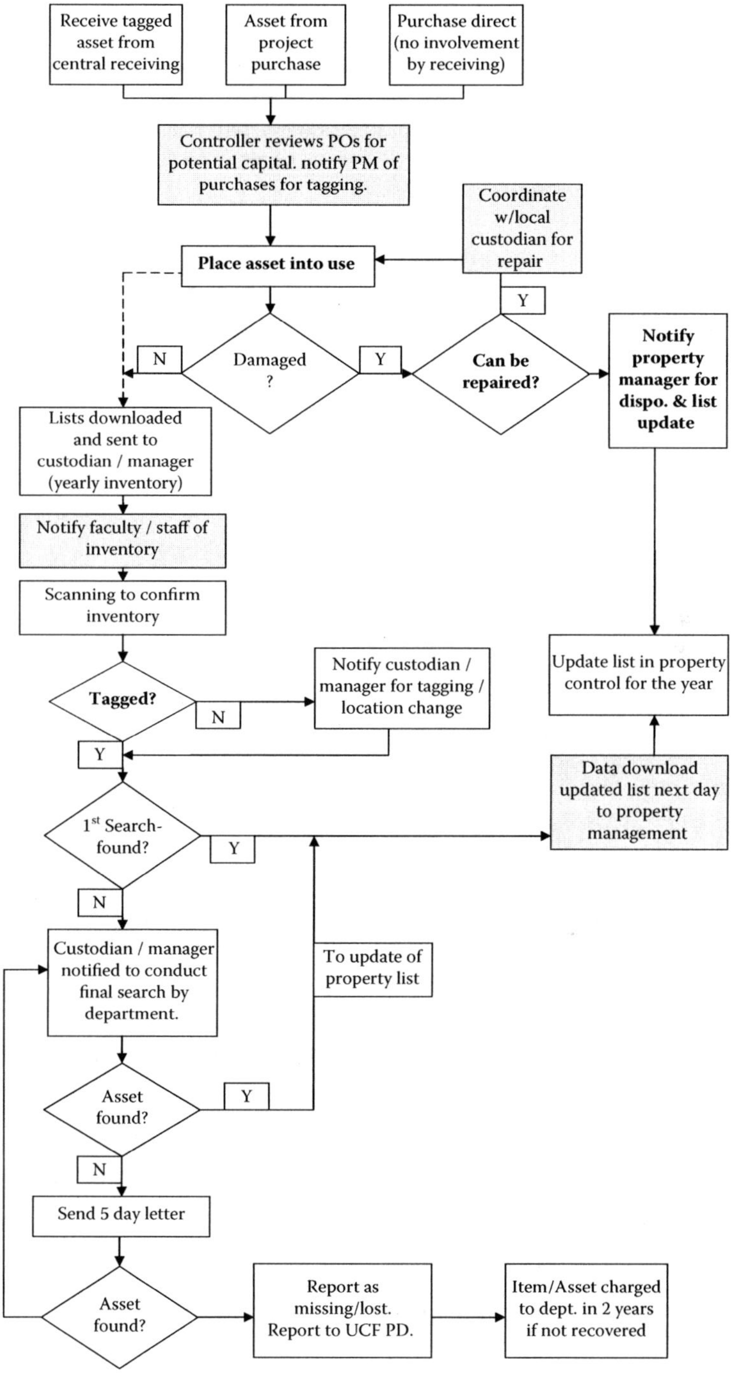

FIGURE 7.26 Future state process flow.

6. Revised VOP Matrix

It is necessary to institute performance targets to establish the level of performance needed for the process to operate well. By utilizing a performance measurement system, such as a balanced scorecard, an organization commits to assessing performance, monitoring performance, course-correcting performance and aligning all employees with key objectives. The metrics corresponding to the CTSs in the Measure phase have been modified upon further investigation and completion of the project although the CTSs themselves have not changed. The updated metrics corresponding to the CTS along with parallel performance targets are summarized in Figure 7.27.

7. Cost/Benefit Analysis

The Analyze phase identified costs that are being sustained by what we call "poor quality." These are the appraisal and failure costs derived from what we know about the costs of labor and acquisition value of items. To reiterate, these are:

- Appraisal costs (first scan) $2761/year
- Failure costs:
  - Internal – (second scan) $2761/year
  - External – (average item loss rate) $66,000/year

Now we estimate what the costs are of implementation of the proposed improvements:

Short-term improvements:

- Faculty education and information, P-card policy reinforcement, quick-links to F&A website for forms—negligible.
- Notification of faculty of upcoming inventory <$500 (for printing and distribution of cards or flyers).
- Implementation of description system through PO changes and photographic database < $1500.
- Use of visible tags for identification on a limited number of items (for 27% of the items comprising 68% of the total value—this is 1300 items). This was not determined at this time. We presently estimate that based on $0.50 to $1.00 per item, this could be in the range $650 to $1300.

Long-term improvements:

- ABC inventory method: Identify and tag the 130 most expensive items (based on 2.7% of the current 4800 items) with RFID. Cost estimate is based on 12 security cameras at $170/each, 36 antennae at $250/each, 9 readers at $150/each, one month of software development at $175/hour, a roll of 2500 passive tags at $100/roll. Total estimate is $40,590.
- Elimination of second scan: No additional cost. Elimination of internal failure cost of $2761.
- Training of scanners for more efficient process: Assuming four part-time scanners at eight-hours each plus two full-time employees at two-hours each, estimated at $325.

CTS	Metrics	Performance targets
• Faculty/staff awareness of process	• Number of emails, memos or flyers sent out to faculty members	• No definite performance. Success would be measured when all the faculty members are made aware of the process
• Documented location of assets	• Number of department transfer forms, off-campus use forms and forms for relief of responsibility of cannibalization recorded per time period	• Number of such forms on record to match what has been moved + what is not found. • Number of items on list = Number of items scanned + records of items relocated + records for items cannibalized + records for items lost/stolen
• Identification of assets	• Number of items without proper description • Assets found with the help of custodian or property manager (final pass) due to poor description	• Zero • Zero
• Efficiency of yearly scanning	• Percentage of items found on the first-pass • Reduction in scanning time before and after implementation of 5S and Lean concepts	• 80% • Scanning time reduced by at least 50%
• Number of items lost • Value of items lost	• Number of lost items before and after RFID tagging • Number of items >$3,000 before and after tagging with visible tags	• Zero • Zero
• Undocumented assets	• Items found but not documented	• Zero
• Efficiency of list update	• Number of times the list is updated	• Left as an open issue at this time. Subject to consideration by F&A department as to feasibility and optimality of improving their update process
• Loss avoidance	• Number (or value) of items missing: Items relocated but not recovered. This value should total the Number(Value) of items lost.	• Zero

FIGURE 7.27 Revised metrics and targets.

If we examine the short-term costs, the most these costs could total would be approximately $3300, and some of these costs are "one-time" start-up costs. Should visible tags not be employed, these costs would be < $2000. It must be noted that potential reduction in loss of items is the primary source of gains. Therefore, any gains in number of items "found" each year must be weighed against whether short-term improvements were responsible and whether visible tags truly provide a deterrent capability against theft.

The long-term improvement being proposed by the Lean Six Sigma team involves the implementation of an RFID security system. The team is proposing to place RFID tags on high-value assets that are valued above $3000.

The system would also require RFID readers as well as a vast amount of programming for the software development. This is due to the fact that an off-the-shelf RFID system does not exist. The total cost of implementation including all hardware and integration has been estimated by a consultant to be around $40,590. This may seem like a fairly large figure, but there was one asset reported as stolen which alone had a value of $18,000. Additionally, it was determined there was an average of $66,000 per year loss of missing items over a 10-year period.

8. Training Plans and Procedures

As a part of the proposed improvement strategies, training is required in at least two areas. Faculty need to be acquainted (or reacquainted) with the process and procedures involved in asset management. Given the level of expertise and preoccupation with other matters, the approach recommended here would be to appeal to the values implied in maintaining the present level of service (no real perceived problem) versus a gradual decline resulting from possible loss of budgetary funds or even loss of grant or research funding as a result of unfavorable financial reporting to potential grantors. Memoranda from the dean's office are important in highlighting the need to uphold and enforce the system. Effectiveness of training will be determined by an increase (or decrease) in recorded forms submitted by faculty for relocation or removal of items, as well as items to be retired and registration of new items for tagging.

The scanners can be trained within one day by acquainting them with the types of assets they are liable to find in their work. Providing a list with better descriptors (size, color, function, brand and model number), along with briefing by the local custodial manager, would enable them to accelerate their learning curve for becoming more familiar with items to be scanned. Effectiveness of the training will be determined by the time taken to scan and record a given number of items.

9. Improve Phase Presentation

The Improve phase presentation can be found in the downloadable instructor materials.

IMPROVE PHASE CASE DISCUSSION

1. **Improve Report**
   1.1 Review the Improve report and brainstorm some areas for improving the report.
   1.2 How did your team ensure the quality of the written report? How did you assign the work to your team members? Did you face any challenges of team members not completing their assigned tasks in a timely manner, and how did you deal with it?

1.3 Did your team face difficult challenges in the Improve phase? How did your team deal with conflict on your team?
1.4 Did your instructor and/or Black Belt or Master Black Belt mentor help your team better learn how to apply the Lean Six Sigma tools in the Improve phase, and how?
1.5 Compare your improve report with the improve report in the book, what are the major differences between your report and the author's report?
1.6 How would you improve your report?

2. Recommendations for Improvement

2.1 How did your team generate ideas for improvement?
2.2 What tools and previous data did you use to extract information for the improvement recommendations?
2.3 How do your recommendations differ from the one's in the book?

3. Revised QFD

3.1 Does the QFD support the alignment with the CTS characteristics?
3.2 How will you assess customer satisfaction?

4. Action Plan

4.1 How did your Six Sigma team identify the timings for when to implement your recommendations?

5. Future State Process Map

5.1 Compare your future state process map to the one in the book. How does it differ?
Is yours better, worse, the same?

6. Revised VOP Matrix

6.1 Does the VOP matrix provide alignment between the CTSs, the recommendations, metrics and target?

7. Costs/Benefit Analysis

7.1 Would you recommend implementing RFID to track assets based on your cost/benefit analysis?

8. Training Plans and Procedures

8.1 How did you determine which procedures should be developed?
8.2 How did you decide what type of training should be done?

9. Improve Phase Presentation

9.1 How did your team decide how many slides/pages to include in your presentation?
9.2 How did your team decide upon the level of detail to include in your presentation?

CONTROL PHASE EXERCISES

1. **Control Report**
   Create a Control phase report, including your findings, results and conclusions of the Control phase.

2. **Control Plan**
   Develop a control plan for each improvement recommendation from the Improve phase report.

3. **Dashboards/Scorecards**
   Create a dashboard or scorecard for tracking and controlling the process.

4. **Control Phase Presentation**
   Prepare a presentation (PowerPoint) from the case study exercises that provides a short (10–15 minutes) oral presentation of the Control phase deliverables and findings.

CONTROL PHASE

1. Control Report

A strong improvement phase needs an appropriate control plan for measuring the impact that these recommendations have provided. These measures are important because they provide scientific proof on whether the recommendations have improved or worsened the system. A control method must be tailored to each one of the recommendations, and it must supply all the necessary documentation to guarantee a successful evaluation.

2. Control Plan

Recommendation #1

Education of faculty. Publish memo and guidelines for fixed assets (FA) site use.

Proposed Control

Response levels from e-mail. Has there been a change in the amount of faculty who now perform all the procedures as determined by property management?

Goal: Obedience levels after memo > Obedience level last year

Verification Method

- Change in levels of faculty obedience/knowledge. Conduct knowledge survey after memos are distributed.
- Increase/decrease in amount of transfers reported, surplus, cannibalization, lost items.

Counter Reactions

- If positive: None but continue sending e-mails with important notices from property management
- If negative: Resend memos, track changes, repeat if negative.

Data Available

Available data regarding the improvement come primarily from the surveys. A new survey (after memos) is necessary to measure the change in percentages. The aim of the control method should be to prove that the awareness level has increased after the memos are delivered.

System Sustain

In order to maintain the benefits provided by this recommendation, it is important to constantly and emphatically remind the faculty of the importance of following the procedures set forth by property management. A single memo most likely will not fix the problem, so there may be opportunities for face-to-face reviews.

Issues

Simultaneous implementation with other suggestions may complicate the ability to measure the exact changes brought upon by this suggestion.

Recommendation #2

Establish ABC inventory analysis. Employ RFID tagging and sensors or cameras to identify more costly items that leave the premises.

Proposed Control

Gauge the number of high-value items lost. Have the new devices reduced the number of high-ticket items that are leaving the premises?

Goal: Number lost after new security devices < Number lost without security system

Cost of items lost after new implementation < Current ten-year cost average of lost items

Verification Method

- Number of high-value items lost
- Combined worth of items lost

Counter Reactions

If positive: Maintain and support implementation

If negative: Reassess improvement, discard current method

Data Available

The lost/missing items list is the only source of data for which we can measure this implementation. We have extracted the most critical information and displayed it below.

Total number of items lost: 262 (since record-keeping began in 1970)

Items lost last fiscal year: 12

Total worth of lost items: $824,447 (since record-keeping began in 1970)

Ten-year cost average of lost items per year: $66,000 (last 10 years)

After the recommendations are implemented, the aforementioned values would need to be recalculated in order for comparisons to be made.

System Sustain

Proper functioning of all system components at all times is critical to implementation success. All the devices should be reported to physical plant to conduct maintenance and ensure appropriate functioning.

Issues

Numbers of high-ticket items lost is very small, only one every couple of years. It may be a long time before the improvement plan can be evaluated.

Recommendation #3

Improve website for faculty use. Provide shortcut to forms.

Proposed Control

Response levels from website; has it increased the number of forms submitted?

Forms submitted to Jose, have they increased?

Goal: Number of forms submitted after web page redesign > number of forms currently submitted

Verification Method

- Web page hits
- Number of forms submitted

Counter Reactions

If positive: None

If negative: Increase website advertisement, improve ease of navigation on the website.

Data Available

We did not gather any information during the Measure phase that serves to measure the improvements brought about by the web page redesign. We suggest performing a quick survey of the amount of forms that Jose processes before the web page redesign.

System Sustain

Web page maintenance is a minimal cost to the CECS, and it would be no higher than the current costs incurred. However, the hypothetical increase in forms submitted might turn into a burden for Jose, and should be studied after the implementation is made.

Issues

Simultaneous implementation with other suggestions may complicate the ability to measure the exact changes related to this suggestion.

Recommendation #4

Elimination of second scan.

Proposed Control

Efficiency levels of the new scanning method. Can we now cover more items in the beginning months? (Thus allow more time to search for missing items before the end of the fiscal year period.)

Goal: Detection date for lost/missing items earlier than current method

% of items scanned in the first three months > % of items scanned in same period last year

Verification Method

– Detection date of a "lost/missing item"
– Percentage of items scanned in the first three months

Counter Reactions

If positive: None
If negative: Return to old method

Data Available

For the current fiscal year we were able to furnish some data directly related to the percentage of items scanned in the first three months.

System Sustain

Undertaking this recommendation requires a change management procedure, and it is important that the scanners are fully convinced of the benefits otherwise the results will not be valid. This thought process must remain fixed.

Issues

Simultaneous implementation with Recommendations #5 and #6 may complicate measure of specific benefits brought upon by this recommendation specifically.

Recommendation #5

Identification of items through better descriptions.

Proposed Control

Location of items with the new descriptions vs. old descriptions. Do items with standard descriptions take less time to locate versus those with a regular description?

Goal: Time to locate asset with new description < Time to locate asset with old description

Verification Method

– Identification time for items with new description standard
– Identification time for items with old description

Counter Reactions

If Positive: None
If Negative: Redesign the standard (add information, contacts, etc).

Data Available

During the Measure phase we attempted to quantify the number of items with poor descriptions by going over the inventory list and tagging those items which we would not be able to recognize. It was a subjective evaluation, but it clearly proved that hundreds of items would be difficult to identify based on the information provided.

We did not conduct a study to determine the amount of time it takes to locate an item with a poor description. Therefore we recommend that after the new descriptions are established, perform a time study for the location of both items with new and old descriptions.

System Sustain

It is important to standardize the descriptions so that those who complete the description section do it consistently across all items, and they should instruct any new employee who performs the same task. We recommend auditing the descriptions at random times to ensure that the proper information was provided; another option would be to employ a system that rejects the purchase if the description is poor (i.e., P-card is shown to Jose for approval).

Issues

The implementation benefits of this system might be overrated if the reduction in identification time is due to more experienced scanners and not better descriptions.

Recommendation #6

Notification to faculty of upcoming inventory efforts (by e-mail or posted flyers).

Proposed Control

Appraise the increased effectiveness in easier access. Can we now scan more items in the beginning months of the fiscal year?

Goal: % of items scanned in the first three months > % of items scanned in same period last year.

Verification Method

– Number of items scanned on first pass.

Counter Reactions

If positive: Continue with notifications.

If negative: Review purpose of notification to faculty. If still negative, return to old method.

Data Available

This recommendation has no relation to any process currently performed by the property managers. However, the desired results are to achieve reduction of the current scanning time.

System Sustain

Sustaining this implementation is the sole responsibility of the respective custodian. The senior property manager should ensure that all his custodians perform this task before the scanners arrive to their locations.

Issues

Simultaneous implementation with Recommendation #4 may complicate measure of the benefits brought upon by this recommendation specifically.

Recommendation #7

Attach highly visible tags for high-value items.

Proposed Control

Study benefits of implementation. Are the tags deterring people from stealing?

Goal: Number of lost/stolen items after implementation < Number of lost/stolen items for previous years.

Verification Method

– Number of lost/stolen items for fiscal year of implementation.

Counter Reactions

If positive: Continue with system, always ensure that tags are visible.

If negative: Reassess tags, discard improvement if necessary.

Data Available

We have the following data available regarding lost/stolen items. They are the same data from Recommendation #2 because they both target a reduction in lost items:

Total number of items lost: 262 (since record-keeping began in 1970).

Items lost last fiscal year: 12.

Total worth of lost items: $824,447 (since record-keeping began in 1970).

Ten-year cost average of lost items per year: $66,000 (last 10 years).

System Sustain

Execution of this recommendation only involves a person to print and place the tags. To maintain this system, it is necessary to ensure that all new items are tagged, as well as having the scanners evaluate the condition of the tags in case of replacement.

Issues

Simultaneous implementation with Recommendation #2 may complicate measure of the benefits brought upon by this recommendation specifically.

Recommendation #8

Conduct a more efficient scanning process.

Proposed Control

Efficiency in building coverage. Does a systematic approach to the scanning of rooms lead to more items identified in the same period of time?

Goal: Percentage of items scanned in the first three months is greater than the percentage of items scanned in same period last year.

Verification Method

– Number of items scanned on first pass.

Counter Reactions

If positive: Continue with system, establish continuous improvement methods.

If negative: Discard new approach, return to old method.

Data Available

There is no data available regarding a quantified measure of their scanning process. However, the improvement results we wish to accomplish with this new approach are the same as Recommendation #6.

System Sustain

If the new approach is successful, it is important that the scanners establish a sense of continuous improvement so that the scanning method can be more finely tuned.

Issues

Simultaneous implementation with Recommendations #4 and #6 may complicate measure of specific benefits brought upon by this recommendation specifically.

Recommendation #9

System-wide use of P-cards.

Proposed Control

Number of new items accounted for. Can we reduce the number of items not properly documented by allowing everyone to use a purchase card?

Goal: 100% use of P-cards for all new items.

Verification Method

– Number of new items with P-cards.

Counter Reactions

– If positive: Continue use.

– If negative: Discard use of P-card.

Data Available

We have no data available regarding the use of P-cards.

System Sustain

System-wide implementation is to be done initially as a trial run. The success level will determine whether the recommendation is permanently implemented. The reason for this trial run is due to the issues mentioned below.

Issues

Purchasers must link the item to be taggable during purchase. OSR needs to give prior approval for purchases under grants, which could be bypassed with P-card purchase.

3. Dashboards/Scorecards

The project score card visually demonstrates the impact of the project's countermeasures and creates or revises the control plan. The FMEA analysis performed in the Analyze phase serves as a scorecard that Murphy can use to implement the recommended actions and once the actions have been taken, the new RPN can be calculated and compared with the old one. This RPN should be significantly lower to suggest an improvement in the process. The FMEA recommended action plan that can be used as a scorecard is shown in Figure 7.28.

Process Step	Potential failure mode	Recommended action(s)	Responsibility	Actions taken and effective date	Sev	Occ	Det	RPN
Receive tagged asset from office	Tag not readable	Ensure good quality barcodes	Property control	TBD	TBD	TBD	TBD	TBD
	Multiple Tags on item		Property control	TBD	TBD	TBD	TBD	TBD
	Damaged tag	Careful handling of items	Property control	TBD	TBD	TBD	TBD	TBD
		Ensure good quality tags	Property control managers		TBD	TBD	TBD	TBD

FIGURE 7.28 Scorecard using FMEA.

4. Control Phase Presentation

The Control phase presentation can be found in the downloadable instructor materials.

CONTROL PHASE CASE DISCUSSION

1. **Control Report**
   - 1.2 Review the Control report and brainstorm some areas for improving the report.
   - 1.3 How did your team ensure the quality of the written report? How did you assign the work to your team members? Did you face any challenges of team members not completing their assigned tasks in a timely manner, and how did you deal with it?
   - 1.3 Did your team face difficult challenges in the Control phase? How did your team deal with conflict on your team?
   - 1.4 Did your instructor and/or Black Belt or Master Black Belt mentor help your team better learn how to apply the Lean Six Sigma tools in the Control phase, and how?
   - 1.6 Compare your Control report to the Control report in the book, what are the major differences between your report and the author's report?
   - 1.7 How would you improve your Control report?

2. **Control Plan**
   - 2.1 How well will your control plan ensure that the improved process will continue to be used by the process owner?
   - 2.2 Are their control charts that could be used to ensure process control?

3. **Dashboards/Scorecards**
   - 3.1 How would your dashboard differ if it was going to be used to present the results of the process to each department, the college or the entire university?

4. **Control Phase Presentation**
   - 4.1 How did your team decide how many slides/pages to include in your presentation?
   - 4.2 How did your team decide upon the level of detail to include in your presentation?

8 High School Advanced Placement Open Access Process Assessment—A Lean Six Sigma Case Study

Marcela Bernardinez, Ethling Hernandez, Lawrence Lanos, Ariel Lazarus, Felix Martinez, and Sandra L. Furterer

CONTENTS

PROJECT OVERVIEW

Sunshine High School (SHS)* is located in the northeastern corner of Orange County, Florida. The school encompasses 95 acres housing 136 permanent and 80 portable

* The high school's name has been changed to provide a generic case study.

classrooms. Other features include a closed-circuit television production studio, a state-of-the-art performing arts center, specialized vocational, and technical laboratories and an agribusiness complex. Athletic facilities include a 5000-seat stadium, a dance studio, two fully equipped weight rooms and a 1900-seat gymnasium.

SHS was completed in 1990, and opened with an enrollment of 1500 students. It is now one of the largest high schools in the Orange County Public School District with over 3500 students and over 340 faculty members. SHS is divided into two campuses. The East Campus consists exclusively of freshman students, and the West Campus consists of sophomores through seniors. The leadership team consists of a principal, three assistant principals, and nine deans.

The principal at the time of the Lean Six Sigma (LSS) project was David Christiansen. He came from Olympia High School, and had been the principal for the past three years. He brought with him an "Extended Learning Opportunities" plan, which he implemented one year later. SHS has earned a state rating of "B" for the past two years. Their gain of 34 points is one of the most significant school-wide gains in the county and the state. This incredible accomplishment is the result of a comprehensive effort by their students, teachers, parents, staff, administration, and community members.

SHS is part of a cooperative educational endeavor with the College Board. This endeavor known as the Advanced Placement (AP) program works to serve three different groups: students who plan to go on to college; schools that would like to offer these advanced opportunities; and colleges that encourage and recognize such achievement. Pursuing AP courses can be very beneficial for students capable of completing college-level courses. What makes AP so great is that not only is there the possibility of earning college credit, but also students gain an edge in college preparation, standout in the college admission process, and can broaden their intellectual horizons.

The student population at SHS is diverse, with 40% of students being Caucasian, 42% Hispanic, 9% African-American, and 9% being Asian or other. Approximately 40% of SHS students are enrolled in the Free and Reduced (F&R) school lunch program, signifying that they are from a lower socioeconomic income group.

SHS is committed to establishing a cooperative and lasting partnership between home, school, and the community to assist students in acquiring the education and qualities that assure a successful and rewarding life. This commitment is evident in their mission and belief statements, which were developed as part of the school improvement process and are posted in every classroom.

Their mission is to advance student achievement for all students with the education necessary to be responsible and successful citizens.

A LSS project team has been assembled to assess the performance of students in AP courses and to assess whether the percentage of minority and low socioeconomic students has become more representative of the student population percentages of these groups. Additionally, a goal of the project is to identify further improvements to the AP open access registration process to improve the percentages of under-represented groups in AP classes, as well as the overall AP experience and student performance.

SHS implemented an AP open access system in the 2004/2005 academic year that would enable a more diverse population of students (as well as more students) taking AP classes.

Preopen Access Process:

The preopen access system was teacher driven. It consisted of a student fulfilling various requirements for teachers to approve their enrollment to the course. Students were required to:

- Be a level-4 or level-5 reader on the Florida Comprehensive Assessment Test (FCAT)
- Fulfill all class prerequisites
- Score more than 80% in the Norm-Referenced Test (NRT)
- Have a minimum 3.5 grade point average (GPA)
- Submit an essay
- Pass an interview
- Have recommendations from five teachers

If a student did not meet any one of the prerequisites, the teacher would be able to override and not allow the student to schedule the class.

Open Access Process

The current open access process is student driven. All students visit a counselor to schedule classes for the following year to begin the process. There are four scenarios outlined as follows:

1. **The student requests to take an AP course and he/she has a strong academic record.**
   During the visit to the counselor, the student may express interest in taking an AP class(es) . The counselor reviews the student's academic documents such as the FCAT, PSAT, or SAT and, if applicable, GPA and reading level. If the student's record shows that he/she has potential to academically succeed in the AP class, the counselor asks if the student knows which AP course he/she would like to take. If the student knows which AP course he/she would like to register in, the counselor will register him/her. If not, the counselor will recommend a General Education course.

2. **The student requests to take an AP course and he/she has a weak academic record.**
   If the counselor finds that the student has performed poorly in the past, the counselor will analyze on a case-by-case basis if the student possesses special abilities that will allow him or her to succeed in the course. Special abilities may be defined as math abilities, familiarity of a second language, etc. If the student possesses special abilities that will help him/her in the selected course, the course is scheduled. If the student does not possess special abilities for the class, the counselor will recommend a non-AP course. If the student feels strongly about the AP class, the counselor will allow him or her to register.

3. **The student does not request an AP course and he/she has a strong academic record.**
   The student's records are reviewed (as with every student). If the counselor finds that the student has a strong academic record, he/she may recommend that the student take an AP course. If the student agrees, the counselor will recommend a General Education AP course to the student. If this course is acceptable to the student, the student is registered for the course.

4. **The student does not request an AP course and he/she has a weak academic record.**
   In this scenario, the counselor will review the student documents and recommend non-AP classes to the student.

DEFINE PHASE EXERCISES

It is recommended that the students work in project teams of four to six students throughout the LSS Case Study.

1. **Define Phase Written Report**
   Prepare a written report from the case study exercises that describes the Define phase activities and key findings.

2. **LSS Project Charter**
   Use the information provided in the Project Overview section above, and the project charter format to develop a project charter for the LSS project.

3. **Stakeholder Analysis**
   Use the information provided in the Project Overview section above, in addition to the stakeholder analysis format, to develop a stakeholder analysis, including stakeholder analysis roles and impact definition, and stakeholder resistance to change.

4. **Team Ground Rules and Roles**
   Develop the project team's ground rules and team members' roles.

5. **Project Plan and Responsibilities Matrix**
   Develop your team's project plan for the DMAIC project. Develop a responsibilities matrix to identify the team members who will be responsible for completing each of the project activities.

6. **SIPOC**
   Use the information provided in the Project Overview section above, to develop a SIPOC of the high-level process.

7. **Team Member Bios**
   Each team member should create a short bio of themselves so that the key customers, stakeholders, project champion, sponsor, Black Belt and/

or Master Black Belt can get to know them and understand the skills and achievements that they bring to the project.

8. **Define Phase Presentation**
   Prepare a presentation (PowerPoint) from the case study exercises that provides a short (10–15 minutes) oral presentation of the Define phase deliverables and findings.

DEFINE PHASE

1. Define Phase Report

Sunshine High School (SHS) is located in the northeastern corner of Orange County, Florida. The school encompasses 95 acres housing 136 permanent and 80 portable classrooms. Other features include a closed-circuit television production studio, a state-of-the-art performing arts center, specialized vocational and technical laboratories and an agribusiness complex. Athletic facilities include a 5000-seat stadium, a dance studio, two fully equipped weight rooms and a 1900-seat gymnasium.

SHS was completed in 1990, and opened with an enrollment of 1500 students. It is now one of the largest high schools in the Orange County Public School District, with more than 3500 students and over 340 faculty members. SHS is divided into two campuses. The East Campus consists exclusively of freshman students, and the West Campus consists of sophomores through seniors. The leadership team consists of a principal, three assistant principals, and nine deans.

The principal at the time of the LSS project was David Christiansen. He came from Olympia High School, and had been the principal for the past three years. He brought with him an "Extended Learning Opportunities" plan, which he implemented one year later. SHS has earned a state rating of "B" for the past two years. Their gain of 34 points is one of the most significant school-wide gains in the county and the state. This incredible accomplishment is the result of a comprehensive effort by their students, teachers, parents, staff, administration, and community members.

SHS is part of a cooperative educational endeavor with the College Board. This endeavor known as the Advanced Placement (AP) program serves three groups: students who plan to go on to college, schools that would like to offer these advanced opportunities, and colleges that encourage and recognize such achievement. Pursuing AP courses can be very beneficial for students capable of completing college-level courses. What makes AP so great is that not only is there the possibility of earning college credit, but also students gain an edge in college preparation, stand-out in the college admission process, and can broaden their intellectual horizons.

The student population at SHS is diverse, with 40% of students being Caucasian, 42% Hispanic, 9% African-American, and 9% Asian or other. Approximately 40% of SHS students are enrolled in the F&R school lunch program, demonstrating a socioeconomic diversity within the study body.

SHS is committed to establishing a cooperative and lasting partnership between home, school, and the community to assist students in acquiring the education and qualities that assure a successful and rewarding life. This commitment is evident in their mission and belief statements, which were developed as part of the school improvement process and which are posted in every classroom.

Their mission is to advance student achievement for all students with the education necessary to be responsible and successful citizens.

2. Lean Six Sigma Project Charter

SHS has recently implemented a new process to allow more open access to all students in AP courses. The school administration wants to assess the impact to the quality of performance and quantity of students across the diverse student body enrolled in the courses. Part of this effort will involve benchmarking best practices of other high schools within Orange County and the state of Florida. Based on meetings with the project champion and sponsor, the project charter has been created for SHS (Figure 8.1).

The goal of the project was to understand and analyze the current selection process for the AP courses to assess the impact to the quality of student performance, as well as to assess whether the percentage of students by race/ethnicity and socioeconomic

Project Name: High School Advanced Placement Open Access Process Assessment.

Problem Statement: Sunshine High School has recently implemented a new process to allow more open access to all students in Advanced Placement (AP) courses. The school administration wants to assess the impact to both the quality of performance and quantity of students across the diverse student body enrolled in the courses.

Customer/Stakeholders: (Internal / External) leadership team, assessment team, students, faculty, counsellors.

What is important to these customers – CTS: AP class grades, AP test scores, student motivation, experience of the teacher teaching AP courses, student attendance, topics covered, and student evaluation of the AP course, percentage of minorities enrolled in the AP courses, the percent of students in the lower socioeconomic groups F&R lunch, number of AP experiences (students taking AP classes).

Goal of the Project: Understand and analyze the current selection process for the AP courses to assess the impact to the quality of student performance as well as to assess whether the percentage of students by race/ethnicity and socio-economic class mirrors the general student body population. The team will also provide recommendations for further improving the AP experience, and further enabling open access to AP courses.

Scope Statement: The project will make use of the student information from those enrolled in AP courses from the academic year prior to AP open access comparing to the academic year after the new AP open access process was implemented. The project will focus on assessing performance of all students enrolled in AP courses, both before open access and in the school year after the open access process was implemented.

Financial and Other Benefit(s): Increase school's funding through improving test scores; Improve school status through school grade; Teacher bonus for each student that achieves a 3 or more in AP exams; College credit awarded to students who earn a 3 or more in the AP exams; Optimize student academic achievements; provide diagnostic tools to assess student performance.

Potential Risks: Availability of resources (people and information); university culture; sensitivity and confidentiality of information.

FIGURE 8.1 Project charter.

class mirrors the general student body population. The team will also provide recommendations for further improving the AP experience, and further enabling open access to AP courses.

The project will make use of the student information from those enrolled in AP courses from the academic year prior to AP open access comparing to the academic year after the new AP open access process was implemented. The project will focus on assessing performance of all students enrolled in AP courses before open access and in the school year after the open access process was implemented.

There are many potential benefits to the school, faculty, and students by enhancing student performance in AP classes. The AP open access process could increase the school's funding through improving test scores; improve school status through maintaining or enhancing the school grade; improve the teachers' bonus for each student who achieves a 3 or more in the AP exams; allow college credit to be awarded to students who earn a 3 or more in the AP exams; optimize student academic achievements through study in advanced courses; and provide diagnostic tools to assess student performance.

3. Stakeholder Analysis

A stakeholder is a person who has interest in our project. Stakeholders are separated into two groups: primary and secondary. Primary stakeholders are those that are ultimately affected by the project and secondary are everyone else that has any kind of involvement. As you can see from the first column in Figure 8.2, the team defined who the stakeholders are as well as separated them into primary and secondary categories. Figure 8.3, shows the stakeholder commitment levels. The commitment level is based on how receptive each stakeholder group has been to meet with and work with the LSS team. Only about half (5 out of 9) of the counselors made time to meet with the team. The guidance counselors are an integral component to the success of the open access process. The initial open access process was thrust upon them and the faculty, so it will be important to gain their commitment by the end of the project.

The administration and assessment team have been extremely supportive and committed to the project. The students tended to be neutral at the beginning of the project but, having at least a moderate commitment to the process, will be important by the end of the project.

4. Team Ground Rules and Roles

The following are ground rules brainstormed by the team members. They will serve to ensure project success and teamwork throughout the project lifetime.

Attitudes

- Be as open as possible, but honor the right of privacy
- Information discussed in the team will remain confidential. With regards to peoples' opinions, what's said here stays here

	Stakeholders	Role	Potential impact or concerns
PRIMARY	Leadership team	The team is composed by the principal, nine deans, and three assistant principals	▪ Improve overall school score (+) ▪ Increase budget (+) ▪ Best practices (+) ▪ Recognition (+)
	Assessment team	The group of faculty and staff that work to assess student progress	▪ Improve overall school score (+) ▪ Increase budget (+) ▪ Staff (+)
SECONDARY	Students	Students enrolled in AP courses	▪ College preparation (+) ▪ College credit (+) ▪ Weighted GPA (+)
	Faculty	Teachers at the university high school	▪ Improve overall school score (+) ▪ Increase budget (+) ▪ Compensation (+)
	Counselors	Guidance counselors at the university high school.	▪ Improve overall school score (+) ▪ Increase budget (+)

FIGURE 8.2 Stakeholder definition.

Stakeholders	Strongly against	Moderate against	Neutral	Moderate support	Strongly support
Leadership team					XO
Assessment team					XO
Students			X	O	XO
Faculty		X			O
Counselors		X			O
X = At start of project					O = By end of project

FIGURE 8.3 Stakeholder commitment.

- Everyone is responsible for the success of the meeting
- Be a team player. Respect each other's ideas. Question and participate
- Respect differences
- Be supportive rather than judgmental
- Practice self-respect and mutual respect
- Criticize only ideas, not people
- Be open to new concepts and to concepts presented in new ways. Keep an open mind. Appreciate other points of view
- Be willing to make mistakes or have a different opinion
- Share your knowledge, experience, time, and talents
- Relax. Be yourself. Be honest

Processes

- Use time wisely, starting on time, returning from breaks and ending meetings promptly

- Publish agenda and outcomes
- Ask for what we need from our facilitator and other group members
- Attend all meetings. Be punctual
- Absenteeism is permitted if scheduled in advance with the leader
- When members miss a meeting, we will share the responsibility for bringing them up to date
- 100% focus and attention while meeting
- Stay focused on the task and the person of the moment
- Communicate before, during, and after the meeting to make sure that action items are properly documented, resolved, and assigned to a responsible individual and given a due date
- Phones or pagers on "stun" (vibrate, instead of ring or beep) during the meetings
- One person talks at a time
- Participate enthusiastically
- Don't interrupt someone talking

5. Project Plan and Responsibilities Matrix

The project plan and responsibilities matrix is shown in Figure 8.4. It identifies the detailed activities and who is responsible for completing each during each phase of the DMAIC problem-solving approach.

6. SIPOC

The SHS Six Sigma team has used a SIPOC diagram to identify the suppliers of the process, process inputs, process outputs, and the customers of those outputs so that the VOC can be captured. The team has identified two high-level process flows represented in the SIPOC diagram. These two processes are the AP registration for the current open access system and the AP registration for the pre-open access system. As shown in the SIPOC diagram, the main suppliers for both processes are the AP advisors. These suppliers are the upstream providers of all the inputs needed for the process to perform properly. The potential AP students and the AP prerequisites constitute the inputs of the process. These are all of the inputs needed for the process to perform properly. The process is the high-level description of all the required steps that are needed for the process to perform properly. After looking at the outputs of the processes, the outputs are represented by registered potential AP students in AP courses. The outputs are all of the final outputs that are produced by the process. Finally, the customers for these two processes include the AP students, the parents of those AP students and the AP faculty. The customers are the downstream users of all of the outputs that are produced by the process. The SIPOC for the AP open access registration process that was implemented to help increase the number and diversity of the students in the AP classes, as well as the SIPOC for the pre-AP open access registration process is shown in Figure 8.5.

Activity	Project champion	Project sponsors	Project black belt	Team leader	Team members
Define phase:					
Form team	X		X	X	X
Kick-off meeting		X	X	X	X
Team roles and ground rules			X	X	X
Define project goals, scope and objectives	X	X	X	X	X
Develop project charter				X	X
Stakeholder analysis				X	X
Report and presentation				X	X
Measure phase				X	X
Process flow charts, Pareto charts, CTS, Key metrics				X	X
Prepare and collect data				X	X
Report and presentation				X	X
Analyze phase:					
Cause and effect diagrams, summary of problems, Summary of data collected, cost/benefit analysis				X	X
Identify improvements				X	X
Report and presentation				X	X
Improve and control phases:					
Improvement plan		X		X	X
Recommendations		X		X	X
Quantification of improvement, revised process flow, metrics				X	X
Training plan				X	X
Final report				X	X
Final presentation				X	X

FIGURE 8.4 Responsibilities matrix with project plan.

7. Team Member Bios

Dr. Sandy Furterer is the assistant department chair in the Industrial Engineering and Management Systems department at the University of Central Florida (UCF). Her teaching and research interests are in quality engineering, engineering management, engineering education, and change management. She has a bachelor's degree and master's degree in industrial engineering from Ohio State, an MBA from Xavier University in Cincinnati, and a PhD in industrial engineering from UCF. Prior to

Open access system – AP registration				
Supplier	**Inputs**	**Process**	**Outputs**	**Customers**
▪ UHS admin ▪ Advising office	▪ Potential AP student ▪ AP requisites	▪ Student has AP potential based on summary of answer report ▪ Potential AP student is invited to take AP courses ▪ Potential AP student visits his/her counselor ▪ Counselor reviews potential AP student's PSAT, GPA, FCAT scores and previous coursework ▪ Counselor completes academic progression plan ▪ Potential AP student meets the requirements ▪ Potential AP student is a level 3 + reader, with exceptions for level-1 and 2 ▪ Potential AP student gets advise on courses to be taken ▪ Potential AP student registers for AP course or courses ▪ Counselor allows student to register ▪ Potential AP student register AP course or courses	▪ Registered potential AP student in AP course or courses	▪ Student ▪ Parent ▪ UHS faculty

FIGURE 8.5 SIPOC.

returning to study for her PhD in 2002, Dr. Furterer was a management consultant specializing in implementing Lean and Quality principles and tools in "white collar" and manufacturing processes. She was a manager of industrial engineering for Mead Data Central (now Lexis Nexis), facilitating improvements in data fabrication and information systems development processes. She also performed information systems analysis for AT&T. Dr. Furterer is an ASQ certified Six Sigma Black Belt (CSSBB) and a certified quality engineer (CQE), as well as a Girl Scout troop leader.

Ethling Hernandez is a master's degree student of the engineering management program in the College of Engineering and Computer Science. She obtained her undergraduate degree in industrial engineering in December 2004 from UCF. She

Pre-Open access system – AP registration				
Supplier	**Inputs**	**Process**	**Outputs**	**Customers**
▪ UHS admin ▪ Advising office	▪ Potential AP Student ▪ AP requisites	▪ Potential AP student is invited to take AP courses ▪ Potential AP student visits his/her counselor ▪ Potential AP student meets the requirements ▪ Potential AP student has at least 3.5 GPA ▪ Potential AP student gets 5 recommendation letters from teachers ▪ Potential AP student is at least level-4 reader ▪ Potential AP student fulfills pre-requisites ▪ Potential AP student passes interview ▪ Potential AP student writes an essay that needs to be accepted ▪ Potential AP student registers for AP course or courses	▪ Registered potential AP student in AP course or courses	▪ Student ▪ Parent ▪ UHS faculty

FIGURE 8.5 (Continued)

is a student member of the Industrial Engineering Society as well as the Society of Hispanic Professional Engineers. Since 2003, Ethling has been a research assistant for the Center for NASA Simulation Research under the mentorship of Dr. Luis Rabelo and Dr. Jose Sepulveda.

Felix Martinez is a graduate student in quality engineering at UCF. He obtained his bachelor's degree in industrial engineering in spring 2005. Felix works as a graduate research assistant in the Housing Constructability Laboratory, where he is leading a project regarding water intrusion in masonry walls. Previous work experience includes a year-long internship with the United Parcel Service, where he helped implement a new package-tracking system and conducted time studies on different personnel.

Ariel Lazarus is studying for her master's degree in quality engineering from UCF. She also received her bachelor's degree in industrial engineering from UCF. She is working for the Industrial Engineering and Management Systems department as a graduate research assistant on the E-Design project. While an undergraduate, she participated in a project for Walt Disney World Distribution Services.

Marcela Bernardinez was born in San Miguel de Tucuman, Argentina, but raised in Venezuela because her parents moved. After she finished high school in Venezuela, she decided to have a new experience, meet new people, find new opportunities, and discover a new world, so she came to the U.S. to study industrial engineering. She has been in the U.S. for six years, and it has been a challenge to arrive at where she is now. Marcela has a bachelor's degree from UCF in industrial engineering and is pursuing her master's in industrial engineering at the same university. In addition, she is a member of the Institute of Industrial Engineers and the Society of Hispanic Professional Engineers. It is Marcela's goal to graduate and become known as an industry expert and earn a respectable management position with responsibility for a major piece of the business.

Lawrence Lanos is working on his master of science degree in industrial engineering, quality track in the Industrial Engineering and Management Systems Department at UCF. He received his bachelor's degree in electrical engineering from the FAMU/FSU College of Engineering at Florida Agricultural and Mechanical University. He had done research in robot design. He has recently received a Green Belt after working on a Six Sigma project where his team had to assess the effectiveness of the current Student Improvement Plan (SIP) developed by SHS in Orlando, Florida.

8. Define Phase Presentation

The Define phase presentation can be found in the downloadable instructor materials.

DEFINE PHASE CASE DISCUSSION

1. Define Phase Written Report

1.1 How did your team ensure the quality of the written report? How did you assign the work to your team members? Did you face any challenges of team members not completing their assigned tasks in a timely manner, and how did you deal with it?

1.2 Did your team face difficult challenges in the Define phase? How did your team deal with conflict on your team?

1.3 Did your instructor and/or Black Belt or Master Black Belt mentor help your team better learn how to apply the LSS tools, and how?

1.4 Did your Define phase report provide a clear vision of the project, why or why not?

2. LSS Project Charter

Review the project charter presented in the Define phase report.

2.1 A problem statement should include a view of what is going on in the business, and when it is occurring. The problem statement should provide data to quantify the problem. Does the problem statement in the Define phase report provide a clear picture of the business problem? Rewrite the problem statement to improve it.

2.2 The goal statement should describe the project team's objective, and be quantifiable, if possible. Rewrite the Define phase report goal statement to improve it.

2.3 Did your project charter's scope differ from the example provided? How did you assess what was a reasonable scope for your project?

3. **Stakeholder Analysis**
   Review the stakeholder analysis in the Define phase report.
   3.1 Are there other stakeholders not identified that should have been?
   3.2 Is it helpful to group the stakeholders into primary and secondary stakeholders? Describe the difference between the primary and secondary stakeholder groups.

4. **Team Ground Rules and Roles**
   4.1 Discuss how your team developed your team's ground rules. How did you reach consensus on the team's ground rules?

5. **Project Plan and Responsibilities Matrix**
   5.1 Discuss how your team developed their project plan and how they assigned resources to the tasks. How did the team determine estimated durations for the work activities?

6. **SIPOC**
   6.1 How did your team develop the SIPOC? Was it difficult to start at a high level, or did the team start at a detailed level and move up to a high-level SIPOC?

7. **Team Member Bios**
   7.1 What was the value in developing the bios, and summarizing your unique skills related to the project? Who receives value from this exercise?

8. **Define Phase Presentation**
   8.1 How did your team decide how many slides/pages to include in your presentation?
   8.2 How did your team decide upon the level of detail to include in your presentation?

MEASURE PHASE EXERCISES

1. **Measure Report**
   Create a Measure phase report, including your findings, results and conclusions of the Measure phase.

2. **Process Maps**
   Create level-1 and level-2 process maps for each of the following processes:
   - Preopen AP registration process
   - Open AP registration process

3. **Operational Definitions**
   Develop an operational definition for each of the identified CTS criteria:

Quality: AP class grades, AP test grades, student motivation, teacher experience, student attendance, topics covered, course evaluation,
Quantity: percentage of minorities enrolled in AP courses, percentage of students of lower socio-economic class enrolled in AP courses, number of AP experiences.

4. **Data Collection Plan**
   Use the data collection plan format to develop a data collection plan that will collect voice of customer (VOC) and voice of process (VOP) data during the Measure phase.

5. **VOC**
   Develop a plan for collecting VOC information through interviews, focus groups or surveys.

6. **VOP Matrix**
   Create a VOP matrix to identify how the CTS, process factors, operational definitions, metrics, and targets relate to each other.

7. **Statistical Analysis and Pareto Charts**
   Create Pareto charts or histograms using the "AP Data.xls" spreadsheet:
   - Of total student population: percentage by race
   - Of total student population: percentage of students in F&R lunch program compared with those who are not in the program
   - Enrollment by class in 2003/2004 and 2004/2005
   - Class size by class in 2003/2004 and 2004/2005
   - Pareto Chart of number of "A" grades received per AP class
   - Pareto Chart of number of "F" grades received per AP class

   Calculate the mean and standard deviation for the following variables for 2003/2004 and 2004/2005:
   - Unweighted GPA
   - Percentage grade in AP course
   - Average grade on AP exam

8. **COPQ**
   Brainstorm potential COPQ for the case study for the following categories:
   - Prevention
   - Appraisal
   - Internal failure
   - External failure

9. **Measure Phase Presentation**
   Prepare a presentation (PowerPoint) from the case study exercises that provides a short (10–15 minutes) oral presentation of the Measure phase deliverables and findings.

MEASURE PHASE

1. Measure Report

The Measure phase is the second phase of the Six Sigma project DMAIC cycle. In this phase of the project, we first defined the current process. To do this, we created a process flow chart of the current AP open access system and also one of their previous AP registration system. The next step in the Measure phase is to address high-leverage opportunities. This is achieved by gaining the VOC. We have obtained the VOC by talking with the leadership team, as well as by conducting interviews with counselors. We will continue to gather more VOC by having focus groups with faculty and students. Once the current process was defined and confirmed by the customer, we were able to determine the CTSs, Key Process Indicators (KPIs) and key metrics. The final step in this phase was to gather initial data and determine current performance of students enrolled in AP courses due to the open access system.

2. Process Maps

A process flow helps to identify the steps that are followed to achieve a result. As specified by the SIPOC, the process flows will deal with how the inputs or students are successfully enrolled into the AP courses or the outputs. There are two defined process flows. The first is the preopen access registration system, which functioned prior to the 2003–2004 school year. The second is the current open access registration system, which has been in effect since the 2004–2005 school year.

Preopen Access Process Flow

The preopen access system was teacher-driven. It consisted of students fulfilling various requirements for teachers to approve their enrollment to the course. Students were required to:

- Be a level-4 or level-5 reader on FCAT
- Fulfill all class prerequisites
- Score > 80% in the NRT (http://www.fcatexplorer.com/)
- Have a minimum 3.5 GPA
- Submit an essay
- Pass an interview
- Have recommendations from five teachers

If a student did not meet any of the prerequisites, the teacher would be able to override and not allow the student to schedule the class.

Open Access Process Flow

The current open access process is student-driven. Because all students visit a counselor to schedule classes for the following year, the process begins with the visit to the counselor's office. There are four scenarios outlined as follows:

1. **The student requests to take an AP course and he/she has a strong academic record**
   During the visit to the counselor, the student may express interest in taking an AP class or classes. The counselor then reviews the student's academic documents, such as the FCAT, PSAT, or SAT, and if applicable, GPA and reading level. If the student's record shows that he/she has potential to succeed in the AP class, academically, the counselor will ask if the student knows which AP course he/she would like to take. If the student knows which AP course he/she would like to register in, the counselor will register him/her. If not, the counselor will recommend a General Education course.

2. **The student requests to take an AP course and he/she has a weak academic record**
   If the counselor finds that the student has performed poorly in the past, the counselor will analyze on a case-by-case basis if the student possesses special abilities that will allow him or her to succeed in the course. Special abilities may be defined as math abilities, familiarity of a second language, etc. If the student possesses special abilities that will help him/her in the selected course, the course is scheduled. If the student does not possess special abilities for the class, the counselor will recommend a non-AP course. If the student feels strongly about the AP class, the counselor will allow him or her to register.

3. **The student does not request an AP course and he/she has a strong academic record**
   The student's records are reviewed (as with every student). If the counselor finds that the student has a strong academic record, he/she may recommend the student take an AP course. If the student agrees, the counselor will recommend a General Education AP course to the student. If this course is acceptable to the student, the student is registered for the course.

4. **The student does not request an AP course and he/she has a weak academic record**
   The counselor will review the student documents and recommend non-AP classes to the student.

The process flows for AP preopen access registration process and open access registration process are shown in Figures 8.6. and 8.7.

3. Operational Definitions of CTSs

Performance Measures

Before the VOC could be captured, the performance measures for evaluating the open access system needed to be defined. The methodology implemented in defining these performance measures included:

- Gaining an understanding of the preopen access system.

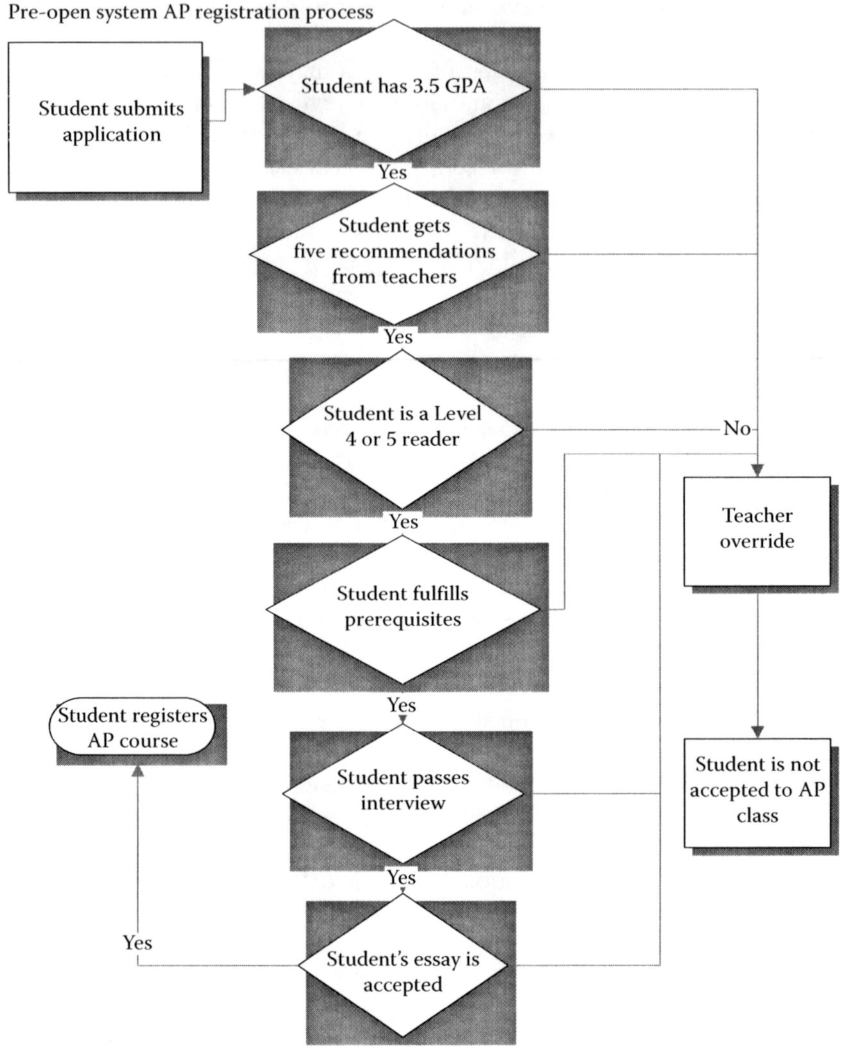

FIGURE 8.6 Preopen access registration process.

- Gaining an understanding of how counselors select students for AP courses in the open access system.
- Brainstorming for performance measures to capture the CTS aspects for evaluating students enrolled in AP courses due to the open access system.

Familiarization with the processes involved extensive research. Basic understanding was gained through meetings with the counselors. Additional clarifications were addressed during client–customer meetings with SHS leadership and the assessment team.

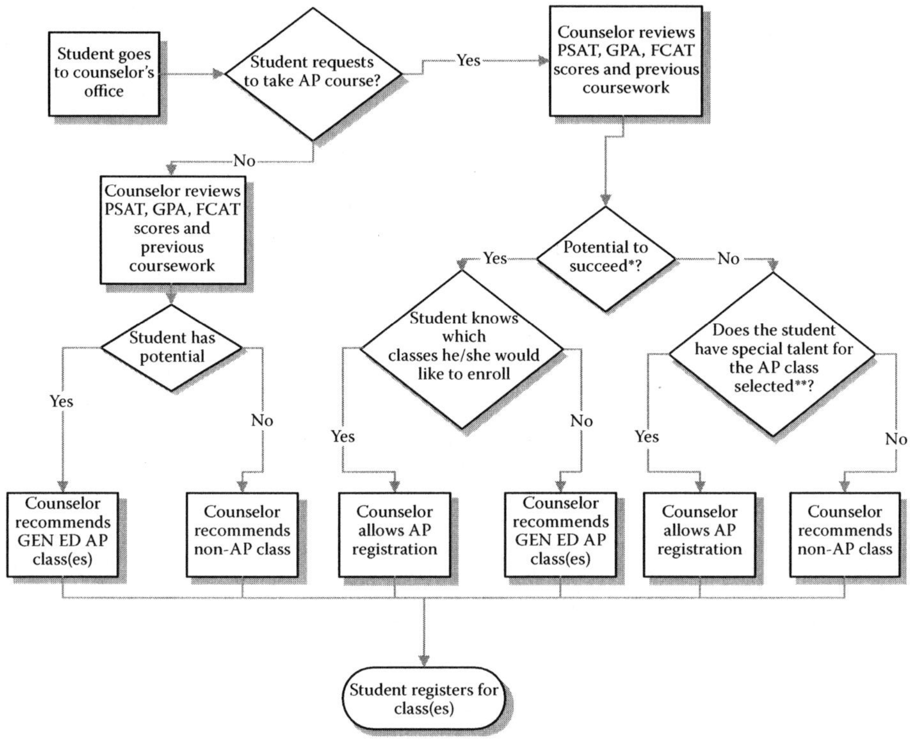

FIGURE 8.7 Open access AP registration process.

CTS

To identify the true quality of the AP experience, the LSS team developed some CTS characteristics. These characteristics provide ways to measure how well the system is functioning. The CTSs were categorized by quality and quantity. The main focus of the open access system was to increase the quantity of students in under-represented groups (race/ethnicity and socio-economic). However, quality of the AP courses is also an important component of customer (student and faculty) satisfaction with the AP courses. CTS characteristics are shown in Figure 8.8.

The quality-oriented CTSs include the AP class grades, AP test scores, student motivation, experience of the teacher teaching AP courses, student attendance, topics covered, and student evaluation of the AP course. Quantitative CTSs were the percentage of minorities enrolled in the AP courses, the percent of students in the lower socioeconomic groups (F&R lunch), number of AP experiences (students taking AP classes).

4. Data Collection Plan

The goal of this project is to determine the effectiveness of the new AP open access system at the SHS and recommend further improvements. Additionally,

	CTS	Description
Quality	AP class grade	Letter grade that they receive in the AP course. This CTS will measure how well the student performed in the course as assessed from the teacher's point of view
	AP test score	The score received on the AP test will show how much material was covered by the teacher and the level of comprehension of the subject by the student
	Student motivation	This assessment, which can only be measured qualitatively will provide an insight in how interested students are in the course that they are taking. Can also potentially be measured by attendance
	Teacher experience	The teacher's experience in teaching AP has a great impact in a student's AP score. Ideally, the only teachers that have taught AP in the past would be the only ones that teach AP or a teacher that has taught AP in the past would mentor new teachers
	Student attendance	Student attendance is necessary to obtain a good grade in any class, particularly in AP. The higher the student attendance, the more likely the student will perform better in the AP class or test
	Topics covered	The number of topics covered should consist of those identified as core curriculum for the AP exam
	Course evaluation	The students should evaluate courses at the end of the semester. This evaluation will help to identify what the problem areas are in the course and how prepared students think they will be for college. This course evaluation can also help have a knowledge base to draw upon
Quantity	% Minorities enrolled	The percent of minority students enrolled in AP classes. Historically, the numbers have been lower than the school's ratio
	% Lower socioeconomic enrolled	The percent of free and reduced students enrolled in AP classes. Historically, the numbers have been lower than the school's ratio
	Number of AP experiences	The total number of AP experiences for the school. This number is equal to the total number of students taking an AP class

FIGURE 8.8 Critical to satisfaction characteristics.

the Lean Six Sigma team needs to determine the performance of minority and socioeconomic students in AP courses for the 2003–2004 (preopen access) and 2004–2005 (open access) academic years. The objective of the data collection is to gain insight into the current open access system performance and to identify areas of improvement.

Data Collection Process

The SHS LSS team focused on three groups (AP student placement counselors, AP faculty, and AP students) to capture the current performance of the SHS open access system. The data collection started early February, lasted for about eight weeks, and was based on information gathered from interviews conducted with the SHS counselors, focus groups conducted with the AP faculty and AP students, and data collected from the SHS database.

The team interviewed five of the nine counselors. The team met with the counselors individually and asked about AP student placement. More detail about the questions asked will be explained in the counselor interview matrix section. The main objective of these interviews was to obtain the VOC and to gain insight into the current AP student placement process from the perspective of each counselor.

The SHS LSS team combined the information obtained from interviews with the customer for the purpose of brainstorming questions for the AP faculty and students. The questions were then revised and shared with the project Black Belt prior to the focus groups. The main purpose of conducting AP faculty and student focus groups was to determine the impact of the new AP open access system on faculty and students. The data collection plan is shown in Figure 8.9.

To ensure standardization during the process of interviewing the AP counselors, a standard procedure was followed:

1. The LSS team requested an appointment with each counselor.
2. The LSS team met with each counselor individually.
3. The LSS team asked each counselor the same questions about the process of placing students into AP classes.
4. The LSS team requested that counselors answer every question so that their opinions can be fully reflected in the study findings.

To ensure standardization during the process of conducting the two focus groups with the AP students and AP faculty, a standard procedure was followed:

1. The LSS team requested a meeting with the two different groups.
2. The LSS team will meet with each group at different times.
3. The LSS team brainstormed questions to be asked to the different groups prior to the meeting.
4. The LSS team shared questions with the project Black Belt.
5. The LSS team met with different groups and conducted the focus groups sessions.

5. VOC

The customer needs are referred to as VOC. Identifying customer needs is the most important part of the LSS project. The LSS team conducted interviews with the customers to identify their needs. Instead of solely relying on the historical data to define these needs, team members met with customers to gain a first-hand understanding of their needs. It was recognized that there were multiple customer voices to consider with this project. The customers included the leadership team, assessment team, counselors, faculty, and students.

Critical to Satisfaction (CTS)	Metric	Data collection mechanism (survey, interview, focus group, etc.)	Analysis mechanism (statistics, statistical tests, etc.)	Sampling plan (sample size, sample frequency)
AP class grade	AP class grade	Student database	T-test	Pre- and Post-open access
AP Test score	AP test score	Student database	T-test	Pre- and post-open access
Student motivation	Interest in taking AP courses	Focus group, interviews with students and teachers, attendance records	Summarize	Focus groups with appropriate participants
Teacher experience	Years of experience teach AP classes	Survey	Data analysis	All teachers teaching AP classes pre-open access and after
Student attendance	AP Class attendance	Student database	Data analysis	Cross-reference of AP students and their attendance
Topics covered	% of AP curriculum topics covered	Course syllabus	Data analysis	Define sample from each type of AP course
Course evaluation	% positive responses	Surveys	Chi-square	Define sample from each type of AP course
% Minorities enrolled	The percent of minorities' students enrolled in AP classes.	Student database	Data analysis	All enrolled
% Lower socioeconomic enrolled	The percent of free and reduced students enrolled in AP classes.	Student database	Data analysis	All enrolled
Number of AP experiences	Total number students taking AP courses	Student database	Data analysis	All enrolled

FIGURE 8.9 Data collection plan.

Summary of Focus Groups

To capture the current AP student placement process after the new open access system, the LSS team interviewed five of the nine AP counselors. The counselor interview matrix (Figure 8.10) shows the different counselors that have been interviewed during the Measure phase of this project. The counselor's population includes one counselor from the freshman campus and four from the senior

Questions/Counselors	Counselor 1	Counselor 2	Counselor 3	Counselor 4	Counselor 5
Grade 9 counselor	No	Yes	No	No	No
Grades 10–12 counselor	Yes	No	Yes	Yes	Yes
What do you look for in students when placing them into AP courses	FCAT, PSAT, transcripts	FCAT sometimes	FCAT, SAT, classroom performance	FCAT, SAT, classroom performance	FCAT, SAT, classroom performance
Which AP courses do you usually recommend	Depending on students' talent and skills. What the student is good at	World history or human geography	Depending on students' talent and skills. What the student is good at	Depending on students' talent and skills. What the student is good at	Depending on students' talent and skills. What the student is good at
Do you look if student meets prerequisites before placing them to AP courses	Yes	N/A	Yes	Yes	Yes
Do you recommend level 1 or 2 students take AP courses	No	Yes	No	No	No
Do you make any exceptions for level 1 or 2 students	Yes	Yes	Yes	Yes	Yes
Do you feel this is a better system for students	Yes	Yes	Yes	Yes	Yes
Who makes the last decision on taking AP courses	Student	Student	Student	Student	Student

FIGURE 8.10 Counselor interview matrix.

campus. The LSS team met with each counselor individually to ask specific questions about the current AP student placement process. After analyzing their responses and comparing the way each one of them places students into AP classes, the Lean Sigma Team can conclude they all follow the same concept. Because it is an open access system, they all allow any student who is interested in taking AP classes to register. The counselor interview matrix shows in more detail the questions asked to the counselors, the different counselors that were interviewed, and their responses.

A focus group was also conducted with the AP faculty. The population of AP faculty interviewed was one teacher from AP Statistics, AP Calculus, AP English Language, AP World History, AP European History, AP Environmental Science, and AP Macro Economics. All teachers were interviewed together, which allowed for the faculty to voice their opinion about the open access system while listening to the ideas and opinions of other faculty. All of these teachers believed that due to the open access system, more students were allowed to enroll in AP courses that were not prepared for the rigor of an AP course. The faculty also believed this led to more students failing and also a lower quality of AP courses because teachers were forced to teach at a slower pace to make sure that everyone was on the same page. There were incidents where a student was enrolled in an AP course and had no idea the course they were enrolled in was an AP course. This suggested to the faculty the counselors were pushing students into AP, which was believed to be the source of most of the problems in the open access system.

A third and final focus group was conducted with five students enrolled in various AP courses. Their view of the open access system was identical to the view of the faculty. They too believed students were being pushed into AP courses and the quality of AP courses had fallen. The students all agreed that there should be some minimal requirements that students must achieve before entering an AP course.

6. VOP Matrix

The VOP matrix is shown in Figure 8.11. It provides an understanding of the alignment between the CTS criteria, to the metrics and targets, as well as the operational definition of how the CTS will be measured.

Most of the metrics can be easily assessed through the database that SHS keeps. For qualitative measures, such as student motivation and course evaluation rating, the team suggests teachers have a meeting with the SHS leadership team to discuss if the student's attitude and motivation is changing within the AP classes.

Although some targets seem very optimistic, the LSS team thinks that through encouragement of students as well as parental and teacher involvement, SHS can reach those levels in the future. It is important then, to decrease the focus on the quantitative CTS, which have generously improved in the last year and shift the focus to the qualitative CTS, which have worsened.

7. Statistical Analysis and Pareto Charts

Because the main objective was to evaluate the impact that the open access system has on minority and low socioeconomic students, charts were created to observe the

CTS	Process factors	Operational definition	Metric	Target	Baseline
Quality					
AP class grades	Teacher experience, students' skills	Letter grade that they receive in the AP course. This CTS will measure how well the student performed in the course as assessed from the teacher's point of view	% of grades above B	100%	03 / 04 = A = 48%; B = 33%; C = 14%; D = 4%; F = 2% 04 / 05 = A = 24%; B = 38%; C = 27%, D = 7%; F = 3%
AP test grades	Teacher experience, students' skills	The score received on the AP test will show how much material was covered by the teacher and the level of comprehension of the subject by the student	% of tests over 3	100%	03/04 = 55% 04/05 = 35%
Student motivation	Environmental factors; quality of teacher and course	This assessment, will provide an insight in how interested students are in the course that they are taking. Could be measured by attendance	Teacher assessment, attendance in AP classes	100% attendance (excluding excused absences)	Not available
Teacher experience	Teacher skills, motivation, enrollment	The teacher's experience in teaching AP has a great impact in a student's AP score	% of teachers with experience teaching AP courses > 1 year	100%	2004/2005 = 43%
Student attendance	Student motivation	Student attendance is necessary to obtain a good grade in any class, particularly in AP	% attendance	100% attendance (excluding excused absences)	Measured by attendance

FIGURE 8.11 Voice of process matrix.

CTS	Process factors	Operational definition	Metric	Target	Baseline
Quality					
Topics covered	Student skills, time available	The number of topics covered should consist of those that are identified as core curriculum for the AP course	Number topics covered	All identified as core topics.	Not available
Course evaluation	Teacher experience, students' skills	The students should evaluate courses at the end of the semester	Course evaluation rating	80% of response in positive ratings	Not available
Quantity					
% Minorities enrolled in AP courses	Open access, encouragement, student motivation	The percent of minorities' students enrolled in AP classes. Historically, the numbers have been lower than the school's ratio	% minorities enrolled in AP courses	Representative of student population	03/04 = 40% 04/05 = 50%
% Lower socio-economic enrolled in AP courses	Open access, encouragement, student motivation	The percent of free and reduced students enrolled in AP classes Historically, the numbers have been lower than the school's ratio	% lower socio-economic students enrolled in AP courses	Representative of student population	03/04 = 22/182= 12% 04/05 = 21% 05/06 = 223/959 = 23%
Number AP experiences	Open access, encouragement, student motivation	The total number of AP experiences for the school. This number is equal to the total number of students in AP classes divided by the number of AP eligible students	Total number students taking at least one AP course	.5 AP courses per eligible students	03/04 = .22 04/50 = .26 05/06 = .42

FIGURE 8.11 (Continued)

number of minority and low socioeconomic students enrolled in AP courses prior to the open access system and after the implementation of the open access system.

In the 2004–2005 school years, various AP classes were added to the curriculum. After all of the changes, 31 AP classes were available for students. The chart in Figure 8.12 was developed to show the difference in enrollment for different AP classes. It is obvious that the more general classes, such as psychology, had a higher enrollment than classes that required a certain aptitude, such as calculus. This serves to confirm that counselors recommended classes that would serve a general curriculum purpose in the case where the student was undecided in which AP class to enroll. As a result of the higher enrollment into AP courses, class size appears to have also increased. In Figure 8.13, class size is compared for school years, 2003–2004 and 2004 to 2005.

The Pareto charts in Figures 8.14 and 8.15 were created as a method for comparing the classes that had the highest percentage of A scores and F scores in the 2004–2005 school years, respectively. The class that had the highest percentage of A scores was psychology, which accounted for more than 20% of the total number of A scores given. On the other end of the spectrum, the class that had the highest percentage of F scores was statistics, followed by English composition, both of which accounted for more than 20% of the F scores given.

The chart in Figure 8.16 shows the distribution of letter grades for different ethnicities in the 2004–2005 academic years, after the open access process was implemented. The chart in Figure 8.17 shows the grade comparisons before open access (2003–2004) and after open access (2004–2005) school year for all AP classes. As a preliminary observation, it can be noted there was a larger percentage of failing grades across all ethnicities after the open access system was put in place. Also, there was a lower percentage of A and B grades overall for all ethnicities.

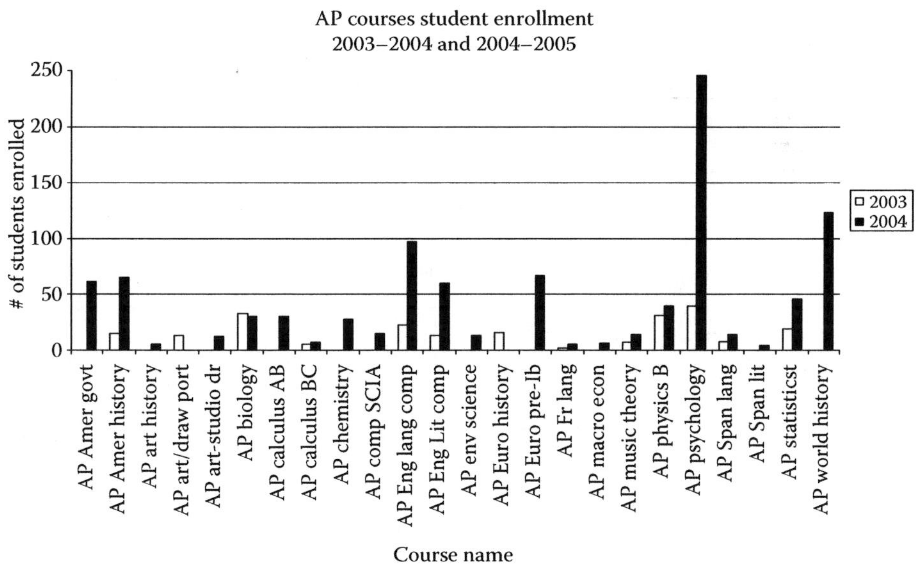

FIGURE 8.12 Enrollment in AP courses: pre- and post-open access.

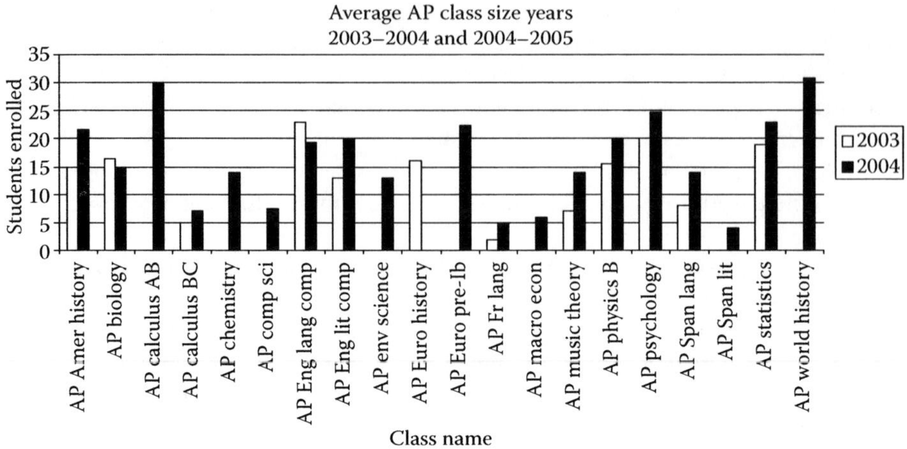

FIGURE 8.13 Average AP class size: pre- and post-AP open access.

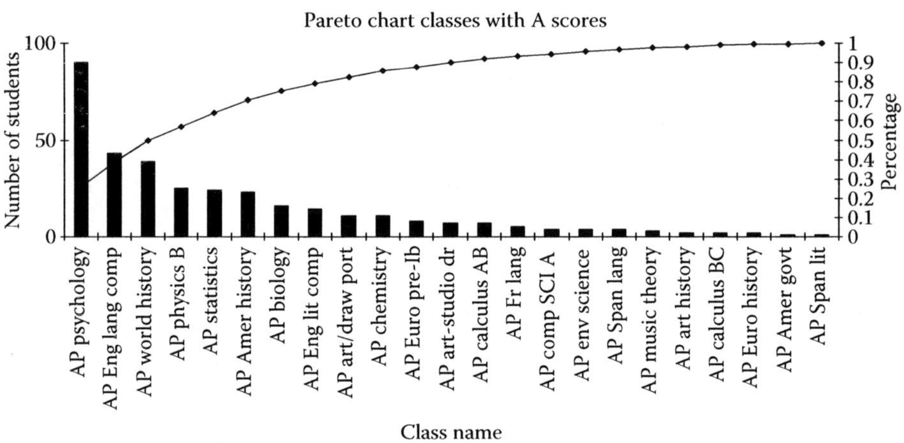

FIGURE 8.14 Pareto chart classes with A scores.

8. COPQ

Quality cost consists of all the costs associated with those school efforts devoted to the open access AP enrollment system, those associated with the efforts to verify that quality is being obtained, and those associated with failures

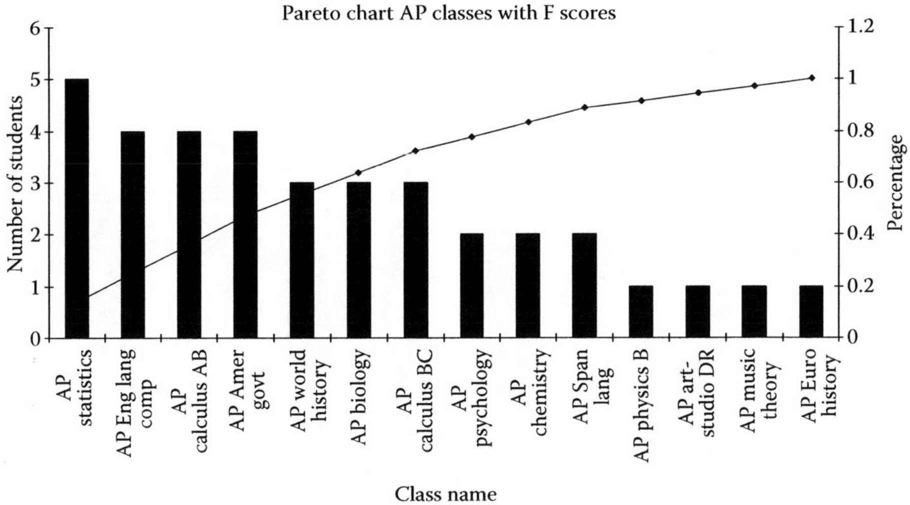

FIGURE 8.15 Pareto chart classes with F scores.

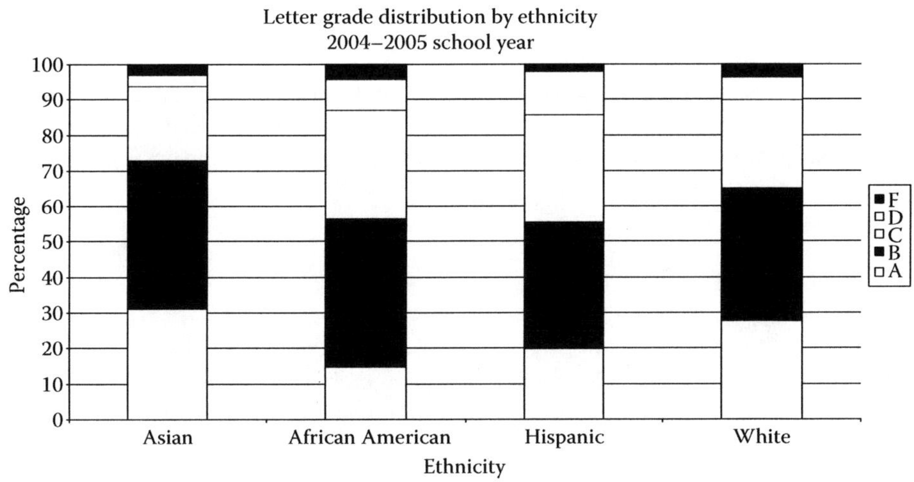

FIGURE 8.16 Letter grade distribution in AP classes by ethnicity after open access (2004–2005).

resulting from the inefficient open access AP enrollment. Quality cost categories are prevention costs, appraisal costs and failure costs (internal and external). The costs of quality described below have been determined for the student and the school.

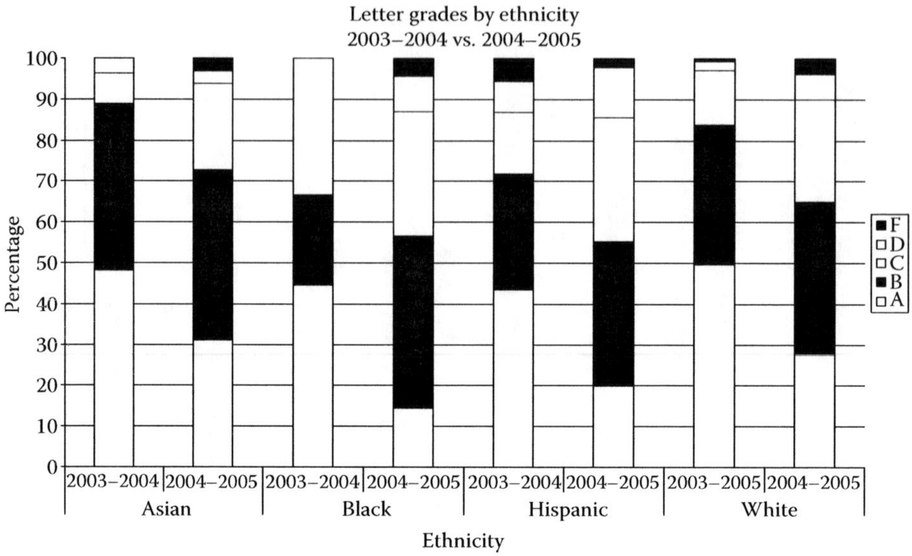

FIGURE 8.17 Letter grades by ethnicity before open access (2003–2004) and after (2004–2005).

Prevention Costs

Prevention costs are those the school incurs before the students are tested or graded. They do not deal with testing but with having the school ready to perform at its best. The expenses are mostly related to faculty and staff. The prevention costs are:

- Students:
  Money spent on a private tutor: 40 hours/year at $20/hour = $800
- School:
  Money spent in training:
  - Teachers: 10 hours at $30/hour for 15 teachers = $4500
  - Counselors: 10 hours at $30/hour = $300
- Money spent in planning quality: Providing resources at school

Total Cost = $5,600.

Appraisal Costs

Appraisal expenditures relate to the assessment and related processes. It involves all processes to assess the current state. Some of them are:

1. Student:
   - Money spent purchasing study guides: $30 per guide

2. School:
   - Money spent on after school tutoring: 5 tutors at $20,000/year = $100,000

- New counselors: 5 at $40,000/year = $200,000
- New experienced teachers = 15 at $40,000/year = $600,000

Total cost = $900,030

Internal Failure Costs

Costs of failures internally are those incurred in diagnosing possible failure causes to identify alternatives for improvement. It also refers to modifying procedures to achieve desired goals. Some failure costs are:

1. Student:
    - Time wasted in class: 200 hours at a job making $6.50/hour = $1300
2. School:
    - Money spent on implementing the AP system: 100 hours at $20/hour = $2000

Total cost =$3300

External Failure Costs

The external failure cost is more subjective because it refers to not perform according to standards. External costs include:

1. Student:
    - Not receiving credit for class toward graduation:
        - One-year delay in graduation = $35,000 earned working
        - Retaking the class = $200
    - Not receiving AP credit for college: one college course of 3 credits at $200/ credit = $600
2. School:
    - Students retaking classes denying room for others: 50 students at $200/ student = $1000

Total Cost = $36,800
Grand Total: $945,730

9. Measure Phase Presentation

The Measure phase presentation can be found in the downloadable instructor materials.

MEASURE PHASE CASE DISCUSSION

1. **Measure Report**
    1.1 Review Measure report and brainstorm some areas for improving the report.
    1.2 How did your team ensure the quality of the written report? How did you assign the work to your team members? Did you face any challenges of team members not completing their assigned tasks in a timely manner, and how did you deal with it?

1.3 Did your team face difficult challenges in the Measure phase? How did your team deal with conflict on your team?
1.4 Did your instructor and/or Black Belt or Master Black Belt mentor help your team better learn how to apply the LSS tools in the Measure phase, and how?
1.5 Did your Measure phase report provide a clear understanding of the VOC and the VOP? Why or why not?

2. Process Maps

2.1 While developing the process maps, how did your team decide how much detail to provide on the level-2 process maps?
2.2 Was it difficult to develop a level-2 from the level-1 process maps? What were the challenges?

3. Operational Definitions

3.1 Review the operational definitions from the Measure phase report, define an operational definition that provides a better metric for assessing some of the quality related metrics.
3.2 Discuss why it may be important to balance the qualitative and quantitative measures.

4. Data Collection Plan

4.1 Incorporate the enhanced operational definition developed in number 3 above into the data collection plan from the Measure phase report.

5. VOC

5.1 How did your team decide how to collect the VOC information?

6. VOP Matrix

6.1 How does the VOP matrix help to tie the CTSs, the operational definitions and the metrics together?

7. Statistical Analysis and Pareto Chart

7.1 What other statistical analysis would you recommend performing?
7.2 Discuss how the Pareto chart provides a priority or focus for what you graphed?
7.3 What conclusions can you draw from the Pareto charts?

8. Cost of Poor Quality

8.1 Would it be easy to quantify, and collect data on the costs of quality that you identified for the case study exercise?

9. Measure Phase Presentation

9.1 How did your team decide how many slides/pages to include in your presentation?

9.2 How did your team decide upon the level of detail to include in your presentation?

ANALYZE PHASE EXERCISES

1. **Analyze Report**
   Create an Analyze phase report, including your findings, results, and conclusions of the Analyze phase.

2. **Cause and Effect Diagram**
   Create a cause and effect diagram for the lower quality of AP courses.

3. **Why-Why Diagram**
   Create a Why-Why diagram for why students are pushed into AP classes.

4. **Waste Analysis**
   Brainstorm potential wastes in the AP open access process.

5. **Correlation Analysis**
   Perform a correlation analysis for the following variables:
   - Student GPA and AP course grade
   - AP course grade and number of AP classes for each student
   - Other variables of interest in the student database

6. **Regression Analysis**
   Perform a regression analysis to try to predict the AP exam grade based on the following independent variables: GPA, AP class grade. Is this a good model?

7. **Histogram, Pareto, Graphical, and Data Analysis**
   Perform a histogram and graphical analysis for the following data from the "AP Data.xls":
   - Percentages by race for students enrolled in AP classes before open access (2003/2004).
   - Percentages by race for students enrolled in AP classes after open access (2004/2005).
   - Percentages of students in the F&R lunch program compared to those not in the program for those students enrolled in AP classes.

8. **Hypothesis Testing, ANOVA**
   Perform the following hypothesis tests:
   - Minority enrollment is the same for 2003/2004 and 2004/2005.
   - The percentage of students who achieved a 3 or better in the AP test in 2003/2004 is the same percentage in 2004/2005.

- Students enrolled in AP for the year 2004/2005 performed equally as those students who were enrolled in 2003/2004

Perform ANOVA to test the following hypotheses:

- F&R lunch program students' performance in AP is equal to non-F&R performance in AP
- Minority students' performance in AP classes is equal to nonminority students' performance in AP classes

9. DPPM/DPMO

Calculate the DPMO and related sigma level for the process, assuming a 1.5 sigma shift, for the following data:

AP grades for 2003 to 2004 (prior to open access):

Opportunities for failure:

- One opportunity per course for final grade average = 1

Defects:

- Number of D or F grades = 12

Units:

- Number of AP course grades = 225

AP grades for 2004 to 2005 (open access):

Opportunities for failure:

- One opportunity per course for final grade average = 1

Defects:

- Number of D or F grades = 106

Units:

- Number of AP course grades = 989

AP exam scores for 2003 to 2004 (prior to open access):

Opportunities for failure:

- One opportunity per course for exam score = 1

Defects:

- Number of 1 or 2 exam scores = 277

Units:

- Number of AP exams taken = 621

AP exam scores for 2004 to 2005 (open access):

Opportunities for failure:

- One opportunity per course for exam score = 1

Defects:

- Number of 1 or 2 exam scores = 835

Units:

- Number of AP course grades = 1287

10. Analyze Phase Presentation
Prepare a presentation (PowerPoint) from the case study exercises that provides a short (10–15 minutes) oral presentation of the Analyze phase deliverables and findings.

ANALYZE PHASE

1. Analyze Report

The Analyze phase takes the VOP and VOC data that have been collected in the Define and Measure phases and analyzes it for patterns, inefficiencies, and root causes. We will use cause and effect analysis to understand the root causes that contribute to the low quality of the AP class experience, low grades and low AP test scores. We will use statistical analysis, hypothesis tests and ANOVA to assess whether the percentages of minorities and lower socioeconomic students have increased in AP classes since the open access system has been put in place in 2003/2004. Identifying the root causes and inefficiencies will help us to identify improvements for the Improve and Control phases.

2. Cause and Effect Diagram

Root cause analysis is a very important process in a LSS project. Data are analyzed in detail and tools are used to determine the root causes of problems and inefficiencies. Too often, data are collected and project team members, champions, or knowledge workers jump to conclusions based on the raw data. LSS and DMAIC prevent this from happening. Several tools are very useful in determining the root causes.

One area of primary improvement was chosen to determine the root causes of the lower quality of the AP classes. While identifying these root causes, it is critical to call upon the expertise and experience of all the people involved in the process to generate as many possible root causes as possible. Therefore, the LSS team conducted brainstorming sessions with three groups consisting of AP counselors, AP students, and AP teachers. During these brainstorming sessions, the LSS team, aided by the three different groups, generated a list of possible root causes, reflecting problems faced by the SHS AP open access system. The outcome of the brainstorming session was a fishbone diagram (cause and effect diagram) that represents the primary area of improvement. This area includes the lower quality of AP open access system. The cause and effect diagram is shown in Figure 8.18.

As seen in the fishbone diagram, five branches were constructed as potential areas where causes exist. The branches were related to the students, communication between stakeholders, counselors, classes, parents, and teachers. From there, the SHS LSS team brainstormed potential causes to the effect. Several root causes were identified as the primary reasons of the lower quality of the new AP open access system. First, by nature of the training and experience, the new AP teachers are not trained formally and the majority of the new AP teachers are inexperienced teaching AP courses. This is evident by the data collected from the student focus group.

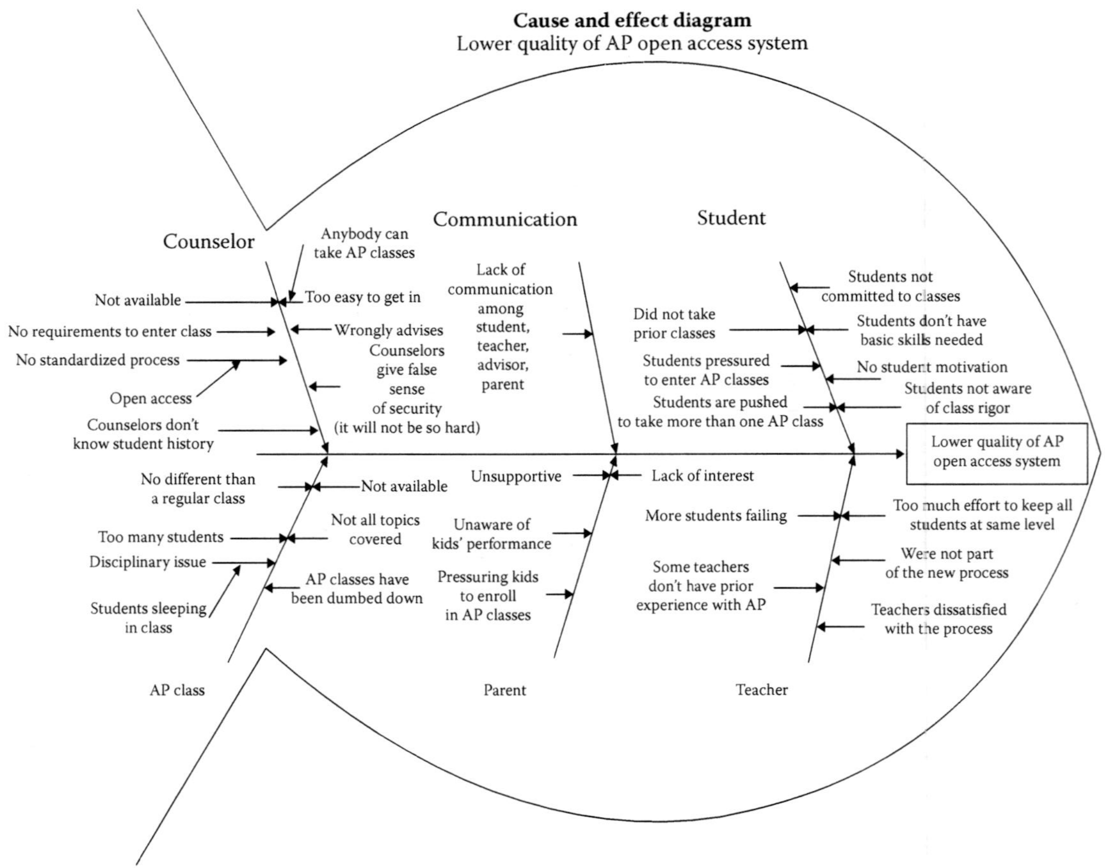

FIGURE 8.18 Cause and effect diagram.

The fact that there are 15 new teachers teaching AP courses, teachers are making a stronger effort to keep all students at the same level, and teachers were not taken into consideration during the new open access system decision making process, could contribute to the lower quality of the AP open access system. Additionally, the fact that students are not committed to the AP classes, students are not aware of the rigor of AP classes, and students are being pressured to enter AP classes without having the basic knowledge for a specific course, contribute to the lower quality of the AP open access system. Other root causes of the lower quality of the AP open access system include parents being unaware of their child's performance in AP classes, counselors not being aware of a student's history outside of paper data, and counselors giving students a false sense of security.

Summary of Problems

After conducting the focus groups with the SHS counselors, AP faculty, and AP students, the Six Sigma team was able to determine some of the major defects or problems of the open access system:

- Students are being pushed into AP classes
- There are no requirements for enrolling in AP courses
- AP classes are too large
- There are inexperienced AP teachers

3. Why-Why Diagram

We used the five Why's and a Why-Why diagram to understand why the students are being pushed into AP classes. The Why-Why diagram is shown in Figure 8.19. The first question is why are students being pushed into AP classes? The counselors and some parents are pushing them into the AP classes. Why are the counselors pushing the students to take AP classes? The administration wants them to increase the percentage of under-represented groups (minorities and lower socioeconomic students). Why? To enhance the school grade and to enhance the students' academic credentials. Why do they want to enhance the school grade? To improve funding and also enhance the school's prestige. Why are the parents pushing the students to take AP classes? To enhance their students' academic credentials. Why? To have better opportunities for scholarships and to get into better colleges, and possibly for prestige.

4. Waste Analysis

The main types of waste in the AP open access process are related to the following wastes:

- Processing: Students do not attend and do not do well, taking teaching resources away from other AP students.
- Defect: Students not doing well in the AP class or not doing well on the AP exam, and not receiving college credit after taking the course; teachers not covering all of the topics that the exam requires.

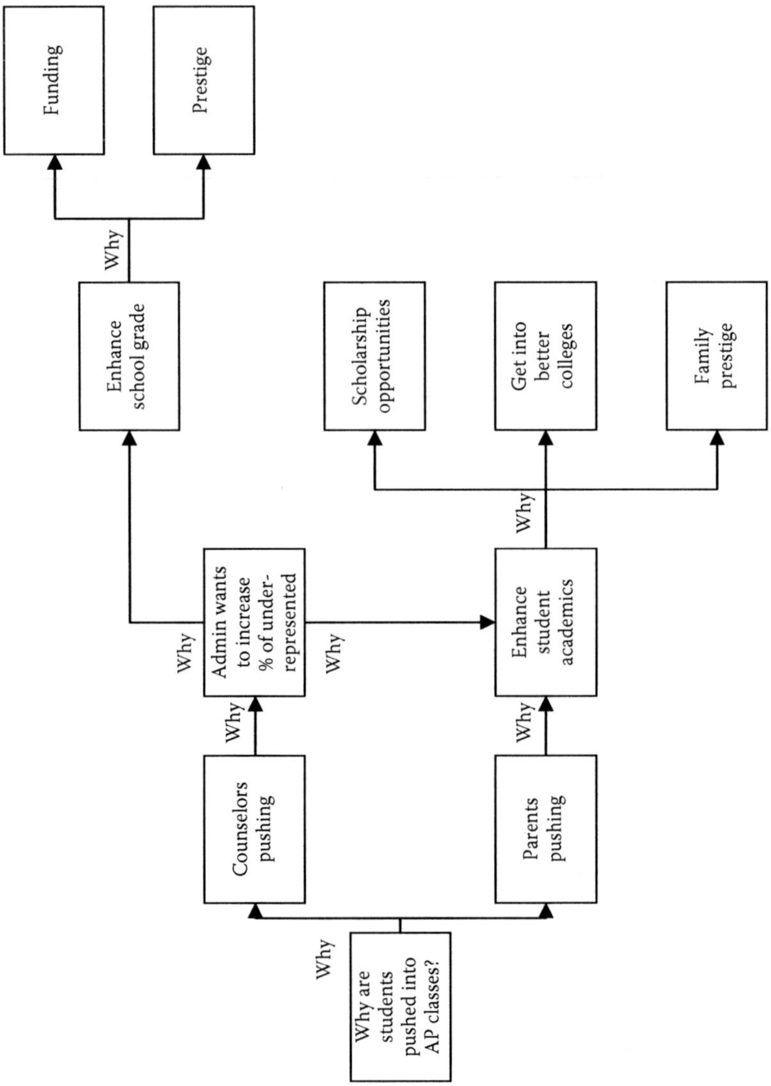

FIGURE 8.19 Why-Why diagram.

- Delay: Students not passing the class and having to take another class to graduate.
- People: Students not being motivated or not attending the class. This impacts the unmotivated student as well as other students in the class that are distracted by them.

5. Correlation Analysis

We performed a correlation analysis to determine if there is a correlation between the unweighted GPA and the AP course grades in academic year 03/04 and 04/05.

There is a fairly strong correlation for both years between the GPA and the grade received in the AP course, as would be expected. The correlation coefficient (r) is 0.554 for 03/04 and 0.571 for 04/05.

We also performed a correlation analysis to determine if the AP course grade and number of AP classes for each student was correlated; r is only 0.125, not representing a correlation between the variables.

6. Regression Analysis

We performed a regression analysis to develop a model that could potentially predict the AP exam score, based on the student's GPA and their performance in the AP class. We did not find a model that predicted the AP exam score very well. The coefficient of determination for 03/04 was 21.9% and 20.1% for 2004/2005.

7. Histogram, Graphical, and Data Analysis

Minority Enrollment before and after Open Access

One of the first issues discussed with the administration at SHS was minority enrollment in AP. It was presumed that after the implementation of open access there would be a more representative distribution of all ethnicities in the AP classroom. Using demographic data it was possible to determine the exact distribution of these ethnicities before and after open access (Figure 8.20).

The current ethnic distribution for the entire school is 43.6% Hispanic, 37.9% White, 10.9% African-American, and 8.4% other. Based on this, we can conclude that the trend from one year to the following was beneficial to all minorities, and it approaches the actual school distribution better than any of the previous years.

Figure 8.20 conceals the fact that the enrollment, as a result of open access, increased significantly, and it would be interesting to see what the increase in enrollment was for each minority as a percentage of their own population. Figure 8.21 is a summary table of the AP enrollment by ethnicities and as a percentage of their own ethnic population.

There was an increase in enrollment for all ethnic races, with the biggest percentage increase seen in the Asian population (8.2–34.7%). The influx of new

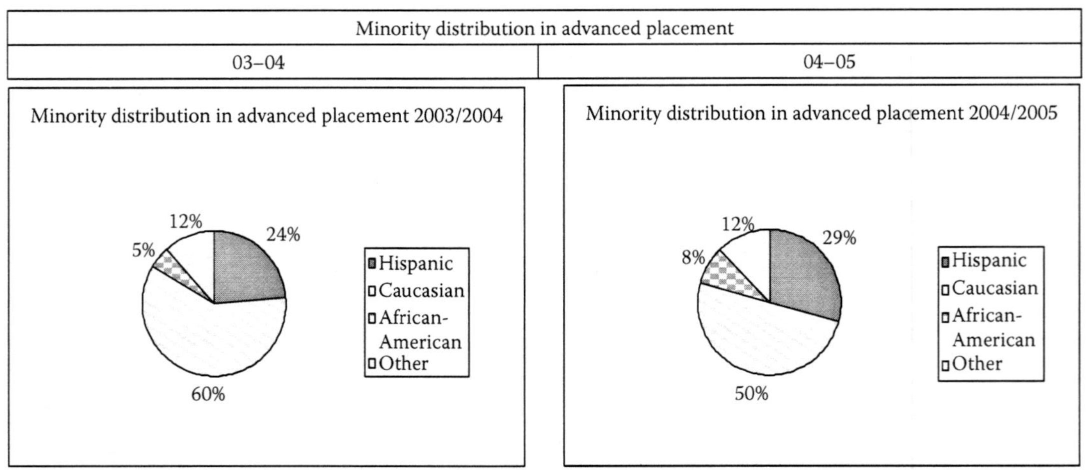

FIGURE 8.20 Minority distribution in advanced placement.

Hispanic				Black			
Year	Enrolled	n	% of H. Pop	Year	Enrolled	n	% of B. Pop
2003	43	1306	3.2%	2003	9	274	3.2%
2004	185	1277	14.5%	2004	53	273	19.4%
Asian				**White**			
Year	Enrolled	n	% of A. Pop	Year	Enrolled	n	% of W. Pop
2003	21	257	8.2%	2003	109	1398	7%
2004	78	225	34.7%	2004	317	1266	25%

FIGURE 8.21 Minority percentages in AP courses.

students into open access completely changed the face of the classroom, it is now a much larger and diverse student body, but the actual impact of this change must be measured to determine the benefits or downfalls of this system. Using a two-proportion test for each of the races, comparing 2003/2004–2004/2005, there was a significant increase in percentage by all races enrolled in AP courses.

The percentage of the lower socio-economic group also increased as a percentage after open access. In the 2003/2004 academic year, the percentage of students that received free or reduced lunch in AP classes was 13%, and increased in 2004/2005 to 21% of the students enrolled in AP courses. Using a two-proportion test with number of F&R lunch students in 2003/2004 of 23 out of a total number of AP students of 181, compared with 2004/2005 of 138 F&R lunch students out of 652 AP students, we conclude that there is a difference between the percentage of F&R lunch students in AP courses between 2003/2004 and 2004/2005. The percentage of students in the lower socioeconomic group increased as a percentage after open access.

8. Hypothesis Testing/ANOVA

Class Performance by Ethnicities

The premise of this analysis is to measure if there are differences in the capabilities of the different ethnicities. In the previous section, it was proven there has been an increase in enrollment across all racial lines, but it is necessary to evaluate how the different groups are performing in class and then take the necessary action to ensure all students are performing on the same level. AP exam scores were obtained for all students in 2004–2005 and an ANOVA test was conducted across the different ethnicities.

The data provided allowed differentiation between a few more ethnic groups (I = Indian, M = Multiracial) that seem to have scored lower than their peers. However, the null hypothesis states that there is no significant difference between the entities. With a *P* value of .125, we fail to reject our null hypothesis; therefore, we cannot conclude that the scores between the different races are different from each other.

F&R Lunch AP Performance

Another division between the student lines besides gender and race is the economic level of their families. SHS has implemented a F&R lunch program for the children of low-income families. It works just like the program title says, offering lunch at a reduced price, or even free of charge to those students in need.

The LSS team was provided with a list of students participating in the F&R program. Using their student Ids, it was possible to identify which of these students were enrolled in AP and compare them to all the other AP students who were not part of the F&R lunch program. Just as in the previous comparison between the races, the null hypothesis is that there is no difference between the two groups.

The results from the ANOVA test showed how the F&R students performed a bit worse than those students who are not in the F&R lunch program. However, with a P-value of 0.708 we fail to reject our null hypothesis, therefore, we cannot conclude the difference between the scores is significantly different.

Learning Gains from Enrollment in AP

Learning gains is a measure obtained from the FCAT exam taken by all the students in their 9th and 10th grade of high school. Although there are limited offerings of Advanced Placement classes for 9th and 10th graders, there are enough students enrolled to justify an analysis.

The results show a significant difference in the average learning gain in FCAT reading between the students who took AP courses and those who did not. The mean Developmental Scale Score (DSS) for students who took at least one AP course is a lot higher for before and after the open access system. A minimum DSS score of 77 is required for a student to be officially recognized as achieving a learning gain, and in 2003–2004 both types of students did show an improvement average. However, for 2004–2005, students that were not enrolled in AP classes only averaged a 56.4, not sufficient for recognition.

It is important to notice how the overall DSS score was lower in 2003–2004 versus 2004–2005, and there seems to be no apparent explanation for this occurrence.

AP Exam Scores in 2003–2004 vs. 2004–2005

Using the data available, it was possible to analyze the performance of those students who took the advanced placement exam before the implementation of open access and compare it with those students who took it afterward.

The test shows the decrease in passing rates from one year to another, dropping considerably from a 55.39% pass rate to 35.12%. Another interpretation of the data proves how a 107% increase in enrollment attributed to a 31% increase in passing scores, which translates into 1 out of every 3.5 **newly** enrolled students actually scoring a 3 or higher on the AP exam.

AP Grades for Students in 2003–2004 vs. 2004–2005

For this analysis, students enrolled in AP during 2003–2004 were tracked the following year. Using their GPA for both years, it was possible to measure any increase or decrease in performance. Although any significant change in performance can be partially attributed to the open access system, it is difficult to specify which specific factor of the open access contributed the most in causing this change. A paired t-test was conducted using the weighted GPA of the same students in 2003–2004 vs. 2004–2005.

It was understood from the teacher focus group the classes had to be "watered down" to accommodate new students. This led us to believe those students enrolled in 2003–2004 would be easily capable of handling the workload and thus maintain or even improve their weighted GPA. However, this was not the case. The *p*-value of zero translates into a significant change in the GPA, meaning those same students in 2003–2004 are actually performing worse after the implementation of open access. Theories as to the cause of this event could be the following, but are not limited to:

- Too many students; less personalized attention
- New teachers; poor knowledge delivery
- Camaraderie; students now surrounded by friends, focused less on class
- A combination of the above

Teacher AP Experience Level 2003–2004 vs. 2004–2005

The quick transition into open access meant many teachers being moved into the AP curriculum with very little training and a short time to prepare. The quick growth of AP at SHS also meant new teachers being hired. Thus, open access started with a mixed pool of experienced teachers and a new set of teachers with little to no experience teaching AP in the classroom. This led to a very simple question: Was there a difference in performance between the two sets of teachers? In this case, we measured performance as the amount of their students who scored higher than a 3 on the AP exam. We are assuming that students were allocated randomly between the teachers, so in theory both teachers' students should perform equally. An ANOVA test was conducted to achieve a conclusion.

Setting up the data for this analysis was fairly simple, all the students from AP teachers in 2003–2004 who were still teaching in 2004–2005 were named "Level-2", and they were compared with all the students from AP teachers who were new to the AP curriculum, who were labeled "Level-1". The mean scores seen in Figure 8.22 represent the average score on the AP exam that their corresponding students received. The difference is evident, and there is no doubt that teachers who have more years of experience are able to teach better, therefore improving the learning experience for the students and reflect it on their AP score.

There were several other hypothesis tests that were performed on the student data related to the AP open access system. The results are shown in Figure 8.23.

9. DPPM/DPMO

Calculating Sigma Levels

The LSS Team analyzed the amount of defects as a fraction of opportunities for error. Our focus is on AP grading in which a defect is defined as a student who obtains a grade of D or lower or a score of 2 or lower on his/her corresponding AP class. Obviously, every student enrolled counts as an opportunity for a defect to occur because they all have the possibility (albeit different likelihoods) of creating a defect.

One-way ANOVA: AP test versus Years Taught

```
Source         DF       SS      MS      F      P
Years Taught    1    82.46   82.46  79.54  0.000
Error         886   918.53    1.04
Total         887  1000.99

S  = 1.018   R-Sq  = 8.24%                R-Sq(adj)    =    8.13%

                                          Individual 95% CIs For Mean
                                          Based on Pooled StDev
Level     N     Mean    StDev             ----+--------+---------+-----
1       251    1.518    0.831            (----*----)
2       637    2.195    1.083                                ( -* -)
                                  ----+---------+---------+--------+------
                                    1.50       1.75      2.00     2.25
Pooled StDev = 1.018
```

Minitab® results.

FIGURE 8.22 ANOVA test of teacher performance as a measure of years teaching.

Sigma Levels for AP Grades

Our first analysis is observing the AP grades obtained by the students during the 2003–2004 school years and comparing it with the 2004–2005 school years.

2003–2004 AP Grades Sigma levels

# of grades obtained (units)	= 225
Defects ("D" or "F" grades)	= 12
Yield → 225–12/	
225 × 100%	= 94.67%

Corresponding Sigma Level The closest value was found to be a sigma of 3.10 for a yield of 94.52%. For 2003–2004, the sigma level for the AP students' grade was 3.1147.

2004–2005 AP Grades Sigma levels

# of grades obtained (units)	= 989
Defects ("D" or "F" grades)	= 106
Yield → 989–106/	
989 × 100%	= 89.28%

Corresponding Sigma Level The closest value was found to be a sigma of 2.7 for a yield of 88.5%. For 2004–2005, the sigma level for the AP students' grade was 2.742.

Figure 8.24 is a graphical representation of the changes that occurred from one year to the next.

After the implementation of open access, the average performance of the AP students decreased as a whole, leading to a noticeable amount of grades

Test no.	Null hypothesis	Data used	Test used	Test result	Conclusions
1	**Minorities performance in AP is equal to nonminority performance in AP**				
1a	Ethnicity vs. AP test score	AP test scores	ANOVA	Fail to Reject	No significant difference in test scores for each ethnicity 03–04, 04–05
1b	Ethnicity vs. grades in each AP Course	AP grades	ANOVA	Fail to Reject	No significant difference in grade in AP courses for each ethnicity 03–04, 04–05
2	**F&R performance in AP is equal to non-F&R performance in AP**				
2a	F&R vs. AP test score	AP test scores	ANOVA	Fail to Reject	No significant difference in test score for low socioeconomic 03–04, 04–05
2b	F&R vs. grades in each AP course	AP grades and lunch codes	ANOVA	Fail to Reject	No significant difference in AP grades for low socioeconomic 03–04, 04–05
3	**Learning gains for students in AP is equal to students not enrolled in AP**				
3a	Learning gains for all AP courses for students in AP vs. not in AP 10th graders. (2004/2005)	FCAT learning gains	ANOVA	Reject	AP students have higher learning gains
3b	Learning gains for all AP courses for students in AP 10th graders. (2004/2005)	FCAT learning gains	ANOVA	Reject	Although there is a difference in gains, on average students achieved gains
4	**Minority enrollment is the same for 2003/2004 and 2004/2005**				

FIGURE 8.23 Hypothesis test results.

Test no.	Null hypothesis	Data used	Test used	Test result	Conclusions
4a	% Increase in Whites	White students, White students enrolled in AP	Test for two proportions	Reject	There is a larger percentage of White students enrolled in AP courses
4b	% Increase in Blacks	Black students, Black students enrolled in AP	Test for two proportions	Reject	There is a larger percentage of Black students enrolled in AP courses
4c	% Increase in Hispanics	Hispanic students, Hispanic students enrolled in AP	Test for two proportions	Reject	There is a larger percentage of Hispanic students enrolled in AP courses
4d	% Increase in Asians	Asian students, Asian students enrolled in AP	Test for two proportions	Reject	There is a larger percentage of Asian students enrolled in AP courses
5	**The percent of students who achieved a 3 or better in the AP test in 2003/2004 is the same percentage in 2004/2005**	AP test scores	Test for two proportions	Reject	Less percentage of students taking the test are earning scores higher than 3
6	**Students enrolled in AP for the year 2004/2005 performed equally as those students who were enrolled in 2003/2004**	AP class grades	Paired T-test	Reject	Students in 2004/2005 had a lower GPA than students in 2003/2004
7	**Years teaching AP courses have no effect in student performance in AP test scores**	Years teaching, AP test scores	ANOVA	Reject	Students who have classes with teachers that have more than 1 year teaching AP perform better in the AP test

FIGURE 8.23 (Continued)

lower than a C. Although it cannot be specified whether the students present in 2003–2004 actually decreased their grades, it can be confidently stated that along with the increase in student enrollment, there was a decrease in the overall class performance.

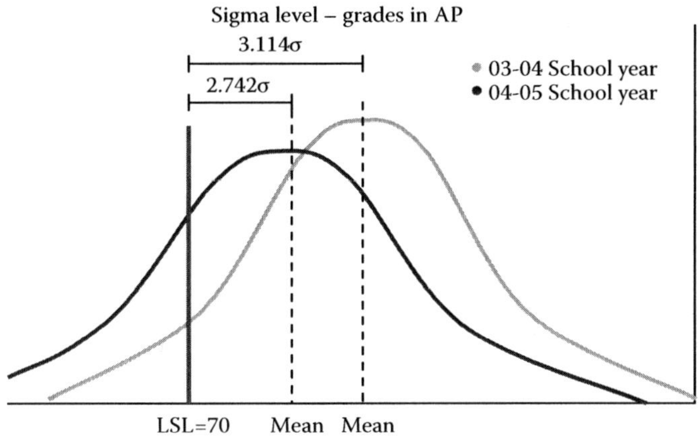

FIGURE 8.24 Sigma levels grades in AP classes.

The open access system has allowed the students to freely enroll in whichever AP classes, however, measurements should be taken so that a student can be properly evaluated to determine whether he/she is capable of handling the workload.

Sigma Levels for AP Scores

The second analysis corresponds to the observation of the AP exam scores obtained by the students during the 2003–2004 school years and compares it with the 2004–2005 school year.

2003–2004 AP Scores Sigma Levels

# of scores obtained (Units) = 621
Defects ("1" or "2" exam score) = 277
Yield → 621–277/621 × 100% = 55.39%

Corresponding Sigma Level The closest value was found to be a sigma of 1.6 for a yield of 54.5%. For 2003–2004, the sigma level for the AP students' score was 1.625.

2004–2005 AP Scores Sigma Levels

# of scores obtained (Units) = 1287
Defects ("1" or "2" exam score) = 835
Yield →1287–835/
1287 × 100% = 35.12%

Corresponding Sigma Level The closest value was found to be a sigma of 1.11 for a yield of 35%. For 2004–2005, the sigma level for the AP students' score was 1.1133.

Figure 8.25 is a graphical representation of the changes that occurred from one year to the next in the AP exam scores.

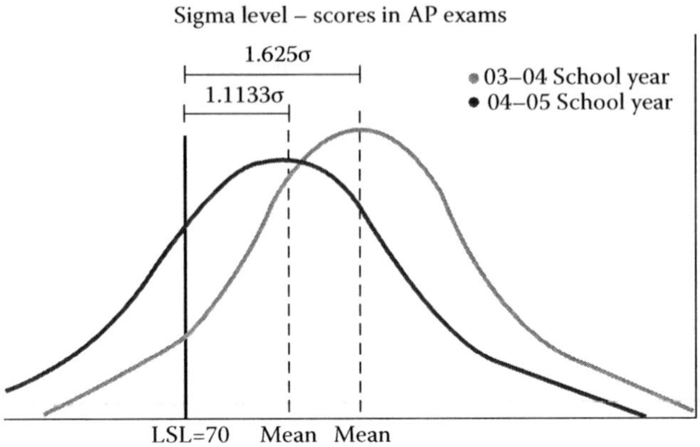

FIGURE 8.25 Sigma level – scores in AP exams.

After the implementation of open access, the average performance of the AP students decreased as a whole, leading to a noticeable amount of grades lower than a C, resulting in a decrease in the overall class performance.

After the implementation of open access, the mean score of the exams taken by the students was lower than in the previous year, thus a bigger percentage of students are getting grades of 2 or lower. Just as in the previous situation, it cannot be specified whether the students present in 2003–2004 actually decreased their performance, but the team can confidently state that along with the increase in student enrollment, there was a decrease in the overall class performance.

The open access system has allowed the students to freely enroll in whichever AP class they desire, but the measurements taken in this phase also strengthen our previous point: a student needs to be properly evaluated to determine whether he/she is capable of handling the workload.

The financial implications of enrolling students in AP are not clearly understood. It is presumed that a high school gets increased funding from the state by increasing the size of their AP curriculum and students enrolled. Also, the state pays $80 to cover the cost of each AP exam taken. There appears to be no out-of-pocket costs for SHS, but there are significant losses to state (whose funding comes from taxpayers) every time a student fails to perform successfully.

10. Analyze Phase Presentation

The Analyze phase presentation can be found in the downloadable instructor materials.

ANALYZE PHASE CASE DISCUSSION

1. **Analyze Report**
   1.1 Review the Analyze report and brainstorm some areas for improving the report.

1.2 How did your team ensure the quality of the written report? How did you assign the work to your team members? Did you face any challenges of team members not completing their assigned tasks in a timely manner, and how did you deal with it?
1.3 Did your team face difficult challenges in the Analyze phase? How did your team deal with conflict on your team?
1.4 Did your instructor and/or Black Belt or Master Black Belt mentor help your team better learn how to apply the LSS tools in the Analyze phase, and how?
1.5 Did your Analyze phase report provide a clear understanding of the root causes of the process. Why or why not?

2. Cause and Effect Diagram

2.1 How did your team determine the root causes, and how did they validate the root causes?

3. Why-Why Diagram

3.1 Was it easier to create the cause and effect diagram, or the Why-Why diagram? Which of the tools was more valuable getting to the root causes?

4. Waste Analysis

4.1 What types of waste were prevalent in the process and why?

5. Correlation Analysis

5.1 Were there any significant variables that were correlated? Do they appear to have a cause and effect relationship, and why?

6. Regression Analysis

6.1 Were you able to identify a model that can predict the grade on the AP exam? Why or why not?

7. Histogram and Graphical Analysis

7.1 What type of distribution does your data appear to be from a graphical analysis?
7.2 Can you test your distribution statistically and determine a likely distribution, what is it?
7.3 Did you have any outliers in your data?

8. Hypothesis Testing and ANOVA

8.1 What were your key findings for your hypothesis tests?
8.2 What conclusions can you make from a practical perspective?
8.3 How might you use these findings in the Improve phase?
8.4 What were your key conclusions in your Analysis of Variance?

9. DPPM/DPMO

9.1 What is your DPPM/DPMO and sigma level. Is there room for improvement, and how did you determine that there is room for improvement?

10. Analyze Phase Presentation

10.1 How did your team decide how many slides/pages to include in your presentation?

10.2 How did your team decide upon the level of detail to include in your presentation?

IMPROVE PHASE EXERCISES

1. Improve Report

Create an Improve phase report, including your findings, results and conclusions of the Improve phase.

2. Recommendations for Improvement

Brainstorm the recommendations for improvement.

3. Revised QFD

Revise or create a QFD house of quality to map the improvement recommendations to the critical to satisfaction characteristics.

4. Action Plan

Create an action plan for demonstrating how you would implement the improvement recommendations.

5. Future State Process Map

Create a future state process map for the following AP open access registration processes.

6. Revised VOP Matrix

Revise your VOP matrix from the Measure phase with updated targets.

7. Training Plans, Procedures

Create a training plan, and a detailed procedure for one of the process steps.

8. Improve Phase Presentation

Prepare a presentation (PowerPoint) from the case study exercises that provides a short (10–15 minutes) oral presentation of the Improve phase deliverables and findings.

IMPROVE PHASE

1. Improve Report

The main focus of this project was to assess whether the percentage of minorities and students in the F&R Lunch Program increased after open access. Early in the Define phase, the team identified that while the quantities looked to improve, there was a perception that the quality of the AP experiences was being negatively impacted. The VOC and the AP grades and exam results has validated this decrease in the quality of the AP courses. The improvement recommendations will mainly be focused at rebalancing the quality with the quantity related to the AP experience in the future.

The SHS LSS team put together multiple recommendations that would improve the overall AP open access system and subsequently the AP academic environment as a whole. The recommendations, and suggested implementation included as a guide for the SHS leadership team to follow when ultimately designing and implementing changes to the system. The recommendations that follow are based on data collected from the three focus groups, interviews with leadership team staff, and analysis of the data collected from the SHS database.

2. Recommendations for Improvement

This list below summarizes the recommendations by this project team.

Recommendation # 1: Develop a more standardized AP enrollment process. This is a written publication that outlines the specific SHS policy on AP placement. The publication would cover all aspects of the AP placement process including the process to enroll in a course, requirements, and students', parents', and teachers' responsibilities. This creates a baseline for the AP placement process which will help to create consistency among counselors when placing students in AP courses.

Recommendation # 2: Set minimum requirements for enrollment into AP classes. This is a written publication outlining the specific requirements students need to meet before enrolling in AP courses. An explanation of each requirement should be in place to ensure an understanding of expectations. This written publication will be reviewed before and during a student's enrollment in AP courses.

Recommendation # 3: Create a contract for students/parents enrolling in an AP course. This contract will be signed by students, parents, and the counselor before students enroll in AP classes. It will ensure students and parents know what is required in AP courses. This contract will describe in a detailed manner, the hours of study the student will have to spend, the workload required by the course, the expectations, the tests the students will have to take at the end of the semester and the benefits of the course.

Recommendation # 4: Establish and encourage parental involvement for students enrolled in AP. This system will allow AP teachers and AP parents to have a closer relationship. This will also allow parents to get more involved in the AP system. The term "parent involvement" is used broadly in this report. It includes several different forms of participation in the AP system and with the schools. Parents can support their children's schooling by attending school functions and responding to school obligations (e.g., parent–teacher conferences). They can become more involved in helping their children improve their AP schoolwork by providing encouragement, arranging for appropriate study time and space, modeling desired behavior (such as reading for pleasure), monitoring homework, and actively tutoring their children at home. Outside the home, parents can serve as advocates for the AP system. They can volunteer to help with school activities or work in the AP classroom. Or they can take an active role in the governance and decision making necessary for planning, developing, and providing an education for the AP students.

Recommendation # 5: Consider keeping AP classes small. Since the implementation of the new open access system and the dramatic increase of students in AP classes, there has not been any assessment on how many students should be placed in each course. AP classes should be kept small so there is more contact between students and teachers. This will increase the performance of the AP students and raise the level of the AP classes. The number of students in AP classes should depend on the teacher. A matrix could also be helpful in determining how many students each teacher is able to handle. The more experience and the better class management skills an AP teacher has, the more students can be enrolled in that particular course.

Recommendation # 6: Set minimum attendance requirements. Right now, there is no minimum attendance requirement for AP students. A system that lets students know what the attendance expectations are when taking AP courses could increase students' performance. This system should explain the consequences when missing classes and consequences should get more severe relative to the number of absences a student has.

Recommendation # 7: Generate a highly detailed class syllabus with a detailed class schedule, workload required, topics to be covered, required books, and assigned homework. Students and teachers will be able to keep track of class progress to make sure they have enough time to cover what is required in the class.

Recommendation # 8: Create a knowledge-sharing program for AP teacher's best practices. This will include the involvement of teachers to discover best practices for effective AP classes. By creating this knowledge-sharing program, teachers will have the opportunity to share their strategies with respect to AP class success with other teachers. Teachers will learn from their colleagues how to work with students, how to complete class topics, and how to apply a variety of management techniques to help students become self-regulated learners. This program could help the entire faculty learn how to increase student motivation, build student–teacher relationships and increase home–school communication. The main purpose of creating this knowledge-sharing program is to enable teachers from the SHS to learn from the experiences, methodologies and achievements of colleagues.

Recommendation # 9: Create a form for students who want to get out of the AP course. This form will have to be signed and approved by the student's teacher, student's parents, student's counselor, and the student. After approval, the student will be able to drop the course. This will include the participation of students, teachers, parents, and counselors. The main purpose is to give students the opportunity to drop the course if everyone involved agrees it is best for the student.

3. Revised QFD

After reaching an agreement over the recommendations that would increase both the quality of the AP courses and the percentage of under-represented students, the LSS team created a QFD house of quality to demonstrate what influence each requirement would have on the CTSs. Additionally, the QFD provides a visual representation of what the interactions between the recommendations will be. The QFD house of quality is shown in Figure 8.26.

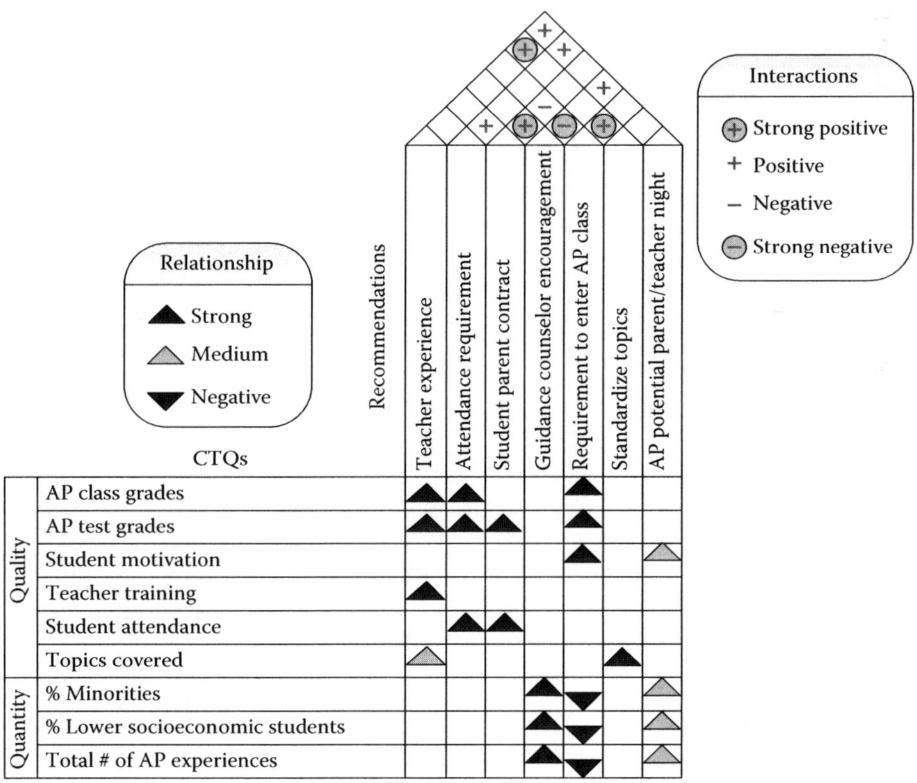

FIGURE 8.26 QFD house of quality.

The QFD shows that the recommendations will target the quality of the AP courses, while others will target increasing the number of students and increasing the percentage of students within certain ethnicities. A notable recommendation is the requirements to be able to register for AP classes. This recommendation will decrease the number of students that enter an AP class but, at the same time, will demonstrate the amount of motivation a student has to have to be in the class. This recommendation will also likely reduce the number of students that enter a class. These requirements to enter an AP class may have negatively interacted with the guidance counselor encouragement to enter AP classes. A counselor will no longer be able to encourage a student that does not meet the minimum requirements to enter a class. However, as stated in the focus groups, the teachers and students both felt that it was necessary to have some kind of requirement prior to enrolling in an AP class.

4. Action Plan

To ensure alignment between the CTSs, the recommendations, and the problems and root causes that they eliminate, we have summarized an alignment matrix in Figure 8.27. We also provide an action plan to organize the improvements into short term and long term recommendations (Figure 8.28).

5. Future State Process Map

We developed a future state process map incorporating the improvement recommendations and providing a new AP open access registration process (Figure 8.29). The recommended AP process flow shown above is a combination of the AP pre-open access system and the new AP open access system. As mentioned in the Measure phase of this report, the pre-open access system was a teacher-driven system, based on students fulfilling various requirements. These requirements included students being a level-4 or -5 reader, fulfilling all class prerequisites, scoring more than 80% in the NRT, having a minimum 3.5 GPA, submitting an essay, passing an interview, and having five teacher recommendations. On the other hand, the new open access system is a student-driven system based on no requirements. This system gave any student the opportunity to take AP courses without considering the student's academic performance, prerequisites, GPA, reading level, or any other requirements in place during the pre-open access system.

After analyzing the results from the different focus groups and all data collected, the SHS Six Sigma team has revised the open access system process flow and is recommending a system based on some requirements. These requirements will not be as rigorous as the pre-open access system, but will only consider students that are motivated to take AP courses. These requirements are as follows:

- Student will have to submit an application to take AP courses with the following information:
  - Letter of recommendation (academic teacher or parent)
  - Essay is submitted for appropriate courses
- Student will have to fulfill prerequisites for the desired AP course

	CTS	Metrics	Problems	Root cause	Recommendation
Quality	AP class grades	Percent of grades above a B	Lower AP grades in 2004/2005 vs. 2003/2004	No requirement to enter AP course Low motivation Low attendance rate	Minimum requirements Attendance requirement
	AP test grades	Percent of test scores over 3	Lower percentage of students receiving a 3,4 and 5 in 2004/2005	Lack of past AP teaching experience Low student attendance Low student motivation	Teacher expertise group Attendance requirement
	Student motivation	Teacher assessment	Lack of student motivation for classes	No requirement to enter AP course Low motivation Low attendance rate	Student/Parent contract Attendance requirement
	Teacher experience	Percent of teachers with experience teaching AP > 1 year	Lack of past experience teaching AP courses	Lack of past AP teaching experience	Teacher expertise group
	Student attendance	Record of student attendance	Low attendance to AP classes	Low student motivation No attendance requirement	Student/Parent contract Attendance requirement
	Topics covered	Number of topics covered	Courses cover less topics	No requirement to enter AP course Lack of past AP teaching experience	Smaller AP classes Class syllabus Teacher expertise group
	Minimum requirements	Percent of requirements met for entering AP class	There are no current requirements to enter AP class	Attempt to increase the number of students enrolled in AP classes	Minimum requirements

FIGURE 8.27 CTS recommendations alignment matrix.

- Student will have an interview with his/her counselor where the reasons why the student is considering taking AP courses will be discussed.

6. Revised VOP Matrix

The VOP matrix and the targets remained the same as it was in the Measure phase. The targets are aggressive, but the team feels that they are attainable. The only

	CTS	Metrics	Problems	Root cause	Recommendation
Quantity	% Minorities	% Minorities enrolled in AP classes	Minorities are not well represented in AP courses	Low encouragement in preopen access system	Guidance counselor encouragement
	% Lower socioeconomic	% Lower socioeconomic students enrolled in AP classes	Lower socioeconomic students are not well represented in AP courses	Low encouragement in preopen access system	Guidance counselor encouragement
	Total AP experiences	Number of students enrolled x classes offered	Total number of AP experiences could be greater	Fewer classes in preopen access system Low encouragement in preopen access system	Standard AP procedures

FIGURE 8.27 (Continued)

Recommendation	Time frame	Owner
Short term recommendations		
Recommendation # 1: Develop a more standardized AP enrollment process	three months	Principal and assessment team
Recommendation # 2: Set minimum requirements for enrollment into AP classes	three months	Principal and assessment team
Recommendation # 3: Create a contract for students/parents enrolling in an AP course	three months	Guidance counselors
Recommendation # 6: Set minimum attendance requirements	three months	Guidance counselors
Recommendation #9: Create a form for students who want to get out of the AP course. This form will have to be signed and approved by the student's teacher, student's parents, student's counselor, and the student. After approval, the student will be able to drop the course	three months	Guidance counselors
Long term recommendations		
Recommendation # 4: Establish and encourage parental involvement for students enrolled in AP	one year	Guidance counselors, principal
Recommendation # 5: Consider keeping AP classes small	one year	Principal
Recommendation # 7: Generate a highly detailed class syllabus	one year	Principal
Recommendation # 8: Create a knowledge-sharing program for AP teacher's best practices	one year	Teachers

FIGURE 8.28 Action plan.

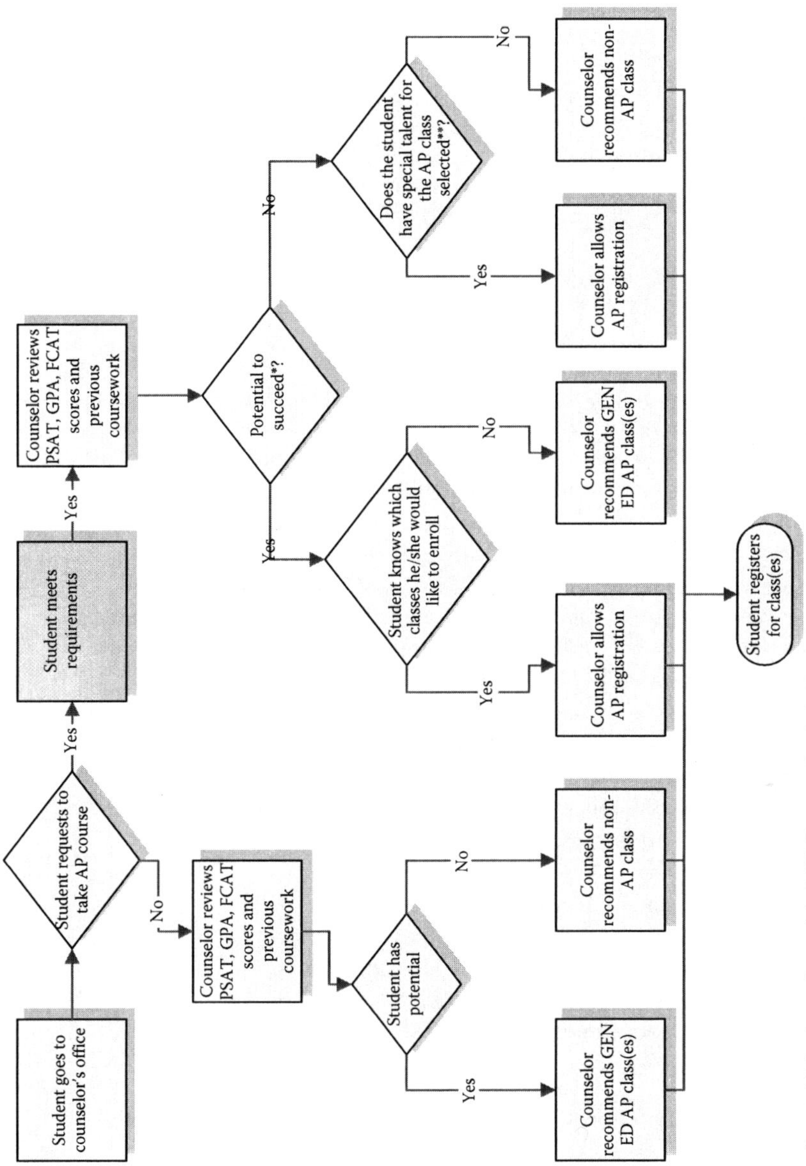

*Potential defined as student with high PSAT score and/or high GPA and/or past pre-AP classes and/or passing FCAT score and/or reading level greater than 2
**Special talents such as math ability, native languages, etc.

FIGURE 8.29 Revised process flow.

distinction at this phase is that we will not measure the CTS for the following, due to the data not being available at this point to define a baseline: student attendance, topics covered, and course evaluations.

7. Training Plans, Procedures

The new process flow will serve as the training procedure, any additional training materials will be developed by the guidance counselors.

8. Improve Phase Presentation

The Improve presentation can be found in the downloadable instructor materials.

IMPROVE PHASE CASE DISCUSSION

1. **Improve Report**
   - 1.1 Review the Improve report and brainstorm some areas for improving the report.
   - 1.2 How did your team ensure the quality of the written report? How did you assign the work to your team members? Did you face any challenges of team members not completing their assigned tasks in a timely manner, and how did you deal with it?
   - 1.3 Did your team face difficult challenges in the Improve phase? How did your team deal with conflict on your team?
   - 1.4 Did your instructor and/or Black Belt or Master Black Belt mentor help your team better learn how to apply the LSS tools in the Improve phase, and how?
   - 1.5 Did your Improve phase report provide a clear understanding of the root causes of the process. Why or why not?
   - 1.6 Compare your improve report with the improve report in the book, what are the major differences between your report and the author's report?
   - 1.7 How would you improve your report?

2. **Recommendations for Improvement**
   - 2.1 How did your team generate ideas for improvement?
   - 2.2 What tools and previous data did you use to extract information for the improvement recommendations?
   - 2.3 How do your recommendations differ from the ones in the book?

3. **Revised QFD**
   - 3.1 Does the QFD support the alignment with the CTS characteristics?
   - 3.2 How will you assess customer satisfaction?

4. **Action Plan**
   - 4.1 How did your Six Sigma team identify the timings for when to implement your recommendations?

5. **Future State Process Map**
   5.1 Compare your future state process map with the one in the book. How does it differ? Is yours better, worse, the same?

6. **Revised VOP Matrix**
   6.1 Does the VOP matrix provide alignment between the CTSs, the recommendations, metrics, and target?

7. **Training plans, procedures**
   7.1 How did you determine which procedures should be developed?
   7.2 How did you decide what type of training should be done?

8. **Improve Phase Presentation**
   8.1 How did your team decide how many slides/pages to include in your presentation?
   8.2 How did your team decide upon the level of detail to include in your presentation?

CONTROL PHASE EXERCISES

1. **Control Report**
   Create a Control phase report, including your findings, results, and conclusions of the Control phase.

2. **Control Plan**
   Develop a control plan for each improvement recommendation from the Improve phase report.

3. **Hypothesis Tests, Analysis of Variance**
   - Compare the percentage of students by ethnicity in AP courses between 2004/2005 and 2005/2006.
   - Compare the percentage of students in the F&R lunch program in AP courses in 2004/2005 and 2005/2006.

4. **Control Charts**
   Create an idea for applying control charts to control the open access AP Registration process.

5. **Replication Opportunities**
   Identify some potential replication opportunities within the high school, and within the school district.

6. **Dashboards/Scorecards**
   Create a dashboard or scorecard for tracking and controlling the AP registration process.

7. **Control Phase Presentation**
   Prepare a presentation (PowerPoint) from the case study exercises that provides a short (10–15 minutes) oral presentation of the Control phase deliverables and findings.

CONTROL PHASE

1. Control Report

The last step in the DMAIC process is the Control phase. The improvement is assessed based on the implemented improvement recommendations, and a control plan developed to ensure that the changes are standardized and controlled. In the case of this project, the SHS LSS team has neither control over improvements to the open access system nor control over implementations. The control plans are included as a guide for the SHS leadership team to follow when ultimately designing and implementing changes to the system.

2. Control Plan

Following are the control plans for each recommendation.

Recommendation #1: Develop a more standardized AP enrollment process.

Control: The goal for this endeavor should be launching the guide/program in the Fall 2006 semester. This gives the AP placement staff and a potential committee the entire summer to develop the publication and the policy that is within it. Representatives that make up this committee should include staff from the AP placement, select experienced AP teachers, counselors, and potentially a Six Sigma consultant that can ensure proper metrics are installed within the framework of the program. In addition, this recommendation would be controlled and evaluated after the fall semester. Counselors will meet with the school principal on a periodic basis and discuss performance of the process. In addition, teacher feedback for the policies created within the guide should be solicited with revisions planned for future editions.

Recommendation # 2: Set minimum requirements for enrolling students in AP classes.

Control: The goal for this endeavor should be launching the guide/program in the Fall 2006 semester. This will help AP placement staff in placing students in AP courses. This recommendation should be controlled and evaluated after the fall semester. Counselors will meet with the school principal who should ensure effectiveness of the implementation.

Recommendation # 3: Create a contract for students/parents enrolling in an AP course.

Control: The goal for this endeavor should be launching the guide/program in the Fall 2006 semester. The contract will be renewed each new semester specifying the amount of work and time needed.

Recommendation # 4: Establish and encourage parental involvement for students enrolled in AP courses.

Control: This system could be implemented in the fall 2006 semester. This will give everyone involved in the development of the contract the summer semester to develop the parental involvement system. This system can be controlled by the student counselor or teacher on a regular basis. Teachers should have a log where they can keep records of the system.

Recommendation # 5: Consider keeping AP classes small.

Control: This recommendation can be controlled every semester. The SHS principal can meet with all AP teachers after each semester to get a better feeling of how well the system is working and get ideas for improvement. Teachers will have the opportunity to let the principal know if they feel their classes are not performing well due to the amount of students.

Recommendation # 6: Set minimum attendance requirements

Control: The attendance requirement matrix could be developed by experienced AP teachers since they have a better idea of the correlation between attendance and performance. This recommendation could be controlled by the SHS principal and the AP teachers at the beginning of each semester.

Recommendation # 7: Generate a highly detailed class syllabus

Control: The syllabus should be implemented and controlled by an expert in the subject matter and should be updated every semester.

Recommendation # 8: Create a knowledge-sharing program for AP teachers' best practices.

Control: A variety of information and communication technologies exist that may be used by the teachers to communicate and share their ideas and inputs on the topic. A knowledge management system for this area could be as simple as a best practices committee that publishes a bi-semester newsletter or a more complicated information technology design that stores best practices in a database. Monitoring and evaluating the teachers' participation in this recommended system would be managed through the school established procedures of teacher reviews. This program can be performed at the end of each school year.

Recommendation # 9: Create a form for students who want to get out of the AP course. This form will have to be signed and approved by the student's teacher, student's parents, student's counselor, and the student. After approval, the student will be able to drop the course.

Control: This system can be implemented during the next school year and could be controlled by students, parents, teachers, and counselors.

3. Hypothesis Testing/ANOVA

Using a two-proportion test with the number of F&R lunch students in 2004/2005 of 138 F&R lunch students out of 652 AP students (21%), compared with

2005/2006 of 210 out of a total number of AP students of 964 (22%), we conclude that there is not a difference between the percentage of F&R lunch students in AP courses between 2004/2005 and 2005/2006. The percentage of students in the lower socio-economic group remained the same as a percentage after open access.

We also wanted to compare the percentage of students in AP classes by ethnicity to compare it with the percentages of the entire school's student population, to see if the AP class percentages are representative of the entire population. Figure 8.30 shows that Black students (9% AP students vs. 10% all students) and Multiracial students (1% in both AP classes and all students) are representative of the entire student population percentages. Asians have 4% more students in AP classes than the student population percentage (11% AP students vs. 7% all students). Whites have a higher percentage of students in AP classes (45%) than the entire student population of 38%. Hispanics have a lower percentage of students in AP classes (34%) compared with 44% in the entire student population. So, there is still more work to be done to align the percentage of students in the AP classes to the entire student population percentages. However, it has vastly improved from the percentages prior to open access AP Registration in 2003/2004 as follows: Asians: 12% AP vs. 8% all students; Blacks: 5% AP vs. 8% all students; Hispanics: 24% AP vs. 40% all students; and White: 60% AP vs. 43% all students).

The percentage of students in AP classes has moved closer to the entire student populations for each ethnicity, increasing if the percentages were lower than the entire student population percent or decreasing if the percentages were higher. The percentages of AP students from 2003/2004 to 2005/2006 were: Asians: 12% to 11%; Blacks: 5% to 9%; Hispanics: 24% to 34%; and Whites: 60% to 45%.

The percentage of students in AP classes for 2005/2006 that are in the F&R lunch program (lower socioeconomic groups) is at 23% compared with the entire student population of 35% of the students being in the F&R lunch program. This is significantly different, with a p-value of 0.000. The percentage of students enrolled in AP classes that were also in the F&R lunch program was 12% in 2003/2004 compared with the overall student population of 28% in 2003/2004. The percentage of students in the F&R lunch program that were taking AP classes (12%) was significantly less than the overall student percent of students in the F&R lunch program (28%), with a p-value of 0.000. The percentage of students has increased since preopen access,

Race	% of Students with AP classes	% of Student population	Significantly different?	p-value
Asian	11%	7%	Yes (4%)	.001
Black	9%	10%	No (−1%)	.362
Hispanic	34%	44%	Yes (−10%)	.000
White	45%	38%	Yes (7%)	.000
Multi	1%	1%	No (0%)	.775

FIGURE 8.30 2005/2006 Ethnicity comparison: AP classes and entire student population.

but it is still below the overall student percentage. Figure 8.31 shows the F&R lunch program percentages for 2005/2006 school year.

4. Control Charts

An idea for applying control charts to control the open access AP registration process could be to track the grades in the AP courses by semester using an individuals and moving range chart across the students enrolled in a particular AP course. Only certain AP courses could be selected to track those that are representative of the subject matter across the AP curriculum, instead of all of the courses. This could be automated and linked to the student grade database and generated automatically, and monitored by the guidance counselors.

5. Replication Opportunities

The open access AP registration program could be replicated in almost any other high school, without the need for major modifications or customizations. This project provided the validation the open access program was a great success in increasing the quantity of the students taking AP courses, and would have value in almost any high school. The quality of the AP courses should now be enhanced and improved.

6. Dashboards/Scorecards

A sample scorecard is shown in Figure 8.32. It consists of the quantitative measures including: percentage of minorities enrolled in AP classes; percent of F&R lunch program students enrolled in AP classes and; the average number of AP classes per student. This can be used after the registration process is complete for the following year in the spring of the prior year to ensure that the school has done a good job at increasing the enrollment of the minority students and the students in the lower socio-economic groups. A similar scorecard could be developed for the qualitative measures related to the qualitative CTSs.

7. Control Phase Presentation

The Control phase presentation can be found in the downloadable instructor materials.

F&R Lunch	% of Students with AP classes	% of Student population	Significantly different?	*p*–value
Yes	23%	35%	Yes (–12%)	.000
No	77%	65%	Yes (12%)	.000

FIGURE 8.31 2005/2006 F&R lunch percentage of students in AP classes compared to the entire student population percentages.

Ethnicity	**Number and % minorities enrolled in AP courses (2005/2006)**	**Number and % minorities in student population**
Asian	101 / 11%	250 / 7%
Black	86 / 9%	354 / 10%
Hispanic	329 / 34%	1568 / 44%
Multi-Racial	10 / 1%	41 / 1%
White	429 / 45%	1346 / 38%
TOTAL	959	3567
In F&R lunch program	Number and % students enrolled in AP (2005/2006)	Number and % in student population
Yes	223 / 23%	993 / 40%
No	736 / 77%	1487 / 60%
Total	959	2480
Number AP experiences	Number AP classes / Number students =	1465 / 3514 = .42

FIGURE 8.32 Scorecard example 2005/2006 data.

CONTROL PHASE CASE DISCUSSION

1. **Control Report**
   1.1 Review the Control report and brainstorm some areas for improving the report.
   1.2 How did your team ensure the quality of the written report? How did you assign the work to your team members? Did you face any challenges of team members not completing their assigned tasks in a timely manner, and how did you deal with it?
   1.3 Did your team face difficult challenges in the Control phase? How did your team deal with conflict on your team?
   1.4 Did your instructor and/or Black Belt or Master Black Belt mentor help your team better learn how to apply the LSS tools in the Control phase, and how?
   1.5 Compare your Control report to the Control report in the book, what are the major differences between your report and the author's report?
   1.6 How would you improve your report?

2. **Control Plan**
   2.1 How well will your Control plan ensure that the improved process will continue to be used by the process owner?

3. Hypothesis Tests, ANOVA

3.1 How did you assess the improvement for the CTS?

4. Control Charts

4.2 Are their additional Control charts that could be used to ensure process control?

5. Replication Opportunities

5.1 How did your team identify additional replication opportunities for the open access AP registration process within the high school, and within the school district?

6. Dashboards/Scorecards

6.1 How would your dashboard differ it is was going to be used to present the results of the open access AP registration process to the school board, or be used across several schools?

7. Control Phase Presentation

7.1 How did your team decide how many slides/pages to include in your presentation?

7.2 How did your team decide upon the level of detail to include in your presentation?

9 Project Charter Review Process Design—A Design for Six Sigma Case Study

Sandra L. Furterer

CONTENTS

PROJECT OVERVIEW

The Information System Division of a major Fortune 50 corporation develops applications to support the business. The division had been reviewing and approving the projects in a cross-divisional weekly meeting with the senior executives. The project charter is developed by the application development team working with the business to understand the scope of the proposed project. The project charter includes a description of the business opportunity, identification of the customers and stakeholders, the goals and objectives of the project, as well as the metrics that assess the successful completion of the project. The project charter also includes identification of the potential risks that could prevent the project from being successfully completed, and the assumptions that are assumed to be true. An initial estimate of the resources and project costs and the hard and soft benefits for doing the project are also assessed. The hard benefits identify financial savings that impact the financial statements, whereas the soft benefits include cost avoidance and intangible benefits to the business for doing the project. The customer signatures signifying buy-in to the project are also included on the project charter. The division's program management office (PMO) provides project management standards, guidance, and training to the division. They have recently decentralized the project charter approval process to the senior vice presidents' (SVP) areas. The review and approval of projects had been performed at a divisional level, looking only at projects that were >1000 hours of effort. If projects were <1000 hours of total effort, they were reviewed by vice presidents (VPs), but not across the SVP area. The goal was to get more visibility of all projects across the entire SVP area. The approval from the customer will be attained, and then the information system division SVP area will review the project charter to identify any cross-area conflicts or overlap and ensure that resources are available to work on the project.

The process and metrics (P&M) team in the SVP's area has been assigned the responsibility of designing a new area council review process to assess the quality of the project charter, and incorporate appropriate metrics to baseline and encourage continuous process improvement. The divisional standards should be maintained to ensure consistency and repeatability of the project chartering process. The stakeholders of the new process include: the management team, who will review and approve the projects within the information system division; the P&M team, who will assess the quality of the project charters and execute the area council review process; the PMO, who provides the divisional standards for reviewing the project charters; the project leaders, who create the project charters; and the business, for whom the project charters are developed.

IDENTIFY PHASE EXERCISES

It is recommended that the students work in project teams of 4 to 6 students throughout the Design for Six Sigma case study.

1. **Identify Report**
   Prepare a written report from the case study exercises that describes the Identify phase activities and key findings.

2. **Design for Six Sigma Project Charter**
   Use the information provided in the Project Overview section above, in addition to the project charter format to develop a project charter for the Design for Six Sigma project.

3. **Stakeholder Analysis**
   Use the information provided in the Project Overview section above, in addition to the stakeholder analysis format, to develop a stakeholder analysis, including stakeholder analysis roles and impact definition, and stakeholder resistance to change.

4. **Team Ground Rules and Roles**
   Develop the project team's ground rules and team members' roles.

5. **Project Plan and Responsibilities Matrix**
   Develop your team's project plan for the DMAIC project. Develop a responsibilities matrix to identify the team members who will be responsible for completing each of the project activities.

6. **Identify Phase Presentation**
   Prepare a presentation (PowerPoint) from the case study exercises that provides a short (10–15 minutes) oral presentation of the Identify phase deliverables and findings.

IDENTIFY PHASE

1. Identify Report

Following is a written report of the Identify phase for the project charter review process design project, including the key deliverables developed as part of the prior exercises. The main purpose of the Identify phase is to understand the opportunity and business that needs a new process to be designed, and to develop a project charter and appropriate scope to design the process. The main activities in the Identify phase are to: (1) develop project charter; (2) perform stakeholder analysis; and (3) develop project plan.

2. Design for Six Sigma Project Charter

The P&M team in the Information System Division's SVP area has been assigned the responsibility of designing a new area council review process to assess the quality of the project charter, and incorporate appropriate metrics to baseline and encourage continuous process improvement. The divisional standards should be maintained to ensure consistency and repeatability of the project chartering process. The stakeholders of the new process include: the management team, who will review and approve the projects within the information system division; the P&M team, who will assess the quality of the project charters and execute the area council review process; the PMO, who provides the divisional standards for reviewing the project charters; the project leaders, who create the project charters; and the business, for whom the project charters are developed.

Following are the sections that comprise the project charter, which defines the problem to be investigated. The project charter is shown in Figure 9.1.

The Information System Division develops applications to support the business. The division's PMO members along with the division's management had been reviewing and approving the projects in a cross-divisional weekly meeting with the senior executives. The project charter is developed by the application development team working with the business to understand the scope of the proposed project. The project charter includes a description of the business opportunity, identification of the customers and stakeholders, the goals and objectives of the project, as well as

Project Name: Project Charter Review Process Design.

Problem Statement: To design a process for the area to review the project charters to determine if the project should move forward to the next phase. Project charters are the project initiation document that identifies a need in the business to perform information systems work.

Customer/Stakeholders: Management team, project leaders, Process & Metrics team, business customers, program management office.

What is important to these customers–critical to satisfaction (CTS): All necessary fields are completed; provide accurate information to make decisions; and review is timely. Obtain approval to continue with the project.

Goal of the project: To provide a process that provides a timely and complete review and decision to continue (or not) with the project.

Scope statement: This process includes the review of the project charters at an area level. Includes project review of the project charter, provides review of the format and content of the project charter, and provides approval of the project charter at appropriate management levels. Link this process to the Quality goals of the organization. This process is just for the identified area.

Financial and other benefit(s): Consistent process, visibility of projects across area to identify overlap and resource sharing.

Project deliverables: Project charter review process; scorecard and metrics with baseline and target goals, and appropriate visibility of reporting requirements.

Potential risks: Being perceived as a bureaucratic instead of value-added process; acceptance and adherence of process of area; and timeliness of review.

FIGURE 9.1 Project charter.

the metrics that assess the successful completion of the project. The project charter also includes identification of the potential risks that could prevent the project from being successfully completed, and the assumptions that are assumed to be true. An initial estimate of the resources and project costs and the hard and soft benefits for doing the project are also assessed. The hard benefits identify financial savings that impact the financial statements, whereas the soft benefits include cost avoidance and intangible benefits to the business for doing the project. The customer signatures signifying buy-in to the project are also included on the project charter. The division's PMO provides project management standards, guidance, and training to the division. They have recently decentralized the project charter approval process to the SVP's areas. The approval from the customer will be attained, and then the information system division SVP's area will review the project charter to identify any cross-area conflicts or overlap and ensure resources are available to work on the project.

Project Name: Project Charter Review Process Design.

Problem Statement: To design a process for the area to review the project charters and determine if the project should move forward to the next phase. Project charters are the project initiation document that identifies a need in the business to perform information systems work.

Customers/Stakeholders: The primary stakeholders are the management team, who will review and approve the project charters, the project leaders, who will develop the project charters, and the P&M team, who will execute the process. The secondary stakeholders are the customers, who the application development teams are developing applications for, and the PMO, who develops and ensures divisional standards are followed.

What Is Important to These Customers (CTS): The management team wants a simple and timely process that provides visibility of the status of the projects that enable the teams to meet the business' information system needs. The project leaders want their projects approved, and want a timely and manageable process. The P&M team wants to implement a simple and measureable process that is of high quality. The business customers want the desired functionality to be delivered in a timely manner. The PMO wants the standards to be followed in a consistent and repeatable manner.

Goal of the Project: To provide a process that provides a timely and complete review and decision to continue (or not) with the project.

Scope Statement: This process includes the review of the project charters at an area level. It includes project review of the project charter, review of the format and content of the project charter, and approval of the project charter at appropriate management levels. It should link this process to the quality goals of the organization. This process is just for the identified area.

Projected Financial and Other Benefits: Consistent process, visibility of projects across area to identify overlap and resource sharing.

Risk Management Matrix: The risk management matrix is shown in Figure 9.2. The main risks are: not having time to get buy-in from the major stakeholders; communication of the new process may not be complete and of high quality; need to consider needed training and rollout; being considered as a bureaucratic rather than a value-added process; and timeliness of the review.

Project Resources: Master Black Belt Mentor: Sandra Furterer. Project Team Members: Carrie Harris, Emily McKenzie, Bridget Corp.

Potential risks	Probability of risk (H/M/L)	Impact of risk (H/M/L)	Risk mitigation strategy
Not having time to get buy in from key stakeholders.	H	H	Create a simple process, which can be enhanced. Identify key stakeholders, and get input quickly.
Communication of the new process is not complete and high quality.	L	M	Identify key stakeholders and create communication and change strategy.
Need to consider needed training and roll out.	H	H	Create training and roll out strategy.
Being perceived as a bureaucratic instead of value added process.	H	H	Alignment with business and project strategies, with value clearly defined. Projects that aren't resourced or aligned shouldn't move forward.
Acceptance and adherence of process.	H	M	Develop change management strategy.
Timeliness of review. Program reviews every two weeks, instead of weekly. Potential maximum impact to project = three weeks.	H	H	Clearly document the process and procedures to help ensure better planning. Contingency process steps may be needed.

FIGURE 9.2 Project risk matrix.

Project Deliverables: Project charter review process; scorecard and metrics with baseline and target goals; and appropriate visibility of reporting requirements.

The business case for this project is that divisional management has made a decision to review information system project charters at the SVP level to provide visibility at the area level. This created an immediate need to design an area council review process within the area to ensure consistency across the division and enable improvement and visibility within the area.

3. Stakeholder Analysis

The PMO has recently decentralized the review of project charters to the areas, resulting in the need for creating a process to review the project charters to ensure they are providing value to the business, and communicating the type of information needed to identify risks and manage projects and resources at an area level. The newly formed P&M team has been assigned the task to design a new area council review process. The team has decided that they will use the Design for Six Sigma tools and IDDOV methodology to ensure a fact-based process is used to design the new process, and to ensure that appropriate measures are incorporated into the process. The primary stakeholders are the management team, who will review and approve the project charters, the project leaders, who will develop the project charters, and the P&M team, who will execute the process. The secondary stakeholders are the customers, who the application development teams are developing applications for, and the PMO, who develops and ensures divisional standards are followed.

Figure 9.3 shows the primary and secondary stakeholders for the process, and their major concerns. Note "+" represents a positive impact or potential improvement, while "–" represents a potential negative impact to the project.

Figure 9.4 shows the commitment level of each major stakeholder group at the beginning of the project.

Stakeholders	Who are they?	Potential impact or concerns	+ / –
Management team	VPs, directors, team managers who manage the development teams.	▪ Simple process ▪ Timely process ▪ Project visibility ▪ Meet business' needs	+ + + +
Project leaders	Application development team leaders.	▪ Approval to continue with the project ▪ Timely process ▪ Manageable process	+ + +
Process and metrics team	Responsible for improving the internal application development life cycle processes, and providing metrics to ensure quality and timeliness of project deliverables.	▪ Simple process ▪ Measurable process ▪ High quality process	+ + +
Business customers	Internal customers who are provided information systems to meet their business needs.	▪ Deliver needed functionality ▪ Delivery in a timely manner	+ +
Program Management Office	Division program management office that provides application development life cycle standards, training and mentoring.	▪ Consistent process is followed	+

FIGURE 9.3 Stakeholder analysis definition.

Stakeholders	Strongly against	Moderate against	Neutral	Moderate support	Strongly support
Management team					XO
Project leaders		X			O
Process and Metrics team					XO
Business customers			XO		
Program management office			X		O
X = At start of project					O = By end of project

FIGURE 9.4 Stakeholder commitment scale.

4. Team Ground Rules

The team adhered to the following ground rules related to working together on the team.

Team Ground Rules

- Be respectful to team members
- Be open minded, share ideas freely
- Provide service to each other, with focus on customers and stakeholders
- Provide excellence to the team
- Respect differences
- Be supportive rather than judgmental
- Be open to new concepts and to concepts presented in new ways. Keep an open mind. Appreciate other's points of view
- Share your knowledge, experience, time, and talents

5. Project Plan and Responsibilities Matrix

The detailed project plan is shown in Figure 9.5, with tasks to be completed, due date, deliverables and resources. It includes the person or people responsible for each activity.

6. Identify Phase Presentation

The Identify phase presentation summarizing the written Identify phase presentation is included in the downloadable instructor materials.

IDENTIFY PHASE CASE DISCUSSION

1. **Identify Report**
   1.1 How did your team ensure the quality of the written report? How did you assign the work to your team members? Did you face any challenges of team members not completing their assigned tasks in a timely manner, and how did you deal with it?
   1.2 Did your team face difficult challenges in the Identify phase? How did your team deal with conflict on your team?
   1.3 Did your instructor and/or Black Belt or Master Black Belt mentor help your team better learn how to apply the Design for Six Sigma tools, and how?
   1.4 Did your Identify phase report provide a clear vision of the project, why or why not?
   1.5 How could you improve your Identify phase report based on the Identify phase report given in the book? How could you improve the Identify phase report in the book?

2. **Design for Six Sigma Project Charter**
   Review the project charter presented in the Identify phase report.

Activity number	Phase/activity	Duration	Predecessor	Resources
1.0	Identify			
1.1	Develop project charter	1 day		Team
1.2	Perform stakeholder analysis	2 days	1.1	Team
1.3	Develop project plan	2 days	1.2	Team
2.0	Define		1.0	
2.1	Collect voice of customer (VOC)	1 day		Team
2.2	Identify CTS measures and targets	14 days	2.1	Team
2.3	Translate VOC into technical requirements	14 days	2.2	Team
2.4	Identify CTS measures and targets	2 days	2.3	Team
3.0	Design		2.0	
3.1	Identify process elements	5 day		Team
3.2	Design process	1 days	3.1	Team
3.3	Identify potential risks and inefficiencies	3 days	3.2	Team
4.0	Optimize		3.0	
4.1	Implement process	60 days		Team
4.2	Assess process capabilities	5 days	4.1	Team
4.3	Optimize design	5 days	4.2	Team
5.0	Validate		4.0	
5.1	Validate process	30 days		Team
5.2	Assess performance, failure modes, and risks	5 days	5.1	Team
5.3	Iterate design and finalize	½ day	5.2	

FIGURE 9.5 Project plan.

2.1 A problem statement should include a view of what is going on in the business, and when it is occurring. The problem statement should provide data to quantify the problem. Does the problem statement in the Identify phase written report provide a clear picture of the business problem? Rewrite the problem statement to improve it.
2.2 The goal statement should describe the project team's objective, and be quantifiable, if possible. Rewrite the Identify phase goal statement to improve it.
2.3 Did your project charter's scope differ from the example provided? How did you assess what was a reasonable scope for your project?

3. Stakeholder Analysis

Review the stakeholder analysis in the Identify phase report.

3.1 Is it necessary to identify the large number of stakeholders as in the example case study?
3.2 Is it helpful to group the stakeholders into primary and secondary stakeholders? Describe the difference between the primary and secondary stakeholder groups.

4. **Team Ground Rules and Roles**
   4.1 Discuss how your team developed your team's ground rules. How did you reach consensus on the team's ground rules?

5. **Project Plan and Responsibilities Matrix**
   5.1 Discuss how your team developed their project plan and how they assigned resources to the tasks. How did the team determine estimated durations for the work activities?

6. **Identify Phase Presentation**
   6.1 How did your team decide how many slides/pages to include in your presentation?
   6.2 How did your team decide upon the level of detail to include in your presentation?

DEFINE PHASE EXERCISES

1. **Define Report**
   Create a Define phase report, including your findings, results, and conclusions of the Define phase.

2. **Data Collection Plan and Voice of Customer (VOC)**
   Develop a data collection plan for collecting VOC and process information to assess the CTS criteria for the project.

3. **CTS Summary**
   Brainstorm ideas to summarize the proposed CTS criteria and prepare a CTS summary and targets.

4. **QFD**
   Develop a QFD house of quality to identify and map the customer requirements to the technical requirements of the process.

5. **Define Phase Presentation**
   Prepare a presentation (PowerPoint) from the case study exercises that provides a short (10–15 minutes) oral presentation of the Define phase deliverables and findings.

DEFINE PHASE

1. Define Report

Following is a written report of the Define phase for the project charter review process design project, including the key deliverables developed as part of the prior exercises. The Define phase of the IDDOV process is designed to gain information on the VOC to understand the needs of the customers and begin translating those customer requirements into the processes' technical elements. The main activities of this phase are to: (1) Collect VOC; (2) Identify CTS measures and targets; and (3) Translate VOC into technical requirements.

2. Data Collection Plan and VOC

The data collection plan is shown in Figure 9.6. It summarizes the potential metrics and how we would collect data to measure the metrics. This will include information on a proposed process and VOC information. VOC data collection consisted of

Critical to satisfaction (CTS)	Metric	Data collection mechanism (survey, interview, focus group, etc.)	Analysis mechanism (statistics, statistical tests, etc.)	Sampling plan (sample size, sample frequency)	Sampling instructions (who, where, when, how)
Timely process	Area council review is held 1st & 3rd Tuesday of month	Track schedule of reviews	Counts of reviews	All reviews	None
High quality Process with metrics	Content quality percentage	Scorecard with content quality criteria and score; stakeholder interviews	Percentage received against grading criteria; control chart	All project charters for each review within SVP area	See scorecard procedures
	Format quality percentage	Scorecard with format criteria and score, stakeholder interviews	Percentage received against grading criteria; control chart	All project charters for each review within SVP area	See scorecard procedures
Accurate information	Content quality percentage	Scorecard with content quality criteria and score	Percentage received against grading criteria; control chart	All project charters for each review within SVP area	See scorecard procedures
Ability to make decisions, go/no go on projects	Percent projects decided on in each meeting	Agenda approval record, stakeholder interviews	Percentage	Each area council project review meeting	See review procedures
Visibility to program/ project relationships	Count of projects related to programs	Scorecard item on format scorecard, stakeholder interviews	Count	All projects reviewed	See scorecard procedures

FIGURE 9.6 Data collection plan.

interviewing the stakeholders to understand what is critical to their satisfaction for the new process, as well as harvesting information on similar processes, and data related to initial design thoughts for the process.

The SVP, the VPs, the directors, program and project leaders, the P&M team, the enterprise, and solution architects who provide cross-area planning of information system blueprints and roadmaps, and the PMO were all interviewed.

The management team (SVP, VPs, and directors) wanted to ensure the programs and project are on budget, have resources, and key sponsors. They also wanted to be able to decide whether they should do the project work or not, and understand how the programs and projects affect other teams within the area and the division. They want to be able to have visibility and knowledge when there is a problem or issue with the project that potentially puts the project at risk of successful completion. The management team wants to be able to understand the project priorities and have the visibility to know if they are working on the right priorities, as well as have a way to periodically review the work being performed in the area.

Some of the questions that they would ask when reviewing projects are as follows:

- What is the scope of processes in the project?
- Should we buy versus build?
- What is the impact to the business?
- Should we outsource any part of the development work?
- Do we have engagement from the business areas?
- What business resources will be required?
- Is there an existing process that they're enhancing?
- Is infrastructure needed?
- Do we have the resources necessary to do the work, or what must be re-prioritized to be able to do this work?

The program and project leaders concerns and critical drivers were to provide resource allocation and management across the programs and projects. The P&M team, who is responsible for the area council review process, wants a simple process that is metrics-based and encourages continuous process improvement for initiating new projects and ensures that there is customer/stakeholder buy-in to the new process.

The enterprise architects were concerned about the ability to be able to see program and project dependencies and assess the impact of adding projects to the business and the information system division. They wanted to provide visibility of program and project changes and periodic updates to the programs and projects. The Architects also wanted to ensure appropriate resource management.

The PMO wants a review that supports and aligns to the divisional standards and the information system development life cycle.

To measure the timely process, we would track that the area council review process is held every first and third Tuesday of the month. To help meet the CTSs for having a high-quality process with metrics and accurate project information, we built a scorecard that would assess format and quality of the content on the project charter. The format-scorecard criteria would ensure that all of the required fields are completed. The content scorecard criteria would assess the quality of the information in the fields against the project charter standard documentation. To assess the ability to make go/no go decisions on the projects, the percentage of projects reviewed and approved or deferred will be tracked at the review meetings. The visibility of projects related to programs can be tracked by counting the number of projects that have a program identified with it.

3. Critical To Satisfaction Summary

The VOC provided insight into the CTS criteria for the project, as summarized below:

- Timely process
- High-quality process with metrics
- Accurate information
- Ability to make decisions, go/no go on projects
- Visibility to program/project relationships

It is important to the project leaders, application development teams, and management that the review process provides a timely review and approval of projects so the teams can get started working on the information systems projects. It is also important to have a high-quality project initiation process and that the metrics designed enable continuous process improvement, and provide project charters that are well-scoped. The new process will also need to enable the ability to make decisions on whether to approve the projects or not, providing information on the business opportunity, goals and objectives of the projects. The process should also provide visibility of the projects and resources required across the area.

4. QFD

The QFD house of quality was used to ensure alignment between the customer and stakeholder needs represented by the CTS criteria and the technical requirements of the process design. After collecting the VOC information that allowed insight into the CTS criteria summarized across all of the stakeholders, the P&M team brainstormed the critical elements to be designed into the new process (technical requirements). The house of quality is shown in Figure 9.7. The P&M team assessed the strength of relationship between the CTS criteria and the design criteria. An importance rating was assigned to each of the CTS, which was then multiplied by the relationship ratings, to derive a relative weighting of the technical requirements. A Pareto chart is shown in Figure 9.8. An area council SharePoint® (Microsoft intranet website software) would be a way to provide a workflow and facilitate the review process. Executive approval is another critical element that should be designed into the new process. Without it there is little chance for the organization to see the value of the reviews, if the management team is not on-board. Detailed procedures and a process map will provide clear definition of the process

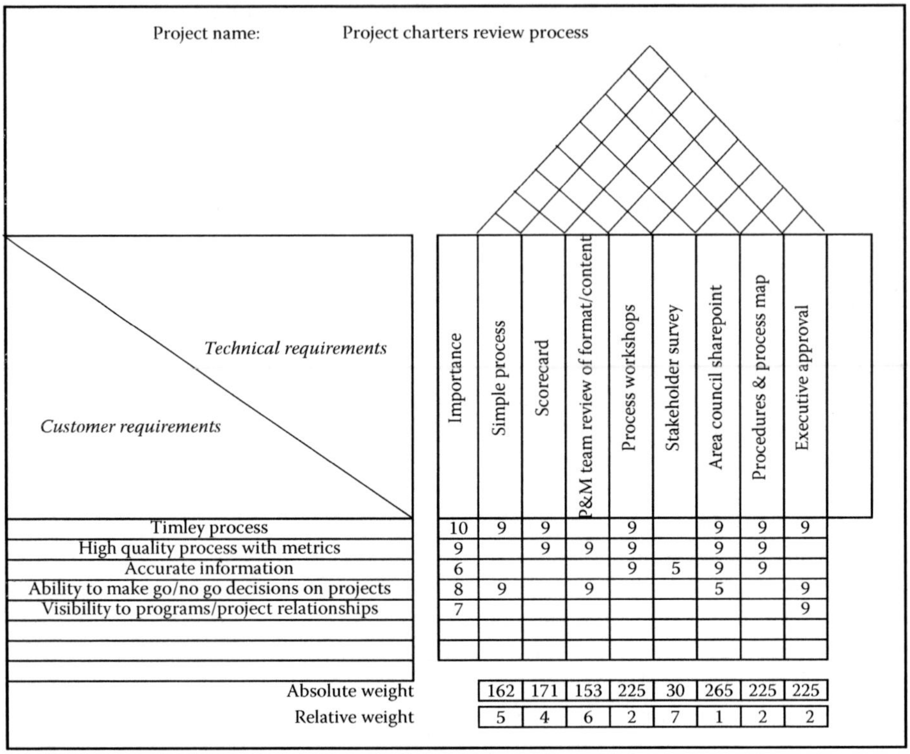

Customer requirements	Importance	Simple process	Scorecard	P&M team review of format/content	Process workshops	Stakeholder survey	Area council sharepoint	Procedures & process map	Executive approval
Timley process	10	9	9		9		9	9	9
High quality process with metrics	9		9	9	9		9	9	
Accurate information	6				9	5	9	9	
Ability to make go/no go decisions on projects	8	9		9			5		9
Visibility to programs/project relationships	7								9
Absolute weight		162	171	153	225	30	265	225	225
Relative weight		5	4	6	2	7	1	2	2

FIGURE 9.7 Quality function deployment house of quality.

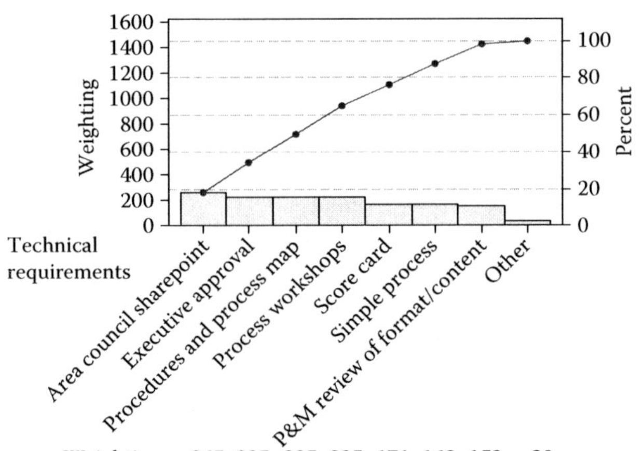

Weighting	265	225	225	225	171	162	153	30
Percent	18.2	15.5	15.5	15.5	11.7	11.1	10.5	2.1
Cum %	18.2	33.7	49.1	64.6	76.3	87.4	97.9	100.0

FIGURE 9.8 Pareto chart of prioritized technical requirements.

and can be used as a training guide, along with workshops to train on the process. A scorecard with definitive criteria for assessing the format and quality of the project charter content is the next most important design criteria, followed by the need for the P&M team to review the project charters. This will ensure consistency of the measurement process. A stakeholder survey is the last design criteria that could be used to validate the new process.

5. Define Phase Presentation

The Define phase presentation summarizing the written Define phase is included in the downloadable instructor materials.

DEFINE PHASE CASE DISCUSSION

1. **Define Report**
    1.1 Review the Define report and brainstorm some areas for improving the report.
    1.2 How did your team ensure the quality of the written report? How did you assign the work to your team members? Did you face any challenges of team members not completing their assigned tasks in a timely manner, and how did you deal with it?
    1.3 Did your team face difficult challenges in the Define phase? How did your team deal with conflict on your team?
    1.4 Did your instructor and/or Black Belt or Master Black Belt mentor help your team better learn how to apply the Design for Six Sigma tools in the Define phase, and how?
    1.5 Did your Define phase report provide a clear understanding of the VOC, why or why not?

2. **Data Collection and VOC**
    2.1 How did you derive the CTS criteria and how would you ensure that they represent the customer and stakeholder needs.

3. **CTS Summary**
    3.1 What would you perceive to be some of the difficulties of collecting VOC information in an interview format?
    3.2 What other ways could you collect the VOC information for this project?

4. **QFD**
    4.1 Why is it important to prioritize the CTS before developing the relationships between the CTSs and the technical requirements?
    4.2 Discuss how the Pareto chart provides the priority for the technical requirements?

5. **Define Phase Presentation**
   5.1 How did your team decide how many slides/pages to include in your presentation?
   5.2 How did your team decide upon the level of detail to include in your presentation?

DESIGN PHASE EXERCISES

1. **Design Report**
   Create a Design phase report, including your findings, results, and conclusions of the Design phase.

2. **Process Map**
   Develop a process map for the process.

3. **Failure Mode Effect Analysis (FMEA)**
   Create a FMEA brainstorming potential failures in the project charter review process.

4. **Process Analysis**
   Prepare a process analysis for the proposed process.

5. **Waste Analysis**
   Perform a waste analysis for the proposed process.

6. **Operational Definitions**
   Develop metrics and operational definitions that relate to the CTSs for the new process.

7. **Design Phase Presentation**
   Prepare a presentation (PowerPoint) from the case study exercises that provides a short (10–15 minutes) oral presentation of the Design phase deliverables and findings.

DESIGN PHASE

1. Design Report

Following is a written report of the Design phase for the project charter review process design, including the key deliverables developed as part of the prior exercises.

The Design phase of the DFSS process is focused on designing a process and the potential failures so they are reduced or eliminated with the potential to achieve a six sigma quality level. The main activities of this phase are as follows: (1) identify process elements; (2) design process; and (3) identify potential risks and inefficiencies.

2. Process Map

The team developed the critical elements to be incorporated into the process as follows:

- Management commitment and review
- Metrics to encourage continuous improvement
- Area review meetings on the first and third Tuesdays
- Development team and VP area reviews project charters before going to the area council
- SharePoint will be used to manage area council workflow and agenda
- Criteria will be set to review certain projects across the entire division
- Skills need to be transferred to project leads to develop high-quality project charters
- Process needs to be simple and based on VOC input

The team designed the new process using the VOC information, and the process elements as a guide. The process map is shown in Figure 9.9. A description of the process follows.

1. **Review Project Charter, Enter into Area Council SharePoint, Enter Scorecard**
   Owners: Development team.
   Purpose: For the development team and the area director to review the project charter and ensure the completeness and content is of high quality.
   Steps:
   1.1 The development team will review their project charter within their team/director area. The initiation scorecard can be used as a guide for the format and content quality levels.
   1.2 The project leader should enter the project information in the SharePoint, with a "Status Initiation" of "pending."
   1.3 The development team should complete the initiation scorecard via the area council SharePoint.

2. **Approve?**
   Owners: Development team, manager/sr. manager, director (as appropriate).
   Purpose: To approve the project charter. This approval includes the format, content, and that the project charter addresses the business needs to be included in the scope of the project charter effort.
   Step:
   2.1 Each director area will define their approval process.

3. **Fix Problem**
   Owners: Development team
   Purpose: To correct any issues identified in the development team and director level review.

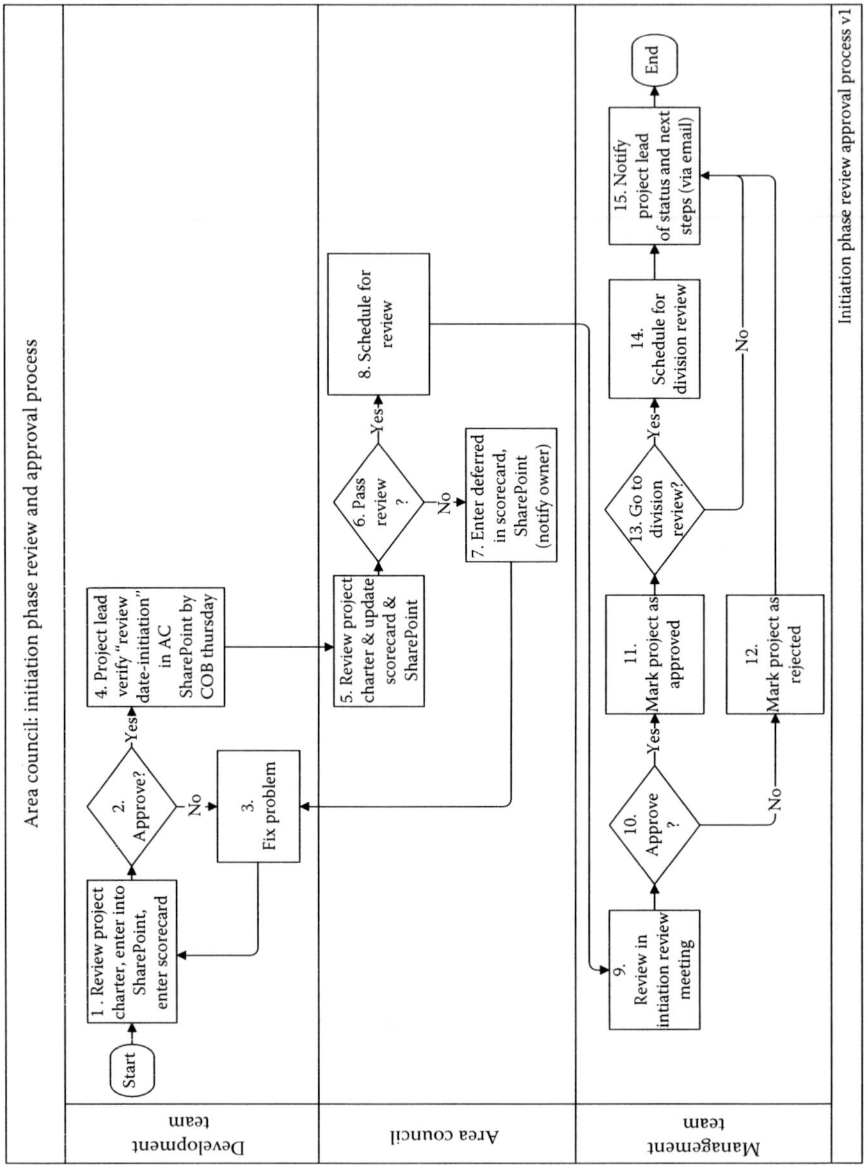

FIGURE 9.9 Process map.

Steps:

3.1 The development team or owner will fix any issues that are identified in the development team and director review of the project charter.

3.2 And resubmit for development team and director approval.

4. Project Leader Verify Review Date: Initiation in Area Council SharePoint by COB Thursday

Owners: Project leader (or development team designated owner).

Purpose: To notify area council that the project charter is ready for the area council format and content review.

Steps:

4.1 Once the project charter is approved by the development team and director, the project leader will verify the date for the "Review Date – Initiation" in the area council SharePoint. This "Review Date – Initiation" should correspond to the supply chain systems area program review dates (currently scheduled as the first and third Tuesday of the month.)

5. Review Project Charter & Update Scorecard and SharePoint

Owners: Area council.

Purpose: To ensure that the format and content are complete.

Steps:

5.1 The area council will review the project charter for format and content, using the project charter scorecard.

5.2 The area council will update the scorecard, and communicate back to the development team as appropriate.

6. Pass Review?

Owners: Area council.

Purpose: To determine if the project charter passes the scorecard criteria.

Step:

6.1 Make decision on initiation scorecard criteria: pass or deferred-pending changes.

7. Enter Deferred in Scorecard, SharePoint (notify owner)

Owners: Area council

Purpose: To notify the project leader of the Development team that there are issues with the project charter to correct.

Steps:

7.1 Enter the decision "deferred pending changes" in the scorecard and SharePoint.

7.2 SharePoint notifies the project lead of the reject and issues.

8. Schedule for Area Council (notify project lead to complete project charter action item)

Owners: Area council.

Purpose: To notify the development team and director that the project charter is scheduled for the area council review and the date that it is scheduled.

Steps:

8.1 Contact the owner (and director) that the project charter is ready to be presented at the area council review, and the date that it is scheduled.

8.2 Add the project charter to the area council (program review) agenda.

8.3 The project lead should complete the action item in Clarity.

9. Review in Area Council

Owners: Management team.

Purpose: To review the project charter and approve or reject the project charter as a project to commence further work.

Steps:

9.1 The area council is to be held the first and third Tuesday of each month from 9 am to 10 am in the True North conference room.

9.2 The project's director (or designee) for the development team will present the project charter at the area council review meeting.

10. Approve?

Owners: Management team.

Purpose: To approve or reject the project charter as a project to commence further work.

Step:

10.1 A decision will be made to "approve" or "reject" the project.

11. Mark Project as Approved in SharePoint & Clarity

Owners: Management team.

Purpose: To update the area council SharePoint and to notify the development team's project leader of the decision.

Steps:

11.1 The decision will be entered into the area council SharePoint (during the program review meeting) and notify the project leader.

11.2 If project was approved, then VP can approve the project in Clarity.

12. Mark Project as Rejected in SharePoint & Clarity

Owners: Management team.

Purpose: To mark the project as rejected in SharePoint & Clarity.

Steps:

12.1 The VP will mark the project as rejected in SharePoint.

12.2 The VP will mark the project as rejected in Clarity.

13. Go to Division's Project Council?

Owners: Management team.

Purpose: To decide whether the project charter should be reviewed at the division's project council.

Step:

13.1 They will decide whether this project charter should be reviewed at the division's project council, typically based upon project size (>1000 hours) and other risk criteria, such as the cross-functional nature of the project, impact to the business, etc.

14. Schedule for Division's Project Council

Owners: Management team.

Purpose: To schedule project charters that need to be reviewed in the division project council meetings.

Steps:

14.1 If it is decided that the project charter will be reviewed at the division's project council, it will be automatically scheduled for division's project council by updating the project council field on the area council SharePoint.

14.2 Division will pull the project charter for the division's project council meetings, based on the area council SharePoint site.

15. Notify Project Leader of Status and Next Steps (via E-mail)

Owners: area council.

Purpose: To notify the project leader of the status of the project and the next steps.

Steps:

15.1 The project leader will receive an email telling him/her whether their project was approved or rejected.

15.2 The email will contain any necessary next steps. For example: if the project is approved, they will be asked to enter their review date for requirements into SharePoint.

3. Failure Mode and Effect Analysis (FMEA)

The team created a FMEA, brainstorming potential failures in the project charter review process. The FMEA is shown in Figure 9.10 with the Pareto chart prioritizing the failure modes by the risk priority number (RPN) shown in Figure 9.11. The highest RPN based on the severity, occurrence and detection, included resources not being available, a project not getting marked as approved, a scorecard not being created, and a project charter not being reviewed by the team prior to being reviewed by the area council team. We identified and incorporated a recommended action into the process and procedures based on the potential failures.

4. Process Analysis

A process value analysis was performed to assess which of the activities provided value to the process. Inherently, the review of the project charter is an inspection step, if the training is done well, the appropriate skills would be transferred to the project charter preparers and a review step would not be necessary. However, some

Process step	Potential failure mode	Potential effects of failure	SEVERITY	Potential causes of failure	OCCURRENCE	Current process controls	DETECTION	RPN	Recommended action
Review project charter, enter into area council SharePoint, Enter scorecard	Submit the project charter without getting it reviewed with their team.	Project charter has errors. Project charter does not explain the problem or identify the scope	5	Lack of training	2	None	9	90	Incorporate director review
	May not identify all errors	Project charter is not high quality	5	Lack of training	10	None	1	50	Scorecard
Fix problem	Preparer may not fix the problem properly	Project charter is not high quality	5	Lack of training	2	None	1	10	Scorecard
Project lead verify review date_initiation in area council SharePoint by COB Thursday	Project lead puts in wrong date.	Project charter does not get reviewed	6	Not reading procedures	2	None	5	60	Training
Review project charter & update scorecard and SharePoint.	Project lead does not create the scorecard	Doesn't catch errors	10	Not reading procedures	10	None	1	100	Verify before review, in procedure
Enter in scorecard, SharePoint (notify owner)	Project lead doesn't correct error	Project charter isn't high quality	10	Lack of engagement	8	None	1	80	Scorecard

FIGURE 9.10 Failure mode and effect analysis.

Schedule for area council (notify project lead to complete project charter action Item)	Reviewer misses the project charter, and doesn't get the project on the agenda.	Project can be delayed	10	Lack of training	1	None	4	40	Training and procedure
Review in area council	Project doesn't get approved	Work so far is wasted	10	Poor scoping, lack of skills, no stakeholder engagement	1	None	1	10	Training
Review in area council	Resources not available	Customer is not satisfied	10	Lack of visibility or budget	8	None	2	160	Reporting
Mark project as approved in SharePoint	Project doesn't get marked as approved.	Project is delayed and customer is not satisfied	10	Mistake	3	None	5	150	Training
Go to ISD project council?	Forget to mark SharePoint for further review	Cross-divisional dependencies may not be identified	5	Mistake	2	None	2	20	Verification step

FIGURE 9.10 (Continued)

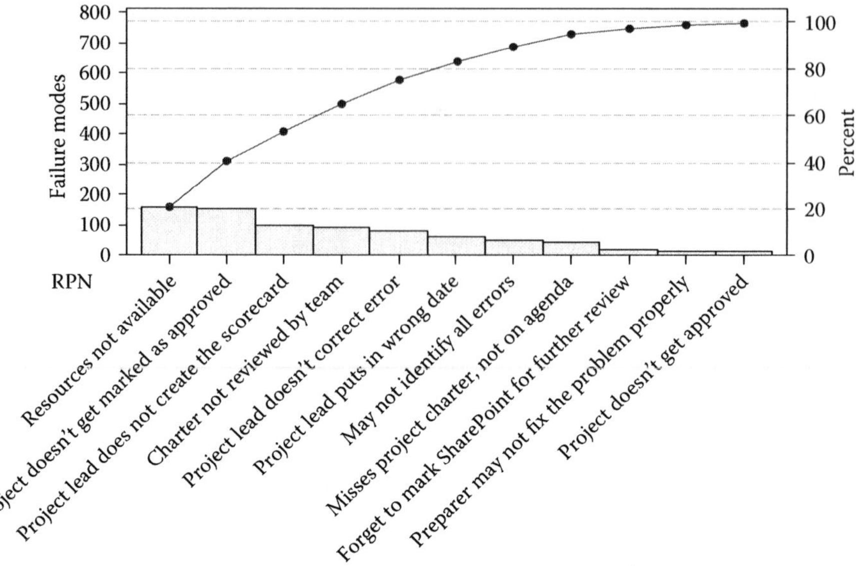

Failure modes	160	150	100	90	80	60	50	40	20	10	10
Percent	20.8	19.5	13.0	11.7	10.4	7.8	6.5	5.2	2.6	1.3	1.3
Cum %	20.8	40.3	53.2	64.9	75.3	83.1	89.6	94.8	97.4	98.7	100.0

FIGURE 9.11 FMEA Pareto chart RPN priority.

of the value of the review is to communicate which projects are being done across the area, and be able to allocate resources across the entire area. The activities in the process that were defined as value-added are the actual decision to approve or reject the project, the area council review held with the SVP, the VPs and directors, and the communication to the project leads of whether the project was approved, deferred, or rejected. The area council review provides value from providing communication of work being performed across the area, and the potential to allocate resources across projects, and find and eliminate any project redundancies. The communication of the project approval to the project leads so that the team can move forward on the project also provides value. Only 25% of the activities add value to the process, with 75% of the activities being nonvalue-added. There is still a great deal of opportunity to incorporate preventive activities and training into the process to further reduce the number of reviews necessary to get a high-quality project charter. The results of the process value analysis combined with the waste analysis results are shown in Figure 9.12.

5. Waste Analysis

A waste analysis was performed on the process. The main types of waste are related to processing embedded in the nature of the process. The project charter review process is being created to provide communication of the work across the entire area and even the entire information systems division, potentially share resources, and to ensure a high-quality project charter. However, there are several levels of review, and the focus of the process should be to incorporate more upfront preventive activities, such as training to reduce the number of reviews necessary to get to a high-quality level. When problems are discovered, this is a defect waste. There are many steps in the process that identify defects, and prevention activities should be incorporated to try to reduce or avoid thc mistakc in thc first placc. The waste analysis identifying the types of waste for each major step in the process is shown in Figure 9.12.

6. Operational Definitions

The potential metrics were developed to help to ensure the CTS criteria could be met. The voice of process (VOP) matrix (Figure 9.13), summarizes and relates the CTS measures, process factors that impact the CTS measures, the operational definition, metrics and proposed targets. The operational definitions describe how you would specifically measure the metrics that relate to the CTS measures. To assess a timely process, we will track that the area council review is held when scheduled, and the

Process step	Value-added	Nonvalue-added	Type of waste
Review project charter, enter into area council SharePoint, enter scorecard		Inspection X	Processing
Fix problem		Defect X	Defect
Project lead verify review Date-initiation in area council SharePoint by COB Thursday		X Inspection	Processing
Review project charter & update scorecard and SharePoint.		X Inspection	Processing
Enter deferred in scorecard, SharePoint (notify owner)		X Inspection	Processing
Schedule for area council (notify project lead to complete project charter action Item)		X Inspection	Processing
Review in area council (communicate value of project)	X		Processing
Mark project as approved, deferred or rejected in SharePoint	X		
Schedule for division's project council		X Inspection	Processing
Go to division's project council?		X Inspection	Processing
Schedule for division's project council		X Inspection	Processing
Notify project lead of status and next steps (via e-mail)	X		

FIGURE 9.12 Process value and waste analysis.

Critical to Satisfaction (CTS)	Process factors	Operational definition	Metric	Target
Timely process	Procedures followed Management commitment Resources available	Area council review is held on the scheduled dates, and projects that are scheduled for the agenda are reviewed during the review.	Area council review is held when scheduled, and projects that are scheduled are reviewed.	100% of projects are reviewed in the identified area council review
High-quality process with metrics	Training Process in place Procedures written, communicated and followed	Scorecard with content quality criteria and score (see scorecard) Scorecard with format criteria and score (see scorecard)	Content quality percentage Format percentage	Content quality: 80% within three months Format: 100% within three months
Accurate information	Training Procedures Relationship with business areas	Scorecard with content and format criteria	Content quality percentage Format percentage	Content quality: 80% within three months Format: 100% within three months
Ability to make decisions, go/no go on projects	Business knowledge of management Quality of project charter	Each project is approved, deferred, or rejected. This would measure the percent approved or rejected, compared to the percent deferred.	Percent of projects approved or rejected the first time (not deferred).	95% (within 3 months of process implementation) of projects that are approved or rejected the first time (0% deferred)
Visibility to program/ project relationships	Program ID is assigned Knowledge of scope of programs and projects	Count of projects that should be related to a program, have the program identified	Count of projects related to programs	80% (within 6 months) of projects that should be related to a program have the program ID.

FIGURE 9.13 VOP matrix.

project charters that are scheduled are reviewed during the session. The target is that 100% of the project charters are reviewed when scheduled.

To assess that a high-quality process with metrics was in place, we developed two initiation scorecards one to assess the format, and the other to assess the quality of the content. The format initiation scorecard verifies every required field is completed. The content initiation scorecard ensures the quality of the content in each field meets the standard criteria identified. We used the standard project criteria provided by the division and used them to create the scorecard for each field of the project charter. For the format, each required field was rated as either complete for 1 point, or as a 0 denoting a missing field. There were a total of 30 required fields, resulting in a total number of 30 points. The format percentage was calculated as the total number of completed fields divided by the total number of points. For example, if a person missed completing four fields, and completed 26 of the fields, their format percent would be 87% (26/30).

For the content scorecard, a Likert type rating scale was used, with a scale from 1 (low quality) to 5 (high quality) for each field. Specific semantic definitions were developed for the ratings of 1, 3, and 5. The 2 and 4 ratings are included to allow a rating between the other ratings when the field entry does not quite meet the next higher rating or the next lowest. There were five points for each field, 1 being the low rating and 5 being the highest rating. There were 12 fields that were assessed for the content. A perfect content project charter would get a total number of content points of 60, for 100% content. If someone received a 3 rating on one of the fields, and 5's on all of the others, a total number of points received on the content scorecard would be 58, for a content percentage of 97% (58/60).

The scorecard criteria will be discussed next. The most important fields on the project charter will be discussed along with the criteria for each.

Business Opportunity

The business opportunity describes the problem, challenge, or opportunity in the business area that initiated the need for the information systems project. We want to ensure that the business opportunity describes the business problem the project is trying to address.

Business Opportunity Scorecard Criteria:

Format: This is a required field and must be entered.

Content: The following criteria were used to assess the business opportunity.

1. Does NOT explain the business problem
2. Somewhat previous answer, but not quite next answer
3. Explains the business problem/uses abbreviations/grammatical errors
4. Somewhat next answer, but not quite previous answer
5. Explains the business problem/impact to business; one paragraph or less; written in business terms; does not reference a solution; factual representation of what the project is to fix, improve, eliminate, or provide; no abbreviations, grammatical errors

Goal

The goal is a statement of how the project will address the identified business problem.

Goal Scorecard Criteria:

Goal: This is a required field and must be entered.

Content: The following criteria were used to assess the goal.

1. Does NOT state how the project addresses the business problem
2. Somewhat previous answer, but not quite next answer
3. Defines how the project addresses the business problem

4. Somewhat next answer, but not quite previous answer
5. Explains the business problem/impact to business; one paragraph or less; written in business terms; does not reference a solution; factual representation of what the project is to fix, improve, eliminate, or provide; no abbreviations, grammatical errors

Objective(s)

The objective is a list of high level bullet points that expand the goal statement and define the boundaries/scope of the project.

Objectives Scorecard Criteria

Format: This is a required field and must be entered.

Content: The following criteria were used to assess the objectives.

1. Does NOT define the scope of the project; task list
2. Somewhat previous answer, but not quite next answer
3. Defines the scope of the project
4. Somewhat next answer, but not quite previous answer
5. Bullet point list; expands upon the goal statement; defines the boundary/scope of the project; descriptive of future desired state; not a list of tasks

Success Criteria

The success criteria identify the end state of the project. The success criteria should be Specific, Measurable, Attainable, Realistic, and Timely (SMART).

Success Criteria Scorecard Criteria:

Format: This is a required field and must be entered.

Content: The following criteria were used to assess the success criteria.

1. Criteria meet 0 of the 5 SMART points
2. Criteria meet 1 of the 5 SMART points
3. Criteria meet 2 of the 5 SMART points
4. Criteria meet 3 of the 5 SMART points
5. Criteria meet 4 of the 5 SMART points
6. Criteria meet all 5 SMART points; completes the statement: this project is successful when…; ties back to objectives.

Risks

Risks identify factors that can negatively impact the outcome of the project.

Risks Scorecard Criteria:

Format: This is a required field and must be entered.

Content: The following criteria were used to assess the risks.

1. No risk factors identified
2. Somewhat previous answer, but not quite next answer
3. Identifies prioritization, resource, or budget risks only
4. Somewhat next answer, but not quite previous answer
5. Identify factors that can negatively impact the outcome of the project

Assumptions

The assumptions are factors considered to be true without demonstration of proof that could impact the outcome of the project.

Assumptions Scorecard Criteria:

Format: This is a required field and must be entered.

Content: The following criteria were used to assess the assumptions.

1. No assumptions identified
2. Somewhat previous answer, but not quite next answer
3. Identifies prioritization, resource, or budget assumptions only
4. Somewhat next answer, but not quite previous answer
5. Identifies factors considered to be true (without demonstration of proof) that could impact the outcome of the project

The concept of the initiation scorecards is to help the project charter authors better understand the criteria for a high-quality project charter, as well as to be used to assess the quality of the charters by the area council. Since the measurement against the scorecard criteria can be somewhat subjective, we use only one person to evaluate the quality of the project charters, until we can train others to consistently score the project charters. We would then perform a gage R&R study to assess the consistency of the measurement system.

7. Design Phase Presentation

Prepare a presentation (PowerPoint) from the case study exercises that provides a short (10 to 15 minutes) oral presentation of the Design phase deliverables and findings.

DESIGN PHASE CASE DISCUSSION

1. **Design Report**
   - 1.1 Review the Design report and brainstorm some areas for improving the report.
   - 1.2 How did your team ensure the quality of the written report? How did you assign the work to your team members? Did you face any challenges of team members not completing their assigned tasks in a timely manner, and how did you deal with it?
   - 1.3 Did your team face difficult challenges in the Design phase? How did your team deal with conflict on your team?
   - 1.4 Did your instructor and/or Black Belt or Master Black Belt mentor help your team better learn how to apply the Design for Six Sigma tools in the Design phase, and how?
   - 1.5 Did your Design phase report provide a clear understanding of the root causes of the process, why or why not?

2. **Process Map**
   2.1 Was it difficult to create a process map for the process, and also the procedures.

3. **FMEA**
   3.1 What other potential failure modes could be identified that were not in the report or in your analysis?
   3.2 How did you determine the recommended actions?

4. **Process Analysis**
   4.1 Discuss how your team defined whether the activities were value-added or nonvalue-added? Was the percentage of value-added activities what you would expect for this type of process and why?

5. **Waste Analysis**
   5.1 What types of waste were prevalent in this process and why?

6. **Operational Definitions**
   6.1 What other metrics could you identify and measure?
   6.2 Was it difficult to clearly define the operational definition?

7. **Design Phase Presentation**
   7.1 How did your team decide how many slides/pages to include in your presentation?
   7.2 How did your team decide upon the level of detail to include in your presentation?

OPTIMIZE PHASE EXERCISES

1. **Optimize Report**
   Create an Optimize phase report, including your findings, results and conclusions of the Optimize phase.

2. **Implementation Plan**
   Develop an implementation plan for the designed process.

3. **Statistical Process Control**
   Develop an example of a control chart that could be used to ensure that the process stays in control.

4. **Process Capability**
   Perform a capability analysis to assess whether the process is capable of meeting the target metrics.

5. **Revised Process Map**
   Revise your process map to incorporate improvements that will further enhance the process.

6. **Training Plans, Procedures**
   Create a training plan, and a detailed procedure for the new process.

7. **Optimize Phase Presentation**
   Prepare a presentation (PowerPoint) from the case study exercises that provides a short (10–15 minutes) oral presentation of the Optimize phase deliverables and findings.

OPTIMIZE PHASE

1. Optimize Report

Following is a written report of the Optimize phase for the project charter review process design project, including the key deliverables developed as part of the prior exercises.

The Optimize phase of the IDDOV process is designed to implement the designed process, and then optimize the design by error proofing and further improving the process by seeing what worked and what did not. The main activities of this phase are as follows: (1) implement process; (2) assess process capabilities; and (3) optimize design.

2. Implementation Plan

The team reviewed the process map and procedures with key stakeholders to ensure it met their needs, and aligned with the divisional standards. They then developed an implementation plan (Figure 9.14). The team also developed a detailed communication plan (Figure 9.15) so they could effectively reach all of the stakeholders so they could understand the new project charter review process. The newly designed process was implemented at the end of February by notifying the entire area through an email with the new process map and detailed procedure. The VPs and directors also communicated the new process to the development teams in their staff meetings and town hall meetings. The first area council was held on March 4. The P&M team gathered input from the stakeholders, and also held some focus groups to understand any issues and to collect improvement ideas regarding the process.

Activity	Responsible	Due date	Stakeholders impacted
Develop communication plan for key stakeholders	Process and metrics team	2/22	All
Distribute new process notice	Process and metrics team	2/29	All
Hold first area council	Process and metrics team	3/4	All
Assess results, and improvement ideas	Process and metrics team	3/18	All
Assess process capability	Process and metrics team	6/17	All
Implement redesigned process	Process and metrics team	7/17	All

FIGURE 9.14 Implementation plan.

Customers/Stakeholders Communication Plan						
	Program	**Initiation**	**Requirements**	**Technical**	**Implementation**	**Post implementation**
VP	- Area program review - SVP staff meeting - VP's staff meetings	- Area council review - SVP staff meeting - VP staff meeting	- Area council review - SVP staff meeting - VP staff meeting	- Area council review - SVP staff meeting - VP staff meeting	- Area council review - SVP staff meeting - VP staff meeting	- Area council review - SVP staff meeting - VP staff meeting
Directors	- Area program review - Director's staff meetings - Program workshops	- Area council Review - Director's staff meetings	- Area council review - Director's staff meetings	- Area council Review - Director's staff meetings	- Area council review - Director's staff meetings	- Area council review - Director's staff meetings
Managers	- Director's staff meetings - Program workshops	- Director's staff meetings	- Director's staff meetings	- Director's staff meetings	- Director's staff meetings	- Director's staff meetings
Development Team:						
Project lead	- Program workshops	- Program/project leader meeting	- Program/project leader meeting	- Program/ project leader meeting	- Program/project leader meeting	- Program/project leader meeting
Program lead	- Program workshops	- Program/project leader meeting	- Program/project leader meeting	- Program/ project leader meeting	- Program/project leader meeting	- Program/project leader meeting

FIGURE 9.15 Communication plan.

Business analysts		BA bi-weekly meeting	BA bi-weekly meeting	E-mail	BA bi-weekly meeting	BA bi-weekly meeting
Technical roles		- Project charter workshops	Gap	E-mail, tech meeting?	E-mail, tech meeting?	E-mail, tech meeting
Division project management office	- Staff meetings - Governance committee - Area council steering committee	- Staff meetings - Governance committee - Area council steering committee	- PMO staff meetings - ISD governance committee - Area council steering committee	- PMO staff meetings - ISD governance committee - Area council steering committee	- PMO staff meetings - ISD governance committee - Area council steering committee	- PMO staff meetings - ISD governance committee - Area council steering committee
Division process engineering	- Staff meetings - Governance committee - Area council steering committee	- Staff meetings - Governance committee - Area council steering committee	- ISDLC staff meetings - ISD governance committee - Area council steering committee	- ISDLC staff meetings - ISD governance committee - Area council steering committee	- ISDLC staff meetings - ISD governance committee - Area council steering committee	- ISDLC staff meetings - ISD governance committee - Area council steering committee
Other areas in division	- Governance committee - Area council steering committee	- Governance committee - Area council steering committee	- BA bi-weekly meeting - ISD governance committee - Area council steering committee	- ISD governance committee - Area council steering committee	- BA bi-weekly meeting - ISD governance committee - Area council steering committee	- BA bi-weekly meeting - ISD governance committee - Area council steering committee

FIGURE 9.15 (Continued)

3. Statistical Process Control

Statistical process control was used to monitor the content and format scorecards by applying a *p*-chart. The control chart for the first three months of format data is shown in Figure 9.16. The format chart for the first three months of content data is shown in Figure 9.17. There were many out of control points, especially in the first month that the review was running.

4. Process Capability

When we implemented the process, we first baselined the process for the scorecard metrics related to the format and content of the project charter. Figure 9.18 shows the

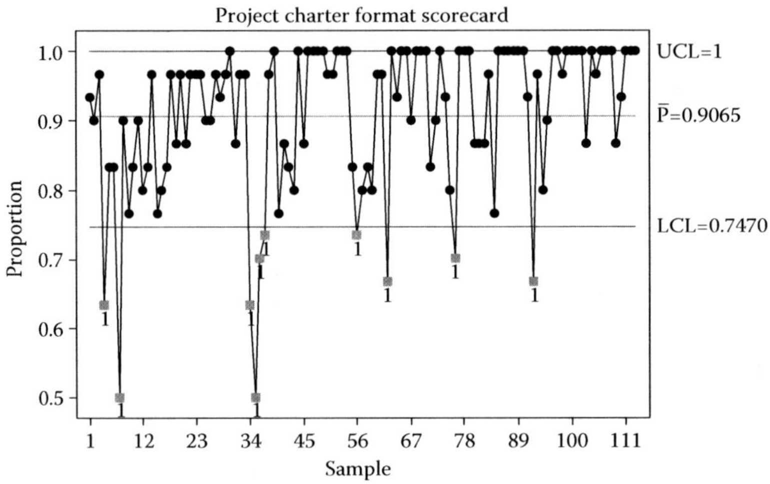

FIGURE 9.16 Format scorecard control chart, with out of control points, date 3/4 to 6/3.

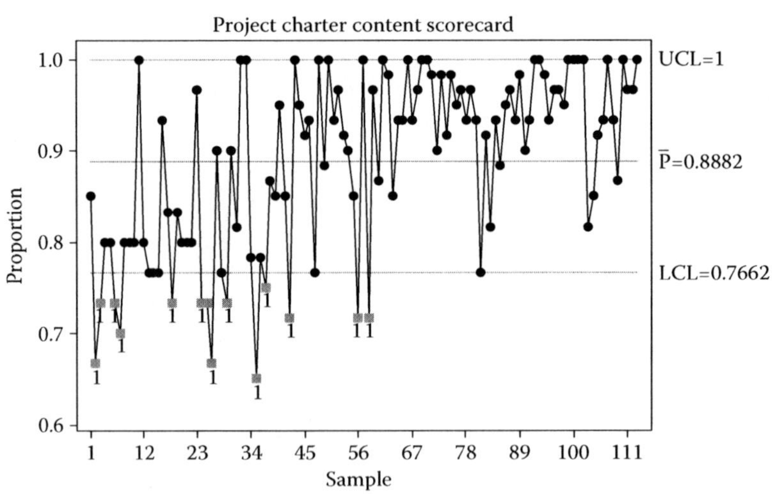

FIGURE 9.17 Content scorecard control chart with out of control points, date 3/4 to 6/3.

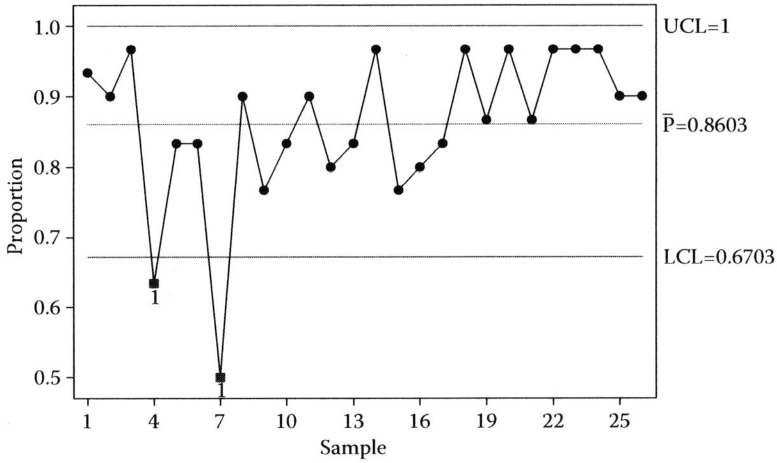

FIGURE 9.18 Baseline format scorecard control chart, date 3/4.

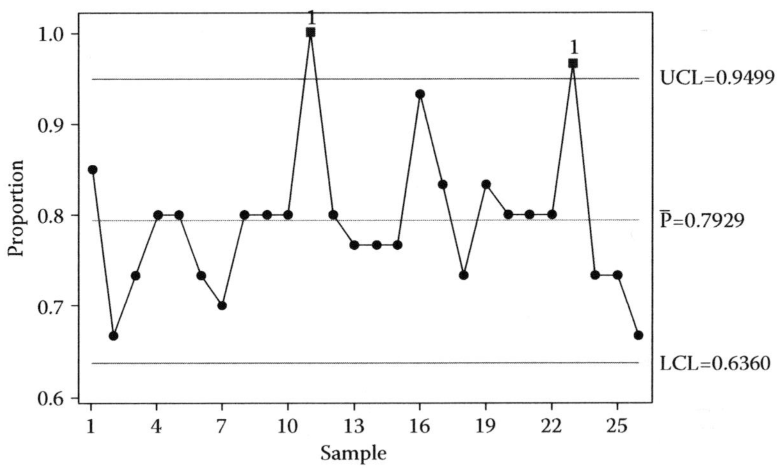

FIGURE 9.19 Baseline content scorecard control chart, date 3/4.

baseline format scorecard percentage of 86%. Figure 9.19 shows the baseline content scorecard percentage of 79%.

The process capability was assessed after three months to have enough data available for an adequate sample size. The initiation scorecard metrics were tracked with each area review to assess improvement from a format and content quality perspective. The format percentage and the content percentage against the scorecard criteria were graphed on *p*-charts. The quality characteristic used for the *p*-charts was percentage of criteria met for the format and content scores, and for each project charter reviewed per session. This data were collected for three months. There were several points that were out of control during each session, when all of the data were placed on a control chart for the first three months' worth of data. Assignable causes were lack of training, or new project leaders were creating the project charters, so these points were removed to calculate the process capability indices.

We calculated the *format* process capability to be 95%, after removing the out of control points (Figure 9.20).

After the assignable causes were removed, we calculated the content process capability to be 96% (Figure 9.21).

The process capability for a *p*-chart is the average *p* value after the process is in control and all of the assignable causes are removed. The process capability for the project charter format is 95% and the process capability for the project charter *content* is 96%. This equates to a sigma level of about 3.2–3.3 sigma, still much room for

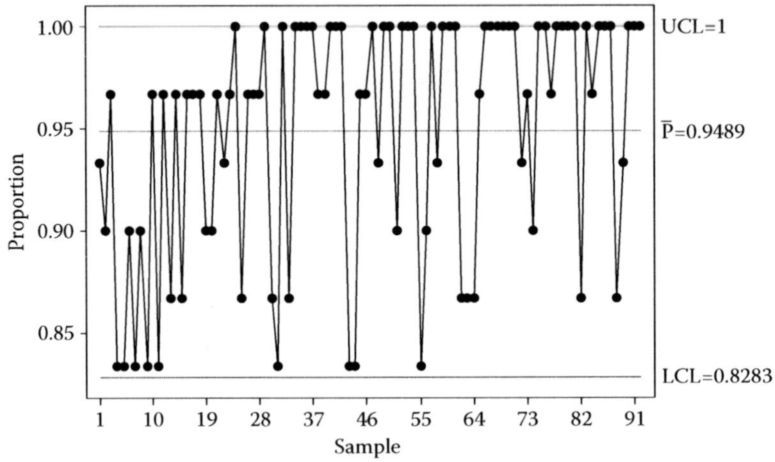

FIGURE 9.20 Format scorecard control chart with assignable causes removed, date 3/4 to 6/3.

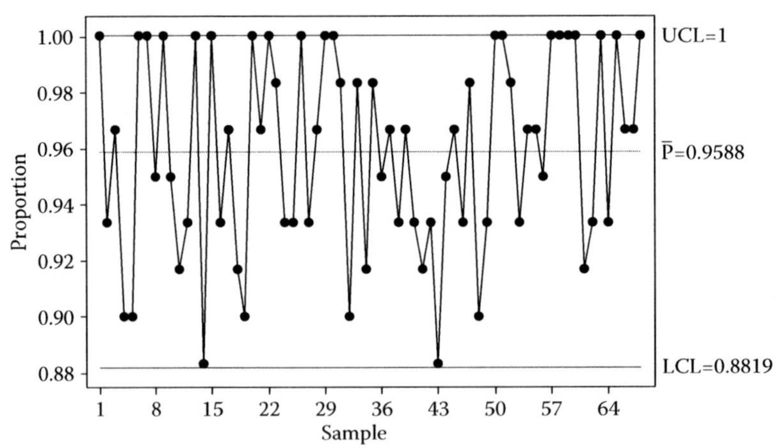

FIGURE 9.21 Content scorecard control chart with assignable causes removed, date 3/4 to 6/3.

improvement if six sigma is our stretch goal. The three-month target for the format was 100%, so we were still shy of the target for filling out all of the required fields on the project charter. The three-month target for the content was 80%, and we have far exceeded the scorecard content target with the process capability of 96%. There is still additional room for improvement related to the project leaders completing all of the required fields.

5. Revised Process Map

We held additional focus groups with the development team stakeholders to understand what worked with the process and what could be improved. We met with the authors of the project charter and project leaders responsible for ensuring that the project charters were reviewed by the area council. There were several elements of the process that the focus group attendees liked as follows:

- The visibility and action items provided by the process and the SharePoint site
- The set deadlines and process consistency
- The ability to have input into the process
- Being able to plan the review schedules better
- The scorecard helps you think through the criteria required on the project charter before sending it on for a review. This comment was given by someone that received a perfect project charter score the first time she ever wrote a project charter.

Some of the improvement ideas from the focus group attendees were:

- Would like to have the scorecard feedback to the authors
- They are not clear on who is supposed to do what
- SharePoint navigation is confusing
- What documents must be attached to the SharePoint?
- Not clear on the review process
- Challenging to coordinate the functional team reviews with the area council review. Timing of the review is difficult (only first and third Tuesdays).

We revised the process to include the following changes:

- In the functional review, we changed the wording from approve to "OK?" to clarify when the project charter is officially approved.
- We combined the format and content scorecard into one document, but kept the ability to report the scores separately. The initial plan was to eventually eliminate the format scorecard when everyone was trained to complete the project charter but, because the format percentage has not reached the target, we combined the two scorecards for ease of entry, but still report on both scores.

- We changed the criteria for the projects that had to also be reviewed in the division's project council to reduce the number of projects that had to be reviewed three different times, down to only twice.
- We moved the due date earlier to better accommodate the volume of project charters that needed to be reviewed.
- We eliminated the need to attach the project charter on the SharePoint site, requiring them to only be uploaded to the project management repository.
- We started to provide the project scorecard feedback directly to the authors. For perfect project charters, we send an email to the project charter author, the project lead, the author's manager, the director, and the VP to share the good news.
- We are tracking the perfect project charters and share those with the management team at the area council review.
- We created a project charter workshop and started training project charter authors to further enhance the quality of the project charters.

A revised process map incorporating many of the improvement recommendations is shown in Figure 9.22.

6. Training Plans, Procedures

We developed the project charter workshops, and started training with the business analysts on the development teams, who create a large number of the project charters. The initial pilot workshop went extremely well. We incorporated suggestions for the workshop to improve the workshop material. We revised the procedures with the revised process ideas.

7. Optimize Phase Presentation

The Optimize phase presentation can be found in the downloadable instructor materials.

OPTIMIZE PHASE CASE DISCUSSION

1. **Optimize Report**
   1.1 Review the Optimize report and brainstorm some areas for improving the report.
   1.2 How did your team ensure the quality of the written report? How did you assign the work to your team members? Did you face any challenges of team members not completing their assigned tasks in a timely manner, and how did you deal with it?
   1.3 Did your team face difficult challenges in the optimize phase? How did your team deal with conflict on your team?
   1.4 Did your instructor and/or Black Belt or Master Black Belt mentor help your team better learn how to apply the Design for Six Sigma tools in the Improve phase, and how?

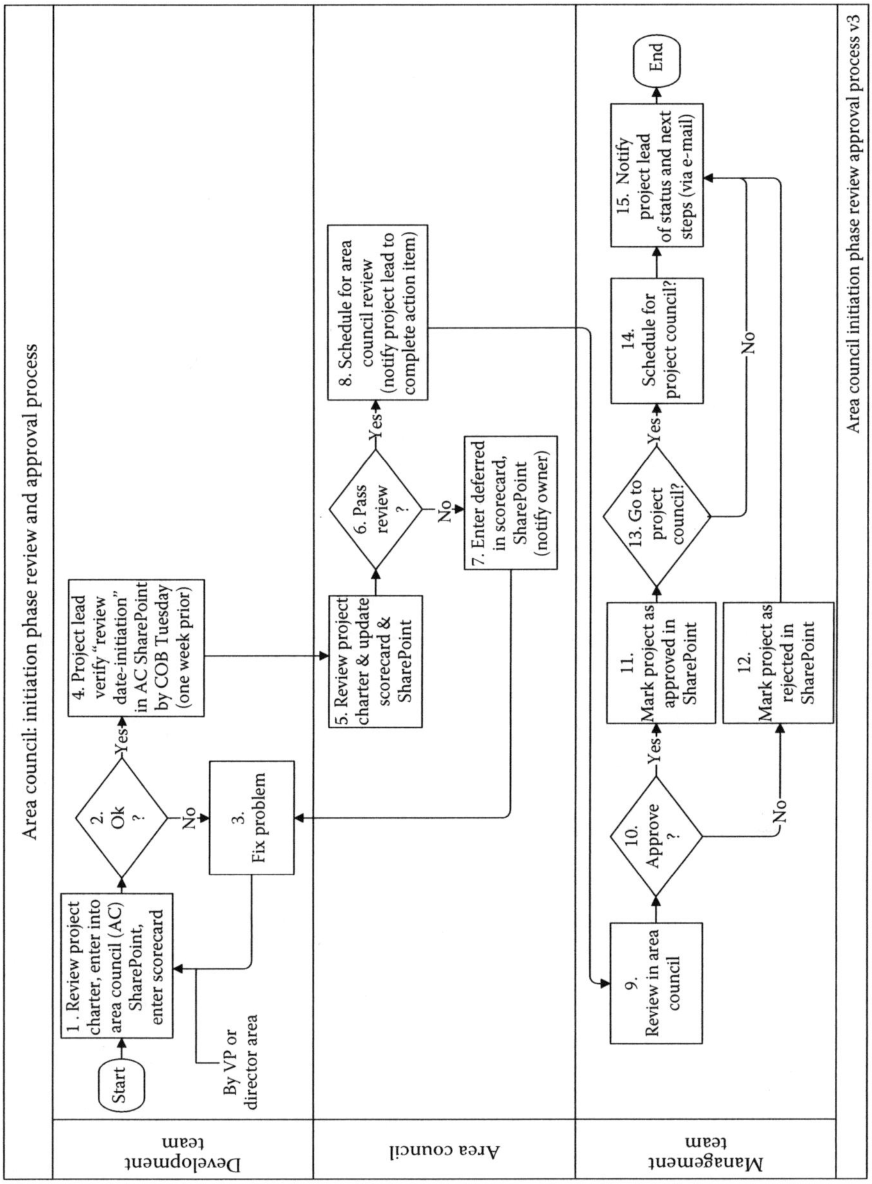

FIGURE 9.22 Revised process map.

1.5 Did your Optimize phase report provide a clear understanding of the new process, why or why not?
1.6 Compare your Optimize report to the Optimize report in the book, what are the major differences between your report and the author's report?
1.7 How would you improve your report?

2. **Implementation Plan**
   2.1 How must the culture be considered in an implementation plan?
   2.2 How must the communication be considered in an implementation plan?
   2.3 How did your Lean Six Sigma team identify the timings for when to implement your recommendations?

3. **Statistical Process Control**
   3.1 How does SPC help us to control the process?

4. **Process Capability**
   4.1 Why is it important to assess process capability?
   4.2 Why is it important to ensure that your process is stable before assessing process capability?

5. **Revised Process Map**
   5.1 Compare your future state process map with the one in the book. How does it differ? Is yours better, worse, the same?

6. **Training Plans and Procedures**
   6.1 How did you determine which procedures should be developed?
   6.2 How did you decide what type of training should be done?

7. **Optimize Phase Presentation**
   7.1 How did your team decide how many slides/pages to include in your presentation?
   7.2 How did your team decide upon the level of detail to include in your presentation?

VALIDATE PHASE EXERCISES

1. **Validate Report**
   Create a Validate phase report, including your findings, results and conclusions of the Validate phase.

2. **Dashboards/Scorecards**
   Create a dashboard or scorecard for tracking and controlling the process.

3. **Mistake Proofing**
   Create a mistake proofing plan to prevent errors from occurring in the process.

4. **Hypothesis Testing/ANOVA**
   Using the data in the "Project Review Data.xls" spreadsheet, perform the appropriate hypothesis test or ANOVA to compare the scorecard quality between the VPs to determine if there is a difference in scorecard quality between the VP areas.

5. **Replication Opportunities**
   Identify some potential replication opportunities within or outside the division to apply the same or a similar process.

6. **Validate Phase Presentation**
   Prepare a presentation (PowerPoint) from the case study exercises that provides a short (10–15 minutes) oral presentation of the Validate phase deliverables and findings.

VALIDATE PHASE

1. Validate Report

Following is a written report of the Validate phase for the project charter review process design project, including the key deliverables developed as part of the prior exercises.

The purpose of the Validate phase of the IDDOV process is to design, develop, and incorporate controls into the improved processes. The main activities of this phase are to: (1) validate process; (2) assess performance, failure modes, and risks; and (3) iterate design and finalize.

2. Dashboards/Scorecards

The dashboard that is reviewed with management at the start of each area council review is shown in Figure 9.23. It shows the initial baseline percent for the format and content scorecard, and the current percentage for the project charters reviewed

Format: All fields complete
- Baseline: 86.3% 96.51% (+10.23%)

Content: Meaningful entries in fields
- Baseline: 79.3% 93.73% (+14.41%)

Number total perfect project charters: 18

FIGURE 9.23 Dashboard.

in the current area council cycle, as well as the overall improvement since the baseline. The dashboard also shows the total number of perfect project charters, those that received 100% on the format and content.

3. Mistake Proofing

To further mistake proof the process we developed the following error-proofing ideas:

- Once the project charter author creates the item on the SharePoint, send them an email if they did not create the scorecard, and encourage them to revise the project charter based on the scorecard feedback. This can help to improve the project charter before the area council review.
- Place a notice on the SharePoint and send an email, to notify everyone of the due date so the authors do not submit the project charters late.
- Provide additional project charter workshop training to prevent project charter errors.
- We asked for a program identifier field be added to the project charter to more easily identify when a project should be associated with a program.
- We added navigational information directions on the SharePoint to reduce confusion identified in the focus group.

4. Hypothesis Testing/ANOVA

Hypothesis Testing between VP Areas

After the first three months of running the process, we wanted to determine if there was a difference in the project charter format and content scores by the VPs areas because our VPs are naturally competitive. We first needed to assess whether the format and content scores were normally distributed to determine which statistical test should be used to compare the scores across the VP areas. If the distributions were normal, we could use an ANOVA test, if not we would need to use a nonparametric test such as Kruskal–Wallis or Mood's median tests.

We performed a normality test in Minitab®, with the null hypothesis being that the data is normal. We received a *p*-value of 0.005 for the both the format and content scores. If *p* is low, the null hypothesis must be rejected. We rejected the null hypothesis, and concluded we did not have a normal distribution for the format or the content scores. The histograms for the data are shown in Figures 9.24 and 9.25.

We next tested whether the variances were equal using the Levene's test for the format and scorecard data. The *p*-value for the format scores was 0.882, and for the content scores was 0.724, so we failed to reject the null hypothesis and concluded the variances are equal. We then performed a Mood median test because it handles outliers better than the Kruskal–Wallis test to test where the median format and content scores are different across the different VP areas. For the format scores, the *p*-value was 0.450, so we failed to reject the null hypothesis and concluded the medians were not significantly different. The medians for each of the VP areas were 29 out of 30 on the format scorecards. The Minitab results are shown in Figure 9.26.

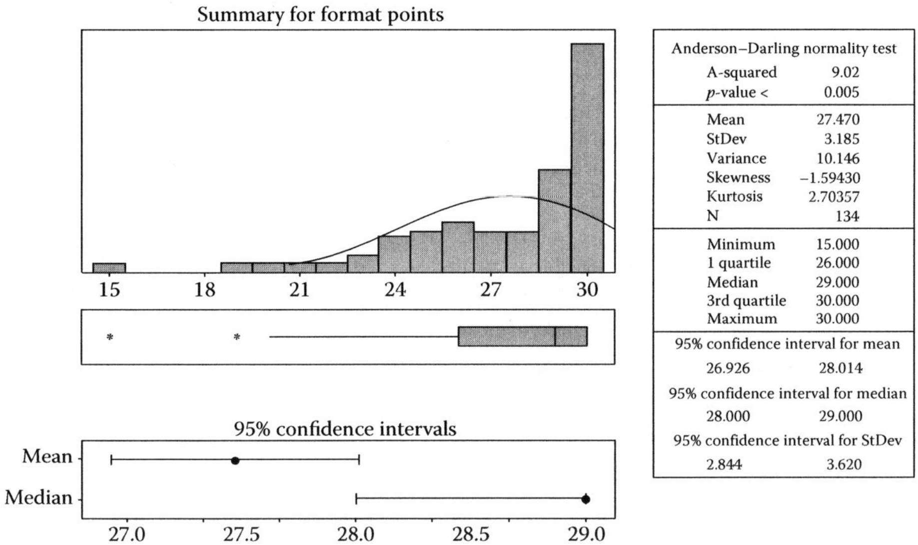

FIGURE 9.24 Histogram of format scores, dates 3/4 to 5/16.

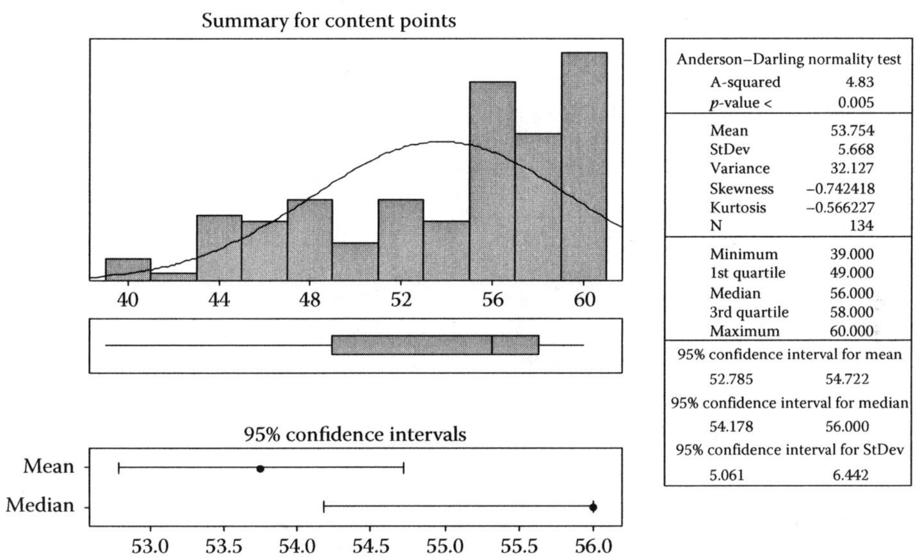

FIGURE 9.25 Histogram of content scores, dates 3/4 to 5/16.

For the content scores, the p-value was 0.228, so we did not reject the null hypothesis and concluded the content scores are not significantly different across the VP areas. The overall median was 56 out of 60 on the content scorecard. The Minitab results are shown in Figure 9.27.

Mood Median Test: Format Points versus VP

```
Mood median test for DOU Format Points
Chi-Square = 2.64    DF = 3    P = 0.450

                                    Individual 95.0% CIs
VP  N<=  N>  Median  Q3-Q1    +---------+---------+---------+------
A    14  13   29.00   4.00      (------------------*------)
B    33  14   29.00   5.00      (------------------*
F     7   5   29.00   4.50   (----------------------*------)
W    29  19   29.00   3.75            (------------*------)
                             +---------+---------+---------+------
                           25.5      27.0      28.5      30.0

Overall median = 29.00
```

FIGURE 9.26 Format scorecard hypothesis test by VP area.

Mood Median Test: Content Points versus VP

```
Mood median test for DOU Content Points
Chi-Square = 4.33    DF = 3    P = 0.228

                                    Individual 95.0% CIs
VP  N<=  N>  Median  Q3-Q1   ------+---------+---------+---------+
A    16  11    56.0    9.0            (-------------*---------)
B    34  13    54.0    9.0         (---------*------)
F     8   4    56.0   10.5   (----------------------*-----------)
W    25  23    56.0   10.8                   (---*-----)
                             ------+---------+---------+---------+
                                 51.0      54.0      57.0      60.0

Overall median = 56.0
```

FIGURE 9.27 Content scorecard hypothesis test by VP area.

Hypothesis Tests from Initial Baseline Results

We wanted to understand if there was an improvement in the format and content scores three months after the process was optimized and implemented.

Format Scorecard

We first tested if the variances between the baseline and the more recent scores were equal with a Levene's test. With a null hypothesis that the variances are equal and a p-value of 0.107, we failed to reject the null hypothesis and concluded the variances are equal. We then performed a Mann–Whitney test to determine if there was a difference between the baseline and later format scores. The null hypothesis was there is no difference between the baseline and the last two months of area council results. The conclusion was the test is significant at 0, so we concluded that there is a difference between the baseline format median score (26.5) and the last two months of results (30.0), showing a significant improvement in the format scorecard results. The Minitab results are shown in Figure 9.28.

Mann–Whitney Test and CI: Format 3 /4, Format 5/6 to 6/17

```
                    N   Median
DOU Format 3 4     26   26.500
DOU Format Rest    59   30.000


Point estimate for ETA1-ETA2 is -3.000
95.1 Percent CI for ETA1-ETA2 is (-4.001,-1.001)
W = 631.0
Test of ETA1 = ETA2 vs ETA1 not = ETA2 is significant at 0.0000
The test is significant at 0.0000 (adjusted for ties)
```

FIGURE 9.28 Statistical test for format scorecard, baseline versus last two months.

Mann–Whitney Test and CI: Content 3/ 4, Content 5 /6 to 6/17

```
                    N   Median
DOU Format 3 4     26   48.000
DOU Format Rest    59   57.000


Point estimate for ETA1-ETA2 is -10.000
95.1 Percent CI for ETA1-ETA2 is (-12.000,-8.000)
W = 496.5
Test of ETA1 = ETA2 vs ETA1 not = ETA2 is significant at 0.0000
The test is significant at 0.0000 (adjusted for ties)
```

FIGURE 9.29 Statistical test for content scorecard, baseline versus last two months.

Content Scorecard

We tested if the variances were equal with a Levene's test. With a null hypothesis that the variances were equal and a p-value of 0.235, we failed to reject the null hypothesis and concluded the variances were equal. We then performed a Mann–Whitney test to assess if there was a difference in the content scores between the baseline period and the last two months. The null hypothesis was there was no difference between the baseline and the last two months of Area Council results.

The test was significant at 0, so we concluded there was a difference between the baseline median content score (48) and the last two months (57), showing a significant improvement in the content scorecard results. The Minitab results are shown in Figure 9.29.

We have optimized and validated our new project charter review process!

5. Replication Opportunities

The concept of incorporating the content and format scorecards would be very effective in any similar process, where there is great value in clearly defining and measuring against specific criteria for qualitative information. This encourages assessing knowledge processes, where knowledge is elicited and presented to gain approval to move forward on an information systems project.

This particular area council review process and procedures were also adopted in the other SVP areas in the division.

6. Validate Phase Presentation

The Validate phase presentation can be found in the downloadable instructor materials.

VALIDATE PHASE CASE DISCUSSION

1. **Validate Report**
   1.1 Review the Validate report and brainstorm some areas for improving the report.
   1.2 How did your team ensure the quality of the written report? How did you assign the work to your team members? Did you face any challenges of team members not completing their assigned tasks in a timely manner, and how did you deal with it?
   1.3 Did your team face difficult challenges in the Validate phase? How did your team deal with conflict on your team?
   1.4 Did your instructor and/or Black Belt or Master Black Belt mentor help your team better learn how to apply the Design for Six Sigma tools in the Validate phase, and how?
   1.5 Compare your Validate report to the Validate report in the book, what are the major differences between your report and the author's report?
   1.6 How would you improve your report?

2. **Dashboards/Scorecards**
   2.1 How would your dashboard differ if it was going to be used to present to just the SVP area or to the entire division?

3. **Mistake Proofing**
   3.1 How well did your team assess the mistake proofing ideas to prevent errors?

4. **Hypothesis Testing/ANOVA**
   4.1 How did you assess the improvement for the CTS?

5. **Replication Opportunities**
   5.1 How did your team identify additional replication opportunities for the process within and outside the information system division?

6. **Validate Phase Presentation**
   6.1 How did your team decide how many slides/pages to include in your presentation?
   6.2 How did your team decide upon the level of detail to include in your presentation?

10 Assessing Lean Six Sigma Project Success—A Case Study Applying a Lean Six Sigma Post Project Assessment Strategy

Sandra L. Furterer

CONTENTS

INTRODUCTION

Lean Six Sigma is an approach focused on improving quality, reducing variation, and defects, while improving profitability in an organization. It is critical to assess the success and effectiveness of the Lean Six Sigma projects so that the organization can understand the impact of the Lean Six Sigma program, and can also gather the lessons learned for subsequent projects. This chapter will present a post-project review strategy that can be utilized by the Lean Six Sigma project teams to assess their performance and the success of their Lean Six Sigma projects. A case study will be presented that applied the post project assessment strategy to three Lean Six Sigma projects. The projects were part of an American Society of Quality (ASQ) Community Good Works Initiative of the ASQ Orlando section 1509 and the University of Central Florida Student Branch. (www.asq.org) Critical learnings were derived from the post-project assessments.

They can be used to improve Lean Six Sigma project success in other organizations' Lean Six Sigma projects, and in improving the ASQ Community Good Works Initiative.

LEAN SIX SIGMA PROJECT ASSESSMENT STRATEGY (LSS PAS)

A Lean Six Sigma project assessment strategy (LSS PAS) that Lean Six Sigma teams can use at the end of their Lean Six Sigma project to identify areas of success, areas of improvement, lessons learned, and to develop improvement strategies was developed to enhance the success of future Lean Six Sigma projects and efforts. Many Lean Six Sigma fledgling efforts in organizations struggle to take hold and fail to entrench the methodology and philosophy into the organization's culture and way of life. The LSS PAS approach is another tool for the learning organization and project teams to enhance and accelerate the impact and the success of their hard work.

The LSS PAS is a five-phase approach, shown in Figure 10.1, and described as follows:

- Phase I: Define Assessment Approach
- Phase II: Develop Assessment Mechanism
- Phase III: Implement Assessments
- Phase IV: Analyze Results, Derive Lessons Learned
- Phase V: Define Improvement Plan

The objectives, activities, and deliverables of each phase of the LSS PAS will be discussed next.

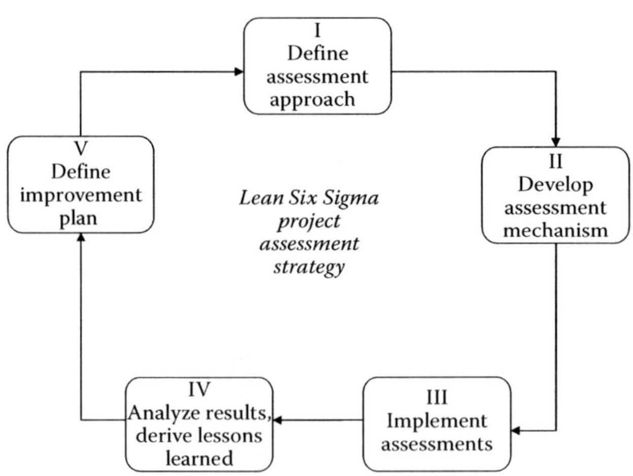

FIGURE 10.1 Lean Six Sigma project assessment strategy.

Phase I: Define Assessment Approach

The objective of the first phase is to define the objectives and criteria for assessing the Lean Six Sigma team's performance, and to obtain leadership buy-in to the assessment approach.

The activities in this phase are:

1. **Define assessment objectives:**
   The main objective of a Lean Six Sigma post-project assessment strategy is to identify areas of improvement for subsequent projects. The team should also understand areas of success so that they are more likely to repeat the areas where they excelled, and also so this knowledge can be shared with other improvement teams.

2. **Obtain Six Sigma leadership buy-in:**
   It is critical that leadership responsible for the Lean Six Sigma program understand and buy-into the need for post-project assessment. They are the people able to share the lessons learned across the organization to help other teams learn the secrets to success. They can remove barriers that impede success of the teams, and can help provide resources for training, team-building, and gaining commitment of project sponsors and team members.

3. **Develop assessment criteria:**
   The assessment criteria help to identify the critical success factors that contribute to Lean Six Sigma project success. Ten areas have been identified that are components of the principles of Lean Six Sigma that can be used as criteria to measure project success:
   - Sponsorship—the level and extent of buy-in and commitment from project sponsors.
   - Project benefits—the value and benefits that the organization, sponsors and team members derive from the Lean Six Sigma projects.
   - Customers and stakeholders—Identifying the primary and secondary customers and stakeholders with respect to the Lean Six Sigma project, and how effective the team is in identifying the voice of the customers and stakeholders.
   - Availability of resources—Identifying and obtaining the appropriate resources for project success. This includes project sponsors, Black Belts, Master Black Belts, team members, training resources, implementation resources, and consultants and experts when necessary.
   - Scope of effort—The scope includes the size and objectives to be met by the project, and whether the scope is appropriate for the resources dedicated to the project and the time available to complete the tasks identified in the project work plan.
   - Deliverables—This area includes the deliverables agreed to by the project team within the project charter developed in the define phase.

It also includes the quality of those deliverables and whether they meet the needs and expectations of the customers and stakeholders.

- Time to complete—This area assesses whether there was appropriate time to complete the project objectives and whether the scheduling of meetings and deliverable due dates was appropriate for the timeframes identified.
- Team synergy—This component addresses how well the team worked together, how well the Black Belts and Master Black Belts mentored and coached the team on the principles and tools, how well the team leaders led the team, and the general synergy of the team.
- Project charter—This area focuses on how well the team defined the project charter, including project objectives, scope, deliverables, business need, costs and benefits, resources needed, work plan, project management and change management approach, and potential risks of project success.
- Value of the Lean Six Sigma approach—This area includes the value that the customers, stakeholders, team members, and the organization received from the Lean Six Sigma projects. This value can be defined by meeting the voice of the customers' expectations; meeting the project objectives; and the costs, benefits and improvements realized by implementing the identified improvements.

The deliverables for Phase I include the objectives of the assessment strategy, a detailed assessment strategy, and documentation of leadership buy-in to the assessment approach.

Phase II: Develop Assessment Mechanism

The objective of this phase is to identify the assessment mechanism(s) that will satisfy the assessment objectives.

The activities to develop the assessment mechanism(s) are:

1. Define assessment mechanism.
   This activity includes defining how the assessment will be performed and what tools will be used to assess the project's success. The author developed a detailed post-project assessment survey tool that incorporates the ten criteria defined above to assess the project. Other avenues could be to interview the project team members using a pre-defined set of questions, or to hold focus groups to understand the areas of success and improvement areas.
2. Develop assessment tools.
   Tools that can be used to assess the Lean Six Sigma project are surveys, focus groups, and interviews.
3. Validate assessment tools.
   Statistical techniques such as variable reduction and factor analysis can be used to reduce the number of questions to streamline the number of

questions, and reduce redundant questions. (Kleinbaum, Kupper, and Muller 1998)

4. Develop sampling and analysis plans.
   This activity includes defining the desired number of respondents, how the survey will be distributed, collected, and analyzed.

The deliverables for Phase II are the assessment tools and sampling and analysis plans.

Phase III: Implement Assessments

The objective of this phase is to implement the assessment tools.

The activities of this phase are to:

1. Distribute the survey, interview respondents or perform focus groups.
   The assessment tool should be distributed to all team members. Confidentiality of individual responses should be ensured.
2. Receive results and encourage participation.
   Many Six Sigma teams have 5–12 participants. It is crucial to encourage all of the team members to participate in the post project review, so that all members provide input into the areas of success and opportunities for improvement. The project champion and sponsors should encourage the participation by writing memos and/or discussing the importance of the post-project review with the team members.

The deliverables for Phase III are the assessment tools, such as the surveys, interview and/or focus group questionnaires.

Phase IV: Analyze Results, Derive Lessons Learned

The objective of this phase is to analyze the results to assess project performance.

The activities for this phase are:

1. Analyze data collected and perform statistical analysis where appropriate.
   The results should be analyzed, using the appropriate descriptive and inferential statistical tools, and summarized in a summary report.
2. Derive lessons learned.
   Lessons learned from the Lean Six Sigma projects can be derived from the survey, interviews or focus groups. If a survey or interviews are used, a focus group can also be used to share the results with the project team so they can provide validation and explanation of the results, as well as any other important lessons learned.
3. Feedback results to leadership and team members.
   The results of the assessment tool should be shared first with the project teams so they can validate the results and provide additional lessons

learned. The project assessment summary should be shared with the leadership, including the Master Black Belt, project sponsors, and champions. The leadership should define mechanisms to share the lessons learned with other project teams so they can incorporate the critical success factors and avoid the areas where other teams could have improved.

The deliverables for Phase IV are a report of the results, and the identified lessons learned to be shared with the leadership, the team members, and other Lean Six Sigma project teams.

Phase V: Define Improvement Plan

The objective of this phase is to develop an improvement plan for future Lean Six Sigma projects.

The activities for this phase are:

1. Develop an improvement plan based on the assessment results
   The results from the assessment tool should be used to develop a best practice improvement plan for each of the assessment criteria.
2. Implement improvements
   The improvements should be applied to subsequent Six Sigma projects.
3. Measure success of the improvements by continuing the assessment strategy for the future Six Sigma projects, starting in Phase I
   Metrics related to the assessment criteria should be measured for all of the Lean Six Sigma teams so that continuous improvement can be a way of life for the Lean Six Sigma program.

The deliverables for Phase V are the improvement plan and on-going metrics to continue measuring success of the Six Sigma projects.

CASE STUDY

The LSS PAS was applied to the ASQ Community Good Works Initiative Lean Six Sigma projects in Orlando, Florida, to assess the teams' performance and success of the three Lean Six Sigma projects.

The Community Good Works Initiative is an outreach program from the national organization of the ASQ. The objectives (www.asq.org) of the ASQ Community Good Works Initiative are to:

1. Stimulate the use of quality practices in the improvement of our communities
2. Create a body of evidence that documents the efficacy of quality in improving communities
3. Improve communities through the use of quality tools and technologies
4. Provide evidence that documents the efficacy of quality in improving communities

5. Engages ASQ members in improvement projects
6. Provide long-term benefit to the community

The three projects were an ASQ Community Good Works Initiative as a collaboration of the ASQ Orlando Section 1509, the University of Central Florida ASQ Student Chapter in the Department of Industrial Engineering and Management Systems (IEMS), and the Harrington Group. The three community-based Lean Six Sigma projects ran from June 2003 through April 2004. The LSS PAS was applied after the completion of the projects to identify areas of success, improvement and to define lessons learned.

The three projects included:

1. Improving a university's distance learning system
2. Developing a needs assessment and governance model for a county's community alliance board
3. Improving the compliance of a nonprofit meal distribution system

The objectives and activities were applied by the teams led by the author.

Phase I: Define Assessment Approach

The objectives of the post-project assessment were to:

1. Assess the performance of the Lean Six Sigma projects
2. Derive lessons learned to be applied on a national basis for the ASQ Community Good Works Initiative.
3. Incorporate improvements into future Lean Six Sigma projects run by ASQ Orlando Section 1509 and the IEMS department at the University of Central Florida

After the teams developed the objectives of the post-project assessment, they obtained Lean Six Sigma leadership buy-in for the assessment from the Master Black Belt, Black Belts, project leaders, and project team members. The previously defined ten assessment criteria were used as the categories for the assessment.

Phase II: Develop Assessment Mechanism

The objective of Phase II was to identify the assessment mechanism that would satisfy the assessment objectives. The author developed a 50-question survey that included the assessment criteria categories. Master Black Belt Frank Voehl, from the Harrington Group, reviewed and validated the survey. The sampling plan was defined, with a goal of receiving 90% response rate on the survey and then holding a focus group with the leadership team to derive lessons learned, using the survey data.

Phase III: Implement Assessment Mechanisms

The objective of Phase III was to implement the survey. The Excel-based survey was distributed via email to each team member, the team leaders, and the Black Belts. Reminder notices were sent to encourage participation.

Phase IV: Analyze Results, Derive Lessons Learned

The objective of this phase was to analyze the results to assess project performance. The author analyzed the data collected and performed descriptive statistical analysis of the survey results. Lessons learned were derived from the results. The results and lessons learned were reviewed with the ASQ Section 1509 leadership team and distributed to all team members. The results were also shared with the ASQ Community Good Works Initiative sponsor.

Phase V: Define Improvement Plan

The objective of Phase V was to develop an improvement plan for future Lean Six Sigma projects. Lessons learned and critical success factors were identified that can be used for future Lean Six Sigma projects. The lessons learned were incorporated into Lean Six Sigma projects being performed as part of a graduate course in the IEMS department at the University of Central Florida. There is a plan to use the post-project assessment survey tool at the completion of the projects.

POST-PROJECT ASSESSMENT SURVEY TOOL

Fifty questions were developed that assessed the performance in each of the ten categories defined in the discussion of the Phase I—Define Assessment Approach, activity three, develop assessment criteria. An agreement scale from one (strongly disagree) to five (strongly agree) was used for the survey questions. The survey also allowed for the respondents to provide additional free-form ideas for improvement, and to identify what they would keep the same for the next project.

CASE STUDY POST-PROJECT ASSESSMENT RESULTS

There was an overall 91% survey response rate across the three Lean Six Sigma teams. The percent of positive ratings (agree, strongly agree) across all of the questions varied by team as shown in Figure 10.2.

The highest rated categories (Figure 10.4) across all of the teams were for: project charter, sponsorship, and value of the Lean Six Sigma approach. The lowest rated categories across all of the teams were: time to complete, availability of resources, and team synergy.

The ratings by category for each team are presented in Figure 10.3. Team 1 is the distance learning team. Team 2 is the community alliance team. Team 3 is the nonprofit meals team.

Team	% Positive ratings
Distance learning	93%
Community alliance	75%
Nonprofit meals	86%
All teams	83%

FIGURE 10.2 Percent positive ratings by team.

For the distance learning team, the highest rated categories were scope of effort, project charter, and team synergy. This team rated the project charter was well defined, the deliverables were well done, there was a reasonable scope and the customers were well defined. Their lowest-rated categories were time to complete, availability of resources and sponsorship.

For the community alliance team, the highest rated categories were the value of the Lean Six Sigma approach, the project charter was well defined, and strong project sponsorship. The lowest rated categories were time to complete, team synergy, and availability of resources.

For the nonprofit meals team, the highest rated categories were project charter, value of Lean Six Sigma approach, and sponsorship. The lowest rated categories were time to complete, scope of effort, and availability of resources.

CASE STUDY CONCLUSIONS: LESSONS LEARNED

Areas of Success

All of the Lean Six Sigma teams did a good job defining their project objectives, customers and the project vision. They also developed the project charters and communicated this information effectively to their customers. The leadership team provided a high level of support for the projects. The value of the Lean Six Sigma approach was highly rated by the teams. Overall, the teams rated their Lean Six Sigma projects as a success. The teams believe that the projects will help them in their profession. Figure 10.5 shows the percentage of positive results across all of the projects. Figure 10.6 shows the overall highest rated categories and the lowest rated categories for the distance learning team.

The community alliance team had a high level of support from their client sponsors. Distance learning had high team synergy where teamwork was encouraged, the team functioned well as a team, the team was receptive to change, and culture change was well managed. The project goals were met on all of the teams. The Black Belts were well trained across all of the teams and their knowledge was appropriate in the community alliance and distance learning teams.

Figures 10.6 through 10.8 show the highest and lowest rated categories for each team.

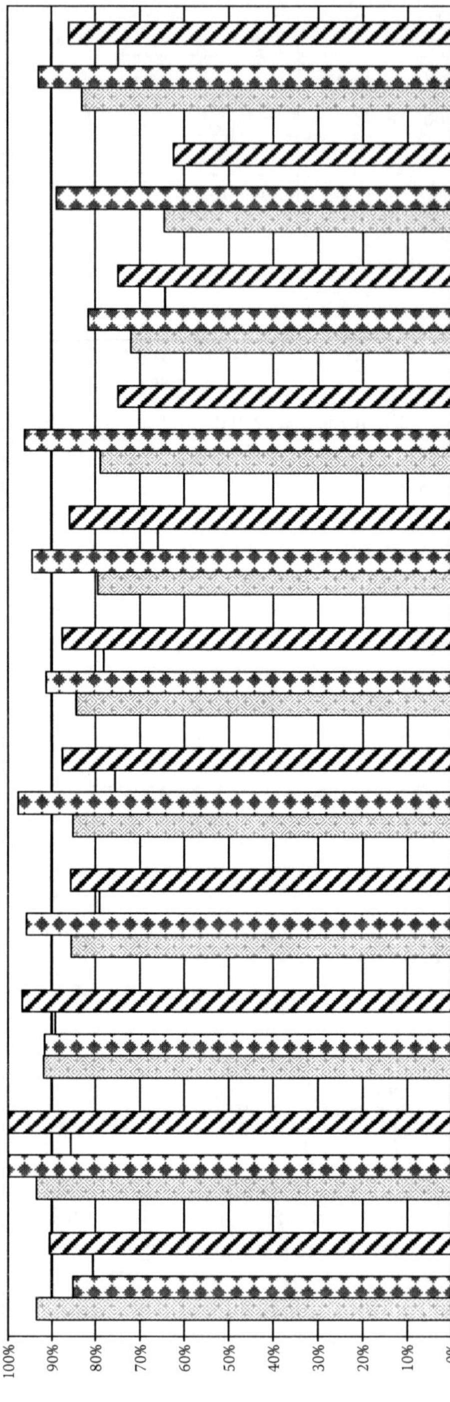

	1.0 Sponsorship:	9.0 Project charter	10.0 Valueof six sigma approach	3.0 Customers	6.0 Deliverables	2.0 Project benefits:	8.0 Team synergy	5.0 Scope of effort	4.0 Availability of resources	7.0 Time to complete	All
All	94%	94%	92%	85%	85%	84%	79%	79%	72%	65%	83%
Dist.	85%	100%	92%	96%	98%	91%	94%	96%	81%	89%	93%
Comm.	81%	86%	89%	79%	76%	78%	66%	70%	64%	50%	75%
Meals	91%	100%	97%	86%	88%	88%	86%	75%	75%	63%	86%

FIGURE 10.3 Post-project assessment results: percent positive ratings by category and team.

Category	Distance learn.	Community	Meals	All
9.0 Project charter	100%	86%	100%	94%
1.0 Sponsorship	85%	81%	91%	94%
10.0 Value of Six Sigma approach	92%	89%	97%	92%
6.0 Deliverables	98%	76%	88%	85%
3.0 Customers	96%	79%	86%	85%
2.0 Project benefits	91%	78%	88%	84%
5.0 Scope of effort	96%	70%	75%	79%
8.0 Team synergy	94%	66%	86%	79%
4.0 Availability of resources	81%	64%	75%	72%
7.0 Time to complete	89%	50%	63%	65%
Overall average	93%	75%	86%	83%

FIGURE 10.4 Case study survey results: percent positive ratings by category.

Highest rated categories	% Positive responses
Project charter	94%
Sponsorship	94%
Value of Six Sigma approach	92%
Lowest rated categories	**% Positive responses**
Team synergy	79%
Availability of resources	72%
Time to complete	65%

FIGURE 10.5 Survey results of all teams combined.

Areas of Improvement

Availability of resources and time to complete the projects were rated low across the teams. There was a lack of a clear reward and recognition system across all three teams. All of the teams needed to improve changing and managing the customers' culture. None of the teams used a clear project work plan with activities, milestones, resources, and timelines to guide their work. Team members' roles and responsibilities were not well defined, nor was feedback appropriate to project team members to

Highest rated categories	% Positive responses
Project charter	100%
Deliverables	98%
Scope of effort/customers	96%
Lowest rated categories	**% Positive responses**
Time to complete	89%
Sponsorship	85%
Availability of resources	81%

FIGURE 10.6 Case study survey results: percent positive ratings by category. Distance learning team.

Highest rated categories	% Positive responses
Value of Six Sigma approach	89%
Project charter	86%
Sponsorship	81%
Lowest rated categories	**% Positive responses**
Team synergy	66%
Availability of resources	64%
Time to complete	50%

FIGURE 10.7 Case study survey results: percent positive ratings by category. Community alliance team.

perform the tasks. The quantity of Black Belts and availability of Black Belts on the nonprofit meals team was lacking. This team needed to better define their internal customers' requirements of the project team. They also needed to improve the client support and obtaining appropriate resources.

The community alliance team did not have high team synergy or function well as a team. The team leaders were not receptive to change, and team empowerment was not encouraged to problem solve or create innovative solutions. Their project governance structure also needed improvement. Community alliance team members needed additional training on the use of Lean Six Sigma tools, and nonprofit meals rated the appropriate use of Lean Six Sigma tools as low. The community alliance project's scope was too large. This team needed to improve measuring project value. The community alliance and nonprofit meals needed to improve

Highest rated categories	% Positive responses
Project charter	100%
Value of six sigma approach	97%
Sponsorship	91%
Lowest rated categories	% Positive responses
Availability of resources	75%
Scope of effort	75%
Time to complete	63%

FIGURE 10.8 Case study survey results % positive ratings by category. Nonprofit meals.

measuring whether the project goals were met. The areas that distance learning struggled with were in obtaining client support and appropriate resources, measuring the value of the Six Sigma project, and improving the community through the use of Lean Six Sigma tools. Distance learning needed to improve focusing on and measuring customer satisfaction, as well as applying a clear problem solving tool.

CONCLUSIONS

The LSS PAS and the survey that was developed and applied are valuable tools to understand areas of improvement, areas where the teams excelled, lessons learned, and whether the Lean Six Sigma projects added value to the clients, based on the perceptions of the team members. Self-assessment is a valuable tool to help the Lean Six Sigma project teams evaluate the success of the Lean Six Sigma projects. The team self-assessment survey could be adapted to allow the customers and stakeholders of the Lean Six Sigma projects to assess the value of the Lean Six Sigma approach and projects, and whether their expectations were met.

This tool can be used to share information with the project sponsors, the organization and other Lean Six Sigma project teams to help them improve the program. The project teams could identify how future project efforts or teams could leverage the areas of excellence and how they could address tactics for improvement for future efforts. The tool can also be used to provide a summary of key project metrics of improvement or satisfaction of customer criteria to capture as part of the assessment so that regardless of the challenges, the organization realizes the benefits. These benefits should be marketed to the organization and the customers to strengthen the future sponsorship.

Following is the Lean Six Sigma Project Assessment Survey.

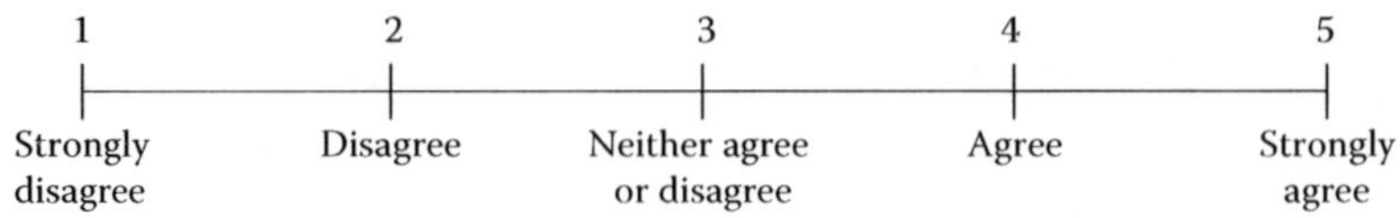

FIGURE 10.9 Survey rating scale.

LEAN SIX SIGMA PROJECT ASSESSMENT SURVEY

Please rate your experience as part of the Lean Six Sigma team(s) by rating the following statements using the scale in Figure 10.9, from 1 to 5. Circle the number on the scale that applies to your response. Only use whole numbers (1, 2, 3, 4, or 5). Thank you for completing this important post-project assessment that will be used to improve the Lean Six Sigma program in the future.

1.0 Sponsorship

1. ASQ Section 1509 leadership supported the projects.
2. Harrington Software Group leadership supported the projects.
3. The client sponsors supported the projects.

2.0 Project Benefits

4. The Lean Six Sigma project sufficiently improved communities through the use of quality tools and technologies.
5. The Lean Six Sigma project had a significant impact on changing the customer's culture.
6. The project goals were successfully met.
7. The project teams' ability to meet the project goals was effectively measured.
8. The project team members' personal and professional goals were successfully met.
9. Customer satisfaction with the Lean Six Sigma project(s) was appropriately measured.
10. My experience on the Lean Six Sigma project was worthwhile.
11. I believe that my experience on the Lean Six Sigma project will help me in my profession.
12. Overall, the Lean Six Sigma project(s) that I was associated with were a success.

3.0 Customers

13. The customer(s) of the Lean Six Sigma project(s) was/were well-defined during the Lean Six Sigma Define Phase.
14. The customer requirements were adequately defined.
15. The customer requirements were well communicated to the project team.
16. The Lean Six Sigma project met the customer's requirements.
17. Customer satisfaction was the project's main goal.

3.1 Stakeholders

18. The project stakeholders were adequately defined
19. The requirements of the project stakeholders were adequately defined

3.2 Internal Customers

20. The requirements of the internal customers (team participants) were adequately defined

4.0 Availability of Resources

21. The Lean Six Sigma project team members' roles and responsibilities were clearly defined
22. A clear project work plan with activities, milestones, identified resources and timelines was used to manage the project
23. The client provided appropriate resources to perform the work

5.0 Scope of Effort

24. The project scope was appropriate
25. The quantity of Black Belts on the project(s) was/were appropriate
26. The availability of the Black Belts to assist/coach team members was appropriate
27. The knowledge of the Black Belts was appropriate
28. The project's governance structure was appropriate
29. The data gathering was fair, open, and honest

6.0 Deliverables

30. A clear problem-solving methodology was applied during the project(s)
31. Training on the use of Lean Six Sigma tools was appropriate
32. The use of Lean Six Sigma tools on the project was appropriate
33. The Lean Six Sigma project sufficiently stimulated the use of quality practices in the improvement of our communities
34. The Lean Six Sigma project team Black Belts were well trained in the Lean Six Sigma tools

7.0 Time to Complete

35. The project length was appropriate
36. The time to complete tasks was appropriate

8.0 Team Synergy

37. The Lean Six Sigma project team was receptive to change
38. The Lean Six Sigma project team leaders were receptive to change
39. The culture and change management was well managed
40. The team was empowered to problem solve and create innovative solutions
41. A clear reward and recognition system existed on the team
42. Team work was encouraged
43. The team functioned well as a team

44. Feedback throughout the project was sufficient for project team members to perform their tasks

9.0 Project Charter

45. There was a clear project vision
46. The project objectives were clearly defined during the Define phase

10.0 Value of Lean Six Sigma Approach

47. The Lean Six Sigma project(s) provided high value to the identified customers
48. The value of the Lean Six Sigma project(s) was well communicated to the project team members
49. The value of the Lean Six Sigma project(s) was well communicated to the customers (clients)
50. The value of the Lean Six Sigma project(s) was effectively measured

Ideas for Improvement

Based on your experience on the Lean Six Sigma team(s), please identify some ideas to improve a team member's experience on the Lean Six Sigma project(s).

Based on your experience on the Lean Six Sigma team(s), what would you keep the same the next time that the Lean Six Sigma project(s) is/are performed. Thank you again for completing this important survey. Best of luck in your endeavors.

ACKNOWLEDGMENT

Portions of this case study were published in Furterer, 2005.

REFERENCES

Furterer, S.L. Assessing Six Sigma project success: A case study applying a Six Sigma post-project assessment strategy. American Society for Quality, Proceedings for Quality Management Division Conference, 2005.

American Society for Quality website. "Community good works initiative." Accessed in 2004 at www.asq.org/goodworks/index.html

Kleinbaum, D., L. Kupper, and E. Muller. *Applied Regression Analysis and Other Multivariable Methods*. 2nd ed. Duxbury Press, Pacific Grove, CA, 1998.

11 The Future and Challenge of Lean Six Sigma

Sandra L. Furterer

This book provided an overview of Lean Six Sigma and the Define-Measure-Analyze-Improve-Control (DMAIC) methodology, the Design for Six Sigma and the Identify-Define-Design-Optimize-Validate (IDDOV) methodology, and real-world service-oriented case studies applying these methods and tools. This last chapter describes a view into the future with the attempt at projecting where Lean Six Sigma will evolve over the next decade.

One of the exciting elements of the Lean Six Sigma evolution has been how many somewhat diverse and at first glance, disparate methods and tools, have come together to provide a more holistic and integrated toolkit for solving extremely complex problems.

The world is getting more complex each day. The problems are getting bigger and more multifaceted, so the tools to solve these problems need to evolve as well.

We can go back to the Evolution of Quality graphic (Figure 2.1) adding the information stream to set the stage for our discussion on the future evolution of Lean Six Sigma (Figure 11.1). The economy in the U.S. and in many other countries is becoming an information and knowledge age. Our economy has evolved from tangible, craft, agricultural, manufacturing economies, to an intangible information, service, and knowledge-based economy. We discussed the progression from Statistical Process Control, which provided control of discrete manufacturing processes, broadening to business process reengineering (BPR) and total quality management (TQM). BPR and TQM incorporated a broader view of the quality management principles and philosophies that needed to be in place to effect change within the cultures that were applying these methods. Six Sigma brought a more structured problem-solving approach, the mentoring and training focus of the belt structure, and a broader toolkit of tools. The Lean side evolved from the Ford production system that provided an assembly process to manufacture and assemble discrete products. Lean and Just-in-Time broadened the spectrum to include more of the supply chain elements. At this same time, information technology was advancing from nonintegrated material requirements planning (MRP) and material resource planning II (MRP II) applications focusing on managing the shop floor and purchasing processes to integrated enterprise resource planning (ERP) and customer relationship management (CRM) information systems that evolved into managing the entire supply chain.

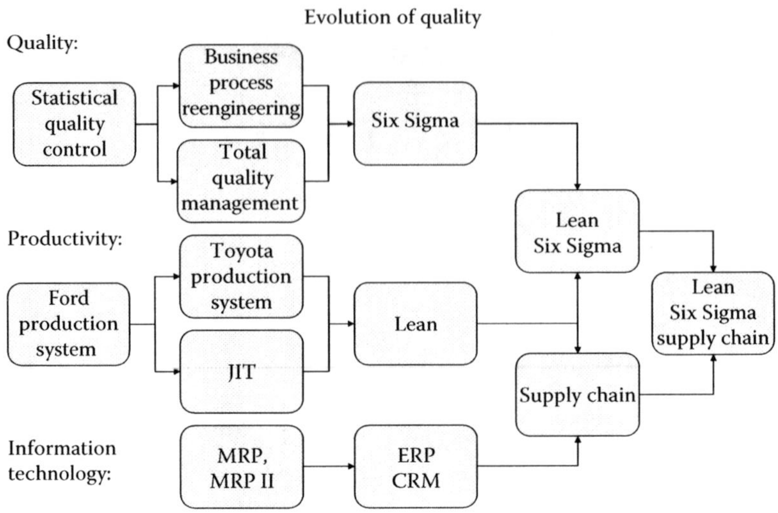

FIGURE 11.1 Evolution of quality to 2008.

The supply chain focus provides an end-to-end view from sourcing of raw material suppliers through the logistics and distribution networks, to the organization and back into the distribution channels to get converted products to the market. The three streams of quality, productivity, and information technology have integrated over the last few years into Lean Six Sigma supply chain, with many variations on the combinations of the names.

For the future, the author sees this integration evolution approach continuing to more tightly couple and integrate the entire enterprise and supply chain, and the philosophies and tools within Lean, Six Sigma and the supply chain areas. As we look at the underlying elements of the three major methodologies, there are many commonalities which the further integration will leverage, as shown below.

Metrics and measurement aligned to business drivers (focusing on reducing costs, improving revenues, and ensuring customer satisfaction): Metrics must continue to evolve to align to business drivers and focus on reducing the cost of doing business, enabling improving revenues by tapping new markets, and ensuring customer satisfaction through measurement and improvement.

Data and information focus to enable flexibility and adaptability to change with changing market and internal conditions: Problem-solving and improvement must enable our organizations to be data- and information-focused to facilitate flexibility and adaptability to adapt with changing market and internal conditions.

Design of our organizations, processes and information systems to support speed to market: Our organizations, processes, and information systems must be designed in the most agile ways to support speed to market. Functional silos must be eliminated to provide a cross-functional view of the businesses that we support.

Customer-focus and engagement: The focus on the customer (not just competitors) will need to grow each day where we more fully engage and align with our customers needs. Mass customization must become a way of life, where we can reach out to each customer to understand their needs and customize our offerings to meet each of their needs in a reachable and cost effective manner.

Data and information supporting the business in alignment with business strategies, leadership, and processes, with supporting information systems and technologies: Data and information must support the business, not exist for its own goals. The place for technology must be identified based on the business strategies and the processes they must support. Leadership will play an even more critical part, and be necessary throughout the organization, not just at the top, but in the middle and at the grassroots of our organizations. Enterprise and business architecture is an emerging framework that fundamentally begins with the needs of the customers and the business to align the strategies, organizations, information, global locations, processes, and timing of all of these elements to enable rapid and controlled change (Boss, Weill, and Robertson 2006).

Empowerment of integrated teams and people working together to common goals: Ultimately, we cannot do any of the above without first considering the needs of our people, empowering them, enabling them with the training, skills, and tools to effectively do their jobs. The central focus of Lean Six Sigma is finding the best way to satisfy customer needs as a never ending process of innovation and improvement. Following are some changes in society and the world economy that Lean Six Sigma must address, and we encourage you to consider how you might take part in the revolution:

- Continued movement in world economies toward service businesses and a world service economy.
- Continued rapid acceleration of technology and making the world "smaller" and more tightly coupled.
- Technology applied to nonvalue-added steps in the business process i.e., inspection (camera imaging), inventory control (radio frequency identification [RFID] real-time tracking of products through the supply chains (global positioning systems [GPS], RFID, Internet, etc.), cashless society (electronic bill payment, debit/credit cards), movement away from the oil-based economy of the world to alternate energy technologies (by necessity), commercialization of nanotechnology.
- Incorporation, acceleration and integration of "green" products, processes, and requirements in product and business process design.
- Continued accelerated increase in customer expectations toward perfection in the products and processes that they utilize.
- Construction and building processes lagging in the technological improvement that the manufacturing and service sector have enjoyed. Lean Six Sigma, automation and mistake-proofing, redesign of the construction management process must and will be employed in this industry that has been labor intensive and dependent.

We challenge the reader and student of Lean Six Sigma to understand the underlying principles embedded in Lean Six Sigma and to be part of the revolution of the tools and philosophies that will continue the evolution of these amazing bodies of knowledge. The success that you experience is in the journey. Create your own world. The best of luck in your Lean Six Sigma endeavors.

REFERENCE

Ross, J., P. Weill, D. Robertson. *Enterprise Architecture as Strategy: Creating a Foundation for Business Execution*, Harvard Business School Press, Boston, 2006.

Appendix A: Financial Process Flows

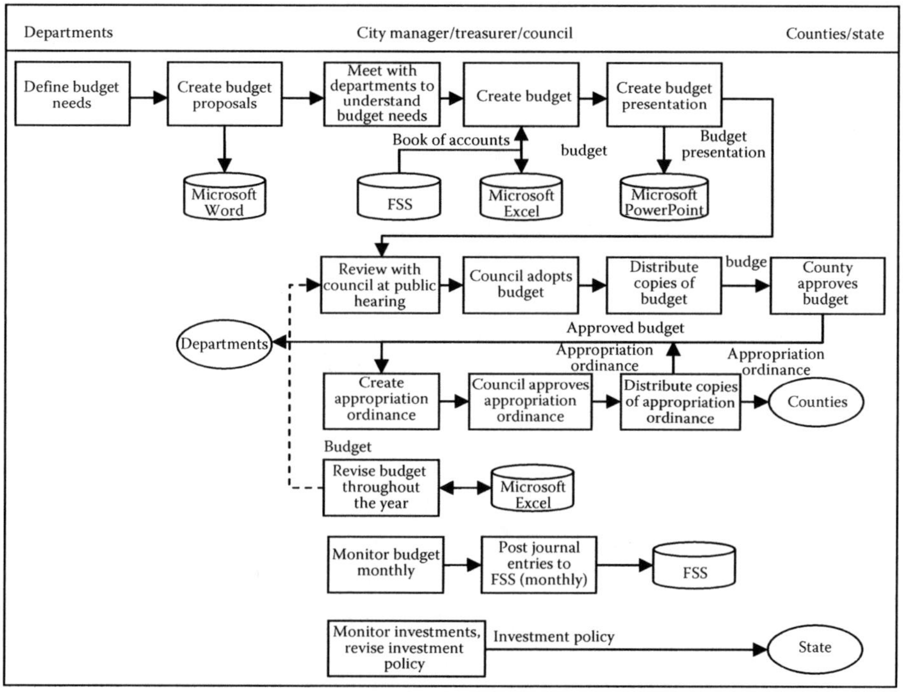

FIGURE A.1 Financial budgeting/investments process flow chart page 1.

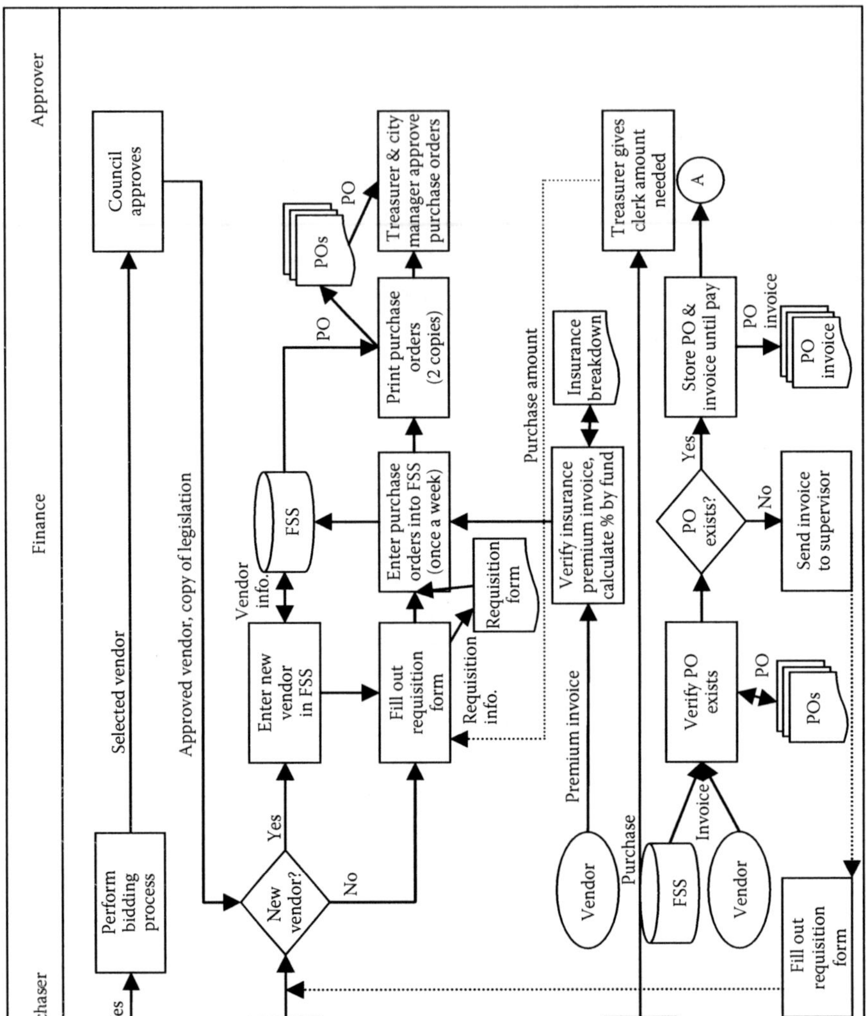

FIGURE A.2 Purchasing/accounts payable process flow chart page 1.

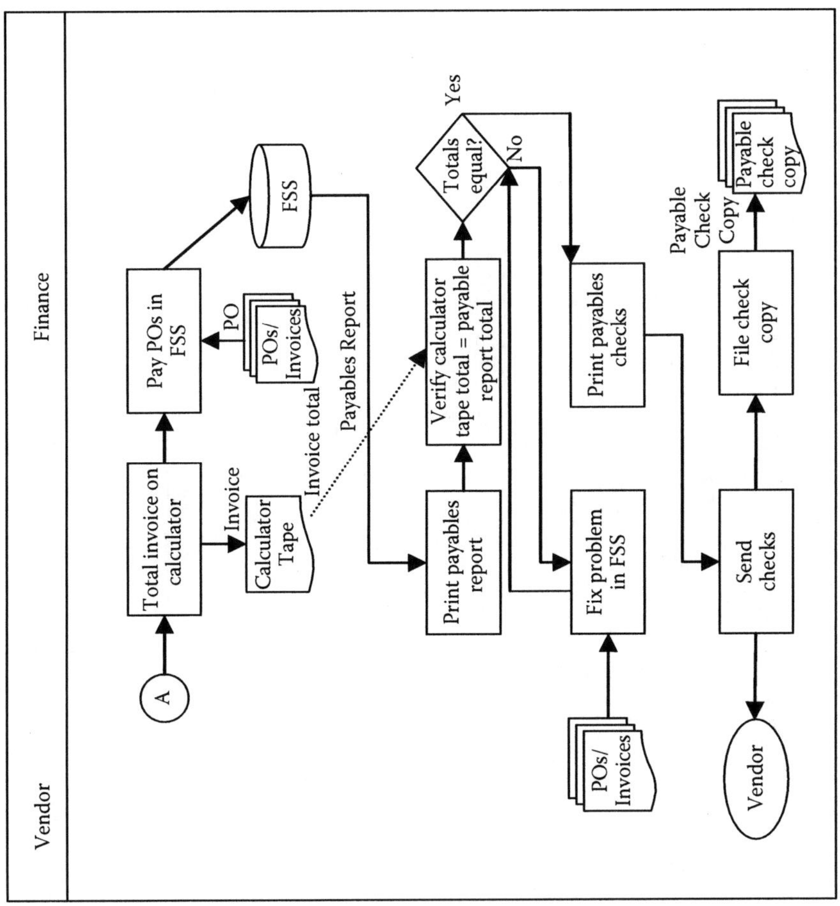

FIGURE A.3 Purchasing/accounts payable process flow chart page 2.

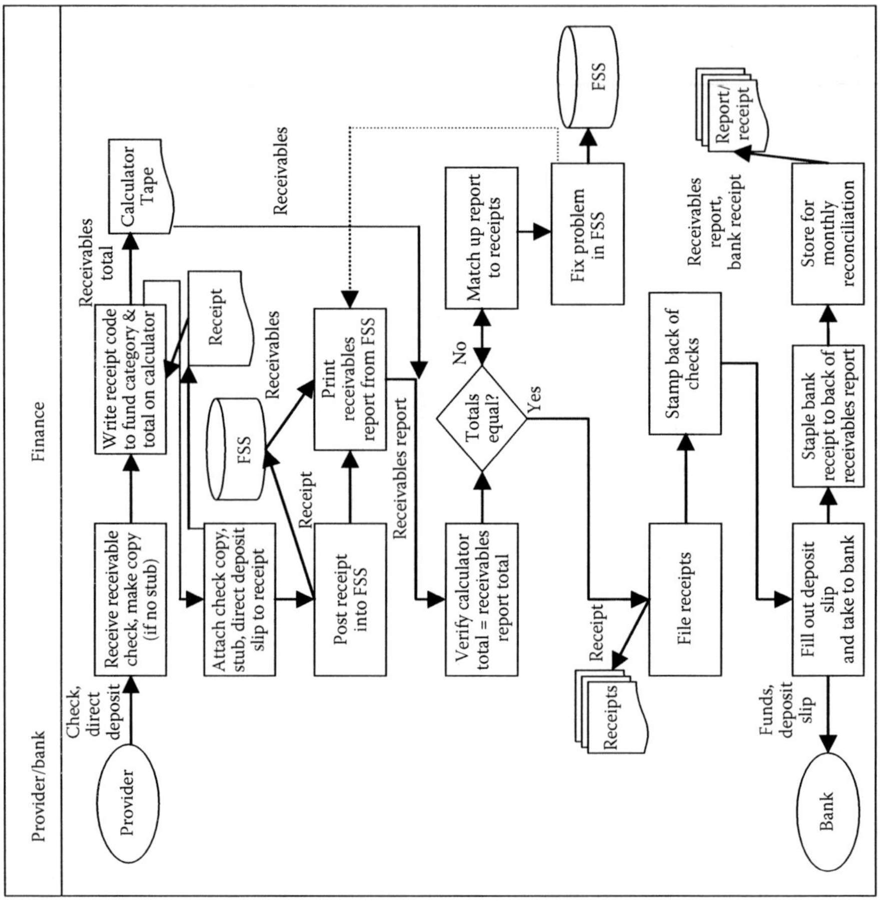

FIGURE A.4 Accounts receivables process flow chart page 1.

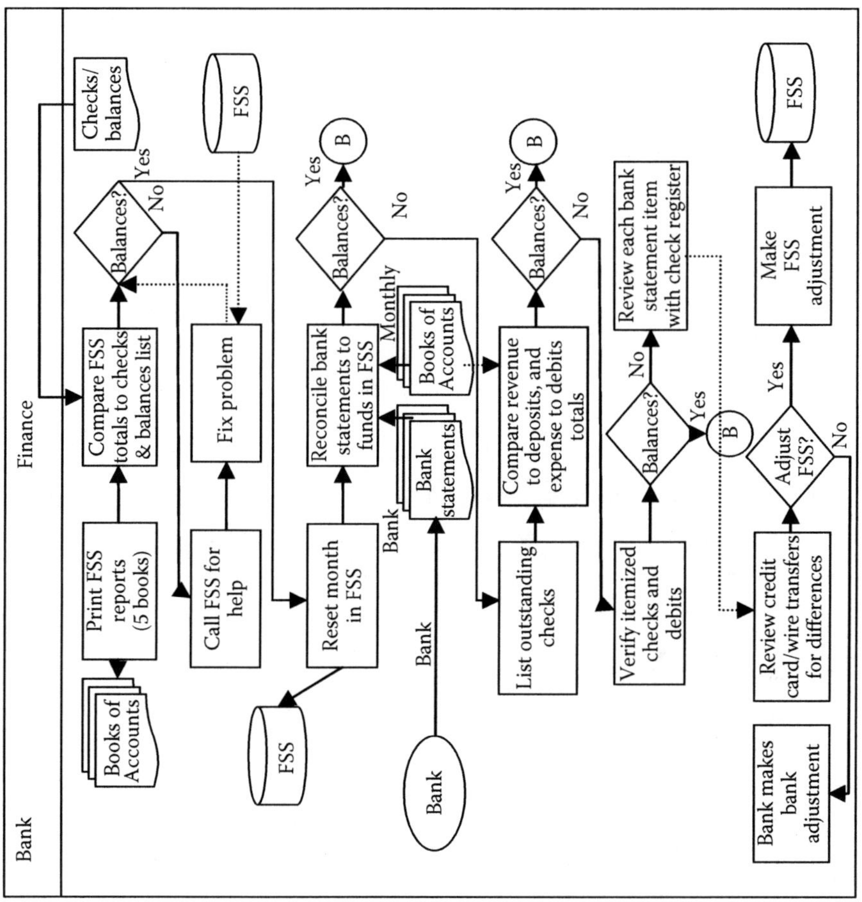

FIGURE A.5 Monthly reconciliation process flow chart page 1.

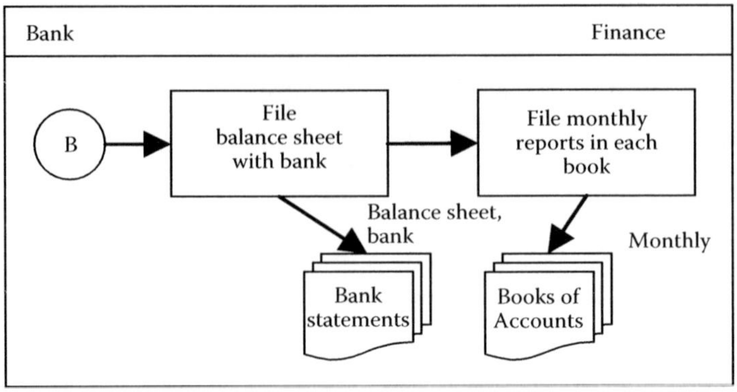

FIGURE A.6 Monthly reconciliation process flow chart page 2.

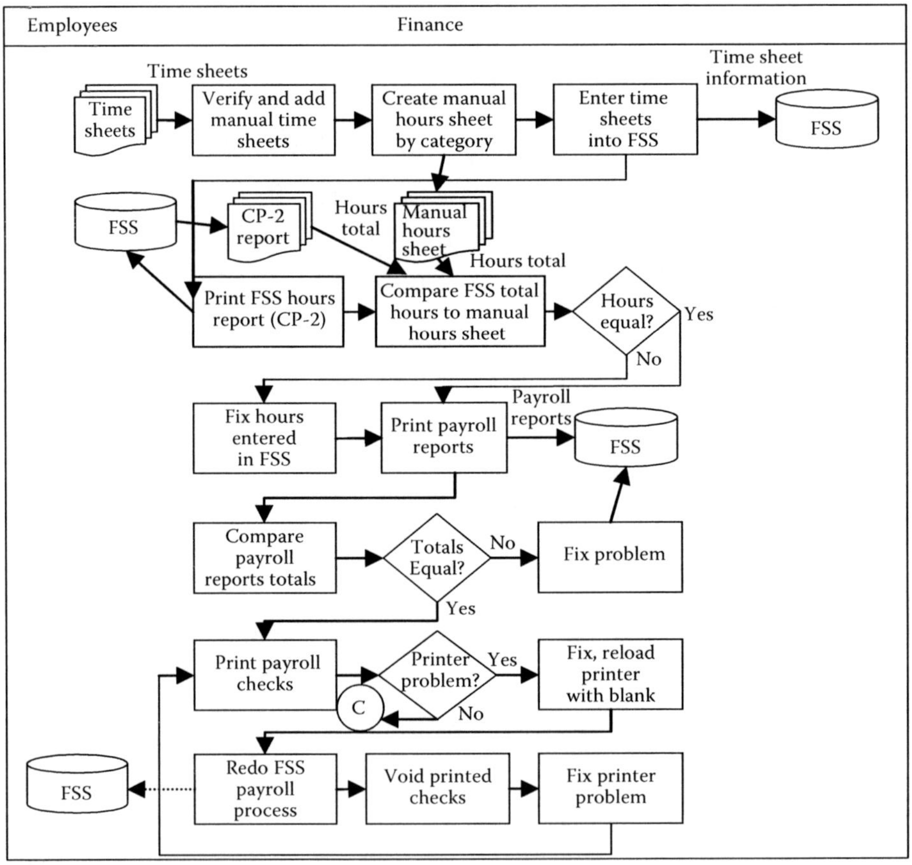

FIGURE A.7 Payroll process flow chart page 1.

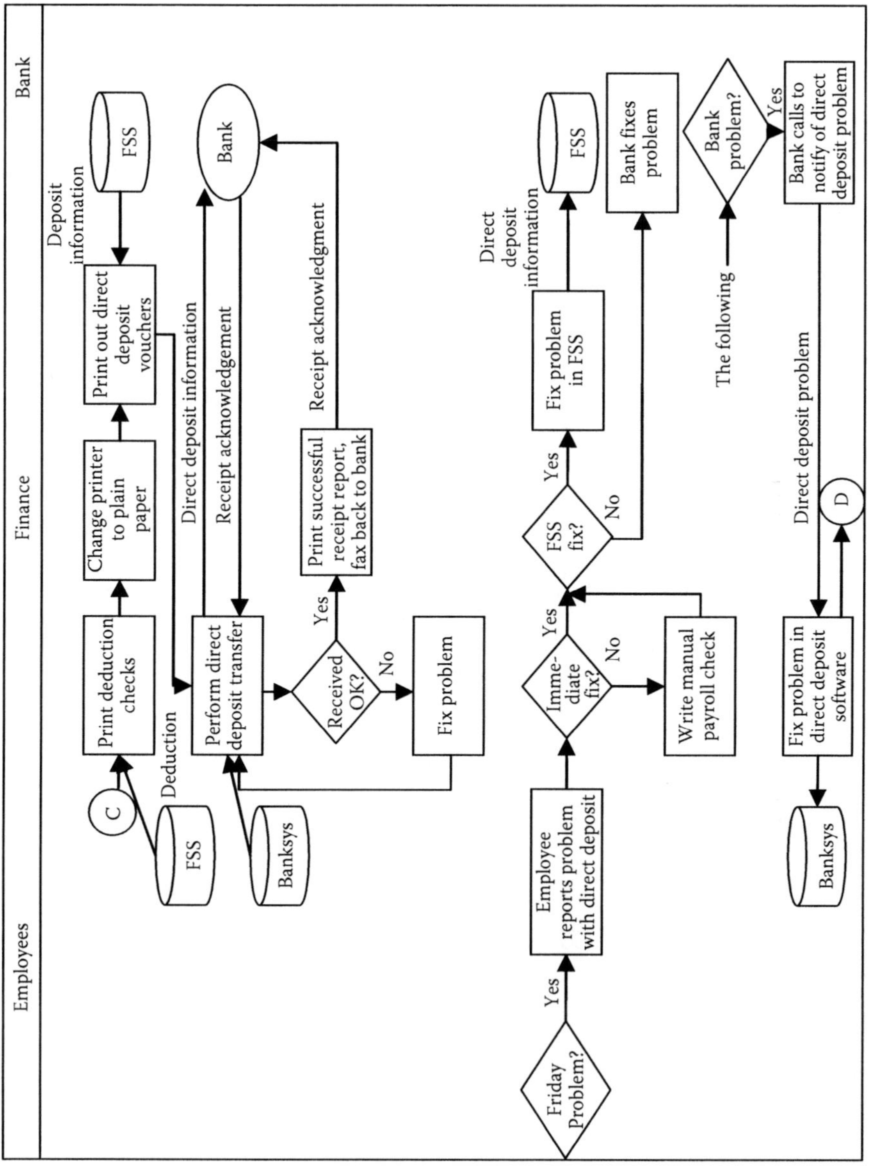

FIGURE A.8 Payroll process flow chart page 2.

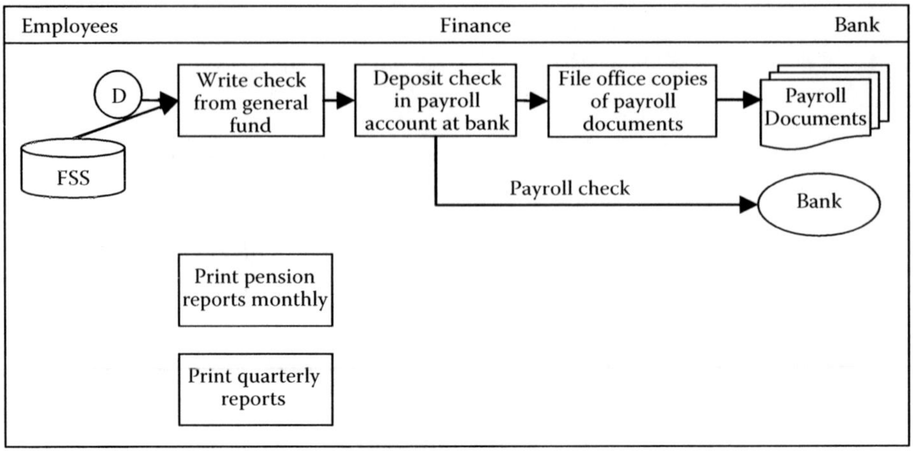

FIGURE A.9 Payroll process flow chart page 3.

Appendix B: Cost Benefit Analysis

The following table summarizes four potential solutions for automating the processing of payroll timesheet hours. The table provides a description of each potential solution, the advantages and disadvantages of each approach, and the estimated costs and benefits for each solution.

TABLE B.1
Cost and Benefit Analysis for Payroll Timesheet Hours Processing

Solution name	Solution 1: Access program	Solution 2: FSI remote payroll	Solution 3: Scanning and OCR	Solution 4: Develop Excel timesheets
Solution description	Create an Access program that contains calculations needed for entry into the financial system's payroll time card program. Verification rules would also be written to verify time sheet data.	Use the Remote Payroll module in the financial information system to enter time sheet data.	Implement a scanning and OCR (optical character recognition) system to scan either manual timesheets or accept Excel spreadsheet time sheet entry input.	Develop Excel timesheets that would standardize the timesheets across all of the departments, enable the departments to enter their own timesheet data, and eliminate the need to verify data off-line with a calculator.
Advantages	• Would provide rule verification and calculations of payroll time data. • Would potentially reduce the manual calculator-based verification processing time.	• Allows entry of time sheet data directly into the format that is accepted by the payroll system. • No custom programming would be needed.	• Allows for input of time sheet data either from manual time sheets or Excel time sheets. • Does not require additional computers for data entry. • Does not require additional data security for remotely located departments.	• Low cost • Enables standardization of the timesheet format and process across city departments • Enables each department to enter their own timesheet data. • Eliminates the off-line calculator verification steps.

(Continued)

TABLE B.1 (Continued)

Solution name	Solution 1: Access program	Solution 2: FSI remote payroll	Solution 3: Scanning and OCR	Solution 4: Develop Excel timesheets
			• Does not require additional computer expertise across city departments. • The OCR and scanning software and hardware has already been implemented in the Income Tax department.	• Would reduce the time needed by the finance clerk to enter and validate the payroll timesheet hours data. • Would provide additional capacity for the finance clerk by eliminating the timesheet entry and validation activities by the finance clerk. • Would enable accountability at the source of the timesheet hours (within each city department) • Short implementation time frame.
Disadvantages	• Requires a certain level of expertise on Access across the city departments for entering and verifying time data. • If the payroll clerk enters the time data, it would potentially reduce data calculation and entry errors but not necessarily reduce data entry time for the payroll clerk.	• Would require a certain level of expertise for the department supervisors or appointees to enter and approve time sheet data in SSI. • Would also require additional computers and data security for remote data entry of time sheet data.	• Does require additional software (OCR and scanning), hardware (scanner) and data security, and software licensing fees. • Requires custom development of OCR and scanning programs. • Requires maintenance and development if time sheet data requirements change.	• Require training by other departments to learn how to use the Excel timesheets. • Would require other departments to buy-in using the Excel-based timesheet process.

(Continued)

TABLE B.1 (Continued)

Solution name	Solution 1: Access program	Solution 2: FSI remote payroll	Solution 3: Scanning and OCR	Solution 4: Develop Excel timesheets
	• Requires custom development of time sheet entry Access programs. • Requires maintenance and development if time sheet data requirements change.		• High cost • Long implementation time frame.	

Index

A

B

C

G

H

I

J

K

L

M